T0328629

Active Disturbance Rejection Control of Dynamic Systems

A Flatness-Based Approach

Active Disturbance Rejection Control of Dynamic Systems

A Flatness-Based Approach

Hebertt Sira-Ramírez
Alberto Luviano-Juárez
Mario Ramírez-Neria
Eric William Zurita-Bustamante

Butterworth-Heinemann
An imprint of Elsevier

Butterworth-Heinemann is an imprint of Elsevier
The Boulevard, Langford Lane, Kidlington, Oxford OX5 1GB, United Kingdom
50 Hampshire Street, 5th Floor, Cambridge, MA 02139, United States

Copyright © 2017 Elsevier Inc. All rights reserved.

No part of this publication may be reproduced or transmitted in any form or by any means, electronic or
mechanical, including photocopying, recording, or any information storage and retrieval system, without
permission in writing from the publisher. Details on how to seek permission, further information about
the Publisher's permissions policies and our arrangements with organizations such as the Copyright
Clearance Center and the Copyright Licensing Agency, can be found at our website:
www.elsevier.com/permissions.

This book and the individual contributions contained in it are protected under copyright by the Publisher
(other than as may be noted herein).

Notices

Knowledge and best practice in this field are constantly changing. As new research and experience
broaden our understanding, changes in research methods, professional practices, or medical treatment
may become necessary.

Practitioners and researchers must always rely on their own experience and knowledge in evaluating and
using any information, methods, compounds, or experiments described herein. In using such information
or methods they should be mindful of their own safety and the safety of others, including parties for
whom they have a professional responsibility.

To the fullest extent of the law, neither the Publisher nor the authors, contributors, or editors, assume any
liability for any injury and/or damage to persons or property as a matter of products liability, negligence
or otherwise, or from any use or operation of any methods, products, instructions, or ideas contained in
the material herein.

Library of Congress Cataloging-in-Publication Data
A catalog record for this book is available from the Library of Congress

British Library Cataloguing-in-Publication Data
A catalogue record for this book is available from the British Library

ISBN: 978-0-12-849868-2

For information on all Butterworth-Heinemann publications
visit our website at https://www.elsevier.com/books-and-journals

Working together
to grow libraries in
developing countries

www.elsevier.com • www.bookaid.org

Publisher: Mara Conner
Acquisition Editor: Sonnini Yura
Editorial Project Manager: Ana Claudia Garcia
Production Project Manager: Anusha Sambamoorthy
Designer: Alan Studholme

Typeset by VTeX

Dedication

This book is dedicated to the memory of my uncle Dr. José Humberto Ocariz-Espinel.

Hebertt Sira-Ramírez

This book is dedicated to Norma, Ma. Inés, Flavio (father), and Flavio (brother) with love and affection.

Alberto Luviano-Juárez

This book is dedicated to my parents Delfin and Ma. Gertrudis, my brother Edmundo, my sister Olivia, and all my family members.

Mario Ramírez-Neria

Dedicated to the memory of my father Élfego Zurita Torres and my second mother Aurea Santiago. To my wife Analleli, my daughters Jana and Sicarú, and my mother Gema Bustamante.

Eric William Zurita-Bustamante

Contents

6. The Challenging Case of Underactuated Systems

A.   Differential Flatness

B.   Generalized Proportional Integral Control

About the Authors

Hebertt Sira-Ramírez

He was born in San Cristóbal, Venezuela. He obtained the Ingeniero Electricista degree from the Universidad de Los Andes (ULA) in 1970. He pursued graduate studies at MIT (Cambridge, USA), from where he obtained the degrees of MSc in Electrical Engineering, 1974, Electrical Engineer, 1974, and the PhD in Electrical Engineering in 1977 under the joint supervision of the late Prof. F.C. Schweppe and the Late Dr. E. Hnyilicza. He was Head of the Control Systems Department at ULA, Vicerrector of the University, and Head of the Graduate School of Automatic Control. He received numerous awards and prizes in his native Venezuela. He retired from the ULA in 1998. He is currently a Titular Researcher at the Mechatronics Section of the Electrical Engineering Department at Cinvestav-IPN in Mexico City.

Alberto Luviano-Juárez

He obtained the bachelor degree in mechatronics engineering from the Instituto Politécnico Nacional (Mexico, 2003), the master of science degree on Automatic Control (2006), and the PhD in Electrical Enginnering (2011) from CINVESTAV-IPN (Mexico). Since 2011, he is with the Graduate and Research Section at Unidad Profesional Interdisciplinaria en Ingeniería y Tecnologías Avanzadas (UPIITA), IPN. His research interests include robust estimation and control in mechatronic systems, robotics, and algebraic methods in the estimation and control of mechatronic systems.

Mario Ramírez-Neria

He received the BSc degree in Mechatronics Engineering from UPIITA-IPN of the Instituto Politécnico Nacional, Mexico City, Mexico (2008), the MSc degree in Electrical Engineering from the Mechatronics Section of the Electrical Engineering Department of CINVESTAV-IPN, Mexico City, Mexico (2011). Currently, he is a PhD candidate at Automatic Control Departament of CINVESTAV-IPN. His current research interests are applications of control theory, active disturbance rejection control, mechanical underactuated systems, robotics, and mechatronic systems.

Eric William Zurita-Bustamante

He received the BSc degree in electronic engineering from the Universidad Tecnológica de la Mixteca, Huajuapan de León, Oaxaca, México, in 2007, the MSc degree in Electronic and Computer Science from the Universidad Tecnológica de la Mixteca, Huajuapan de León, Oaxaca, México, in 2010. Currently, he is a PhD student at Electrical Departament of CINVESTAV-IPN. His research interests include development and implementation of hardware architectures on reconfigurable logic for power electronics, automatic control, and mechatronic systems.

Preface

Active Disturbance Rejection Control (ADRC) is a mature research and application field of automatic control systems design in a variety of dynamical controlled systems. Its conceptual inception dates back to the 19th century. The teaching of some fundamentally pertinent issues to modern ADRC was already popular in the Soviet Union throughout the last half of the 20th century and its consolidation, as an efficient and innovative controller design tool, begun since the 1980s and all throughout the passing years of the 21st century. One of the key issues and concerns associated with ADRC, during the years, has been the lack of a widely acceptable mathematical basis for the theoretical foundations of the subject. The intuitive aspects, allowing for direct applications and implementations, have reached a stage of development that may be termed as remarkable. Slowly, however, the needed rigorous basis has begun to emerge, with great impetus, in relatively recent publications, putting the whole frame into a wider acceptability and enhanced possibilities for practical developments to be used in academic laboratories and in industrial environments all over the planet.

This book is an outgrowth of the work of the authors during a considerable span of time. Our intention in writing this book has been manifold. We have aimed at bringing ADRC to the academically oriented researchers, graduate students, and practitioners of the field of automatic control in industrial establishments. We tried to produce a book that also allows newcomers to the area of control systems to be able to handle the essential features of the ADRC technique while providing relations with neighboring developed areas that greatly facilitate the design problem from the outset. We have, thus, strived to bring areas of nonlinear control theory such as: differential flatness, generalized proportional integral control, tangent linearization-based control schemes, sliding mode control, nonlinear time-delayed systems and fractional-order systems, within the perspective and realm of ADRC. The style is that of presenting illustrative, worked out, textbook and laboratory examples in a concise, systematic manner, so that the reader finds a chance to be convinced of the breadth and reaches of the technique and may attempt, himself or herself, to use it in their own field of work, or development, with some confidence.

The authors have immensely benefitted of their individual relations with many people doing joint work with them. They take this opportunity to express their deepest appreciation to those who faithfully embarked on the job of making the theory work in challenging areas of academic application. For us, it has been a remarkable experience, and a great source of joy, to have been able to complete this work, to work together, and to see this project in final form. Only the years to come will be able to prove if the material presented here is useful to people in different continents of our beloved Earth. We certainly hope so.

Hebertt Sira-Ramírez
Mexico City
December 2016

Acknowledgments

Hebertt Sira-Ramírez wishes to dedicate this book to the memory of his uncle Dr. Humberto Ocariz-Espinel for his affection, encouragement, and wonderful teachings throughout his life. He would also like to express the superb contributions of his wife Maria Elena Gozaine-Mendoza in all fields of his life. He is indebted to Dr. Rubén Garrido-Moctezuma of Cinvestav for his enthusiasm, mature criticism, and fruitful joint work over the last years. Dr. Jesús Linares-Flores from the Universidad Tecnológica de la Mixteca (Huajuapan de Leon) has been a persistent source of inspiration, fruitful and challenging cooperation throughout the years. His former students at Cinvestav, Dr. John Cortés-Romero, Dr. Carlos García-Rodríguez, and MSc Felipe Gonzalez-Montañez, were pivotal in enthusiasm, technical skills, and insights in early stages of this work. He thanks Prof. Michel Fliess from CNRS (France) and Drs. Rafael Morales and Vicente Feliu-Battle from the Universidad de Castilla-La Mancha (Spain) for a long and most enjoyable academic interaction. The kind cooperation of Dr. Marco Oliver-Salazar from Cenidet (Cuernavaca) and Dr. Jesús Leyva-Ramos of IPICYT (San Luis Potosí) is thankfully acknowledged. The author has enjoyed open, profitable, and lively discussions on ADRC with Drs. Gerardo Silva-Navarro and Hugo Rodríguez-Cortes from Cinvestav (Mexico City) and with Prof. Enrico Canuto from the Politecnico di Torino (Italy). Dr. Martín Velasco Villa, Dr. Alejandro Rodríguez, and Dr. Rafael Castro from Cinvestav have pursued joint ADRC ventures with the author, resulting in successful implementations. He thanks them for their patience, understanding, and contributions. He wishes to acknowledge the privilege of interesting conversations, joint work, and generous invitations of Prof. Ziquiang Gao from Cleveland State University (USA) and Prof. Congzhi Huang from the North China Electric Power University (China). Dr. Sira-Ramírez recognizes the creativity, enthusiasm, generosity, and effectiveness of his colleagues coauthoring this book.

Alberto Luviano-Juárez dedicates his work in this book to Norma Beatriz Lozada Castillo for her support, wisdom, love, advice, and kind understanding even in the most critical moments. He would like to dedicate his work to his parents and brother: Ma. Ines Juárez García, Flavio Luviano Palacios, and Flavio Luviano Juárez. He would like to express his sincere gratitude to Griselda Sanchez Otero, a great boss and better friend. He owes recognition to his former students, colleagues, and his study fellows from SEPI-UPIITA, from the Department of Automatic Control at Cinvestav, and from the Mechatronics section, especially to John Alexander Cortés Romero, Oscar Octavio Gutiérrez Frías, Alejandro García González, and Jorge Isaac Chairez Oria. He would like to thank Cecilia Castillo and Manuel Lozada for all their support and Rubén Cruz Rosales for his unconditional friendship. Finally, Alberto would like to thank Professor Hebertt Sira-Ramírez, Mario Ramírez-Neria, and Erik William Zurita-Bustamante for giving him the opportunity of being part of this project and supporting him along all the process.

Mario Ramírez-Neria expresses his deepest gratitude to his family members: his parents Delfin Ramırez-Flores and Marıa Gertrudis Neria-Garcıa, his brother Edmundo Ramirez-Neria, and his sister Olivia Ramirez-Neria. Without their love, patience, encouragement, and sacrifices, he would not have completed this work. He wishes to dedicate this book to all of them. He would like to add these words: "I only can offer all my affection, love, and work."

Eric William Zurita-Bustamante dedicates this work to his dear wife Analleli Jimenez Duran for her love, affection, and patience in the development of this work. To his daughters Jana and Sicaru for the unconditional love, kisses, hugs, and laughs that they give me everyday. Eric expresses his total gratitude to his mother Gema Bustamante and his beloved father Elfego Zurita, who passed away and always supported his projects. He wishes to thank his brothers Didyer, Moctezuma, and Quetzalcoatl for their love and support. A very special thanks goes to professor Dr. Jesus Linares Flores for his advices, support, and especially for his friendship. Eric greatly appreciates the support of Dr. Hebertt Sira-Ramrez, Dr. Alberto Luviano, and Dr. Mario Ramírez during the development of this project.

Hebertt Sira-Ramírez, Alberto Luviáno-Juárez, Mario Ramírez-Neria, Eric William Zurita-Bustamante
Mexico City
December 2016

Chapter 1

Introduction

Chapter Points

- A brief historical review of contributions to Active Disturbance Rejection Control (ADRC) putting in perspective the utilization of differential flatness as a fundamental tool to systematize and substantially facilitate, in planning and design issues, nonlinear control problems.
- The similarities and differences with recent trends in automatic control, such as Model-Free Control, Intelligent PID, and other fruitful areas, is also examined with the desire to stimulate future interactions and cross fertilization of ideas among researchers and practitioners in the unexplored areas of the diverse disciplines that have nonempty intersections with ADRC.
- This introduction contains a brief description of the contents of each one of the chapters constituting this book.

1.1 A BRIEF HISTORICAL PERSPECTIVE ON ACTIVE DISTURBANCE REJECTION CONTROL

Active Disturbance Rejection Control has been studied since the 19th century. The main idea is to simplify the plant description so as to group all disturbances, foreign and internal, and all unknown or ignored quantities and expressions into a single disturbance term and proceed to estimate the effects of this disturbance, in some accurate manner, and devise the means to cancel its effects, using the gathered estimate as part of the feedback control action. The method is simple, elegant, and efficient in a vast number of situations. The disturbance is, generally speaking, unstructured, i.e., it may depend on internal variables of the system in an either unknown or an ignored manner, or else, it may also depend on external phenomena affecting the system. In any case, disturbances are assumed to be uncertain, unknown, unpredictable, or unmodeled. As stated, we distinguish the nature of the disturbances in two classes and refer to them as endogenous disturbances, which are dependent upon internal variables (states, outputs, even control inputs, unmodeled parasitics, and nonlinearities), and exogenous disturbances, which are generated by the environment, by other systems interacting with the plant under study, or represented by the effects of willingly applied

Active Disturbance Rejection Control of Dynamic Systems. http://dx.doi.org/10.1016/B978-0-12-849868-2.00001-0
Copyright © 2017 Elsevier Inc. All rights reserved.

1

disturbing tests. The main virtue and, perhaps, the most questionable viewpoint of ADRC consists in treating the two kinds of disturbances in a common, unified framework, which makes them indistinguishable from each other and essentially of the same disturbing nature. It is the lumped, overall, effects of disturbances that really concern the designer using the ADRC methodology. ADRC then vastly and indiscriminately simplifies the system dynamics into an input–output relation in which the most essential structural aspects of the system dynamics are retained, namely, the order of the system and, possibly, the factors escorting the control input variables. In some research trends, these control input gains are also replaced, at the design stage, by simple, elementary constant, control gains, relegating the true modulated influence of the inputs to the overall unknown disturbance term. In other instances and trends, currently known as model-free control, even the known, or suspected, order of the plant is ignored on purpose and specifically set to be as low as possible (order one or, at most, order two). ADRC is then aimed at reducing a complex process to a possibly linear perturbed plant affected by a total disturbance term, which is easy to control by means of a linear output feedback control law. ADRC advocates and practitioners enjoy a severe, and probably justified, criticisms from the mathematically oriented automatic control communities. More recently, groups and individual control scientists actively doing research in this area have produced different mathematical justifications of the procedures commonly exercised by the ADRC users in plant simplifications, disturbance estimation options, and feedback-based cancellation of uncertain disturbances (see [28], [29], [30]).

The French engineer and mathematician Jean Victor Poncelet seems to have been the first person to suggest in 1829 [1] the so-called invariance principle *"the disturbances must be used to generate an activating signal which will tend to cancel the effect of the same disturbance,"* which was formally established by Professor G.V. Schipanov in 1939 [2]. The invariance principle became popular in control theory in the former Soviet Union [3]. In those early days, the approach was fundamental linear, and the analysis circumscribed to systems described by transfer functions in the complex variable domain. The invariance principle, or Poncelet principle, was reinforced with the so-called feedback principle. Fundamentals contributions in this area were given by V.S. Kulebakin in 1948 [4], by B.N. Petrov in 1955 [5], and by L. Finkelstein in 1960 [6]. In America, the first developments in this direction were initiated by F.G. Shinskey in 1963 [7], and L.E. McBride in 1965 [8].

In the 1960s, the methodology was approached within the state-space representation of systems. In this context, the input disturbance was treated as a fictitious state, to be estimated with the help of a suitable state observer, known later on as the Unknown Input Disturbance Observer (UIDO). Naturally, the estimation process was aimed at providing the feedback control action with a disturbance canceling capability. The initial developments were proposed by C.D. Johnson [9] in the 1970s. The method was known as Disturbance Accommodation Control, and a discrete version was studied in [10]. The adaptive control oriented version of disturbance accommodation was presented in [11]. In the 1990s, the late Professor Jingqing Han, who studied at Moscow State University and was an ex-student of professor Schipanov, proposed for both the dynamic uncertainties and the unknown external disturbances, the Extended State Observer (ESO). The combined effect of both unknown dynamics and external disturbances is treated as a fictitious auxiliary state, estimated using a suitable extra dimension accounted for in a Luenbergeer-type state observer. The obtained estimate is subsequently canceled out, reducing a complex nonlinear time-varying control problem into a simple linear time-invariant one [12]. Han's contributions at the time formed the backbone of the new design paradigm, Active Disturbance Rejection Control (ADRC) [13]. The methodology was successfully applied in industrial implementations [14], [15] and in academic laboratory environments. Prof. Han's ideas where introduced into the English literature in [16] and [17]. Recent contributions by Professor Zhiqiang Gao, coworkers, and students have nourished ADRC with academic applications [18], [19] from various technological engineering areas [20], [21], [22], [23], [24], [25], [26], [27].

Some rigorous mathematical features of the ADRC approach are thoroughly discussed in the works of Zheng, Zhao, and Guo [28–30]. The recent book [31] includes some of these contributions for a wide class of dynamical systems. Recently, ADRC has been extended to a class of distributed parameter systems concerning the output feedback control of an antistable wave equation and of systems subject to mismatched uncertainties (see [32], [33]).

Generalized Proportional Integral (GPI) observers are shown to naturally act as Extended State Estimators. The idea was used in the context of sliding-mode observers for flexible robotic systems in [34]. The nonsliding-mode control version appears in [35], as applied to chaotic-system synchronization. The main idea in this approach is to implement a linear observer, which naturally incorporated iterated output error integral injections for attenuating the effects, on the estimation error dynamics, of exogenous and state-dependent perturbation input signals affecting the input–output model of the plant. The variety of reported results for other applications is increasing, including the control of power converters, as in [36], for robotic systems [37], [38], [39], [40], in electric machines [41], [42], underactuated systems [43,44], in feedback levitation systems [45], and in fault tolerant control [46], among others.

A comprehensive recent survey of the theory and the many applications of the ADRC methodology is contained in the article by Madoński and Herman [47].

Model-Free Control: An Algebraic Counterpart of Active Disturbance Rejection Control

The idea of model-free control dates back to 2006 [48], where the idea of black box identification was approached in the context of algebraic identification techniques [49], [50]. The idea was to avoid dealing with a complete mathematical model and, instead, using a simpler phenomenological model of the form

$$y^{(n)} = F + \alpha u \tag{1.1}$$

The actual model is, thus, replaced by a simpler model of order $n = 1$ or $n = 2$, whereas F and α are to be estimated via algebraic estimation techniques (see [51]) complemented, for $n = 1$, by a proportional or proportional-integral controller or, for $n = 2$, by a proportional-integral-derivative controller. This approach, naturally conforms to the idea of relating a complex control problem with popular industrially oriented control design techniques, such as P, PI, and PID compensation. This is also the practical motivation, in one word, behind ADRC control (see [13]). The robust solution of the control problem for $n = 2$ was termed intelligent PID control (iPID) [52], [53]. The phenomenological plant model was addressed as the ultra-local model [54], [55], and the generalization of the possible solutions was the origin of the so-called model-free control [56]. The intelligent PID, intelligent PI control, and the variations of the intelligent controllers (see [54] for a complete set of intelligent controllers) depend on the accurate state estimation, closely related to the development of robust algebraic differentiators (see [57], [58], [59], [60]), and the trajectory tracking control relies on the flatness viewpoint (this is the major difference with respect to classical ADRC). Though the novelty of this technique, there is a growing set of applications where model free has been successfully applied such as automotive and traffic control [61], [62], [63], [64], [65], [66], biomedical systems [67], agricultural [68], electromagnetic engineering [69], [70], power systems [71], [72], [73], hysteretic actuators [74], time delay systems [75], humanoid robot control [76], aerial robotics [77], variable structure systems [78], [79], etc.

1.1.1   Contents of This Book

Chapter 2 exposes the Active Disturbance Rejection Control methodology in its most general form, within the realm of nonlinear dynamic systems. The need for disturbance estimation–disturbance cancellation, or direct disturbance cancellation, without using observers, is portrayed as part of the fundamental procedure to grant robustness on a linear controller performance. The controller is based on a substantially simplified model of the system. In this regard, flatness plays a crucial role since the simplified system model adopts the form of a set of controlled pure integration systems with additive bounded disturbances. Illustrative examples are presented along with an online evaluation of an integral square error performance index based on the stabilization or output reference trajectory tracking error.

Chapter 3 develops and illustrates two alternative feedback techniques for Active Disturbance Rejection Control in differentially flat systems: The GPI extended state observer and the flat filtering approach (also referred to as the classical compensation networks approach or, simply, the robust GPI control approach).

An Extended State Observer based linear output feedback controller is developed for trajectory tracking problems defined on a class of disturbed differentially flat systems. A linear active disturbance rejection control based on a high-gain linear disturbance observer is proposed and implemented on a variety of challenging cases of study. We show that the state-dependent disturbance can be approximately, but arbitrarily closely, estimated through a linear high-gain observer, called a Generalized Proportional Integral (GPI) observer, which contains a linear combination of a sufficient number of extra iterated integrals of the output estimation error.

On the other hand, the concept of Flat filtering is introduced as a reinterpretation of GPI Control in the form of Classical Compensation Networks (CCNs). Flat filters allow the output regulation of any linear controllable system through a suitable linear combination of the measurable internal states of the filter. Here, it is shown that such a classical control related tool is also capable of efficiently handling control tasks on perturbed exactly linearizable nonlinear systems, i.e. flat systems, including the effects of unknown nonlinearities and unmodeled dynamics. Some study cases are provided to show the effectiveness of the control schemes.

In Chapter 4, we explore some extensions of ADRC control to classes of systems of general interest. We first present the integral reconstruction of states in multivariable systems, just to complete the integral reconstruction-based GPI control schemes, proposed in previous chapters. The extension is presented in the realm of observer-free ADRC, previously used only for the control of simple chains of integration systems. The chapter also addresses how to enhance the robustness and effectiveness in Sliding-Mode control schemes by using extended disturbance observers. This proposal evades the typical need for adaptation of switching "amplitudes," representing binary control values, characterizing typical control input limits in sliding-mode control. We then move to present ADRC results to discrete-time systems in the context of delta transforms, a successful technique to deal with the efficient control of sampled data dynamical systems [80], [81]. The chapter also includes an excursion into the ADRC control of systems exhibiting input delays. The topic is also addressed, in more detail, via a specific laboratory example in Chapter 5. The chapter closes with an illustrative example, pointing toward an extension of ADRC control to a class of fractional-order systems.

Chapter 5 is devoted to presenting some miscellaneous applications of ADRC control. The topics range from robotic systems, nonlinear manipulators of one and two degrees of freedom, controlled by nonlinear motors, to control of Thomson's jumping ring. The idea is to present the wide range of applicability of ADRC schemes and the underlying ease of actual implementation in laboratory environments. Some examples are dealt with using Flat Filtering Controllers, a dual alternative to observer-based ADRC controllers. The obtained results are shown to also extend to a class of control-input delayed systems.

The experimental results included, dealing with several complex nonlinear systems, confirm the effectiveness of Active Disturbance Rejection Control and its dual variation, flat filtering control for dealing with systems exhibiting significant exogenous uncertainties and endogenous uncertainties in the form of unmodeled nonlinearities.

Chapter 6 centers around the ADRC control of underactuated systems, which, in particular, belong to the class of nondifferentially flat systems. These cases are treated in the context of trajectory tracking tasks, just to illustrate the effectiveness of ADRC in the control of nonlinear systems via their controllable tangent linearizations around a given equilibrium point. The trajectories may take the operating point significantly far away from the linearization point, and the neglected nonlinearities may be substantially excited by the features of the desired reference trajectory. In spite of these classical obstacles to tangent linearization-based control, the ADRC scheme yields a quite robust and precise closed-loop performance.

Appendix A is devoted to revisit the concept of flatness. Flatness is a rather ubiquitous property of controlled dynamic systems, which offers several unique features that contribute to simplify the controller design task, the trajectory planning problem, to directly perform offline analysis related to manifolds of equilibria, effects of inputs and states restrictions, and to examine the feasibility of controllers in the context of uncertainties. Flatness trivializes offline computations and controller design issues.

Appendix B provides a reinterpretation of GPI control in terms of CCNs. This allows, in turn, for the conception of a "flat filtering" device, which constitutes a tool for output feedback control design in linear controllable systems. A controllable linear system exhibits a natural flat output (best known as the Brunovsky output), from which the system is also trivially observable. Flat filtering is based on the fact that a GPI controller is viewed as a dynamical linear system that exhibits, as a natural flat output, a filtered version of the plant output signal. This property is particularly helpful in the design of efficient output feedback stabilization schemes and in solving output reference trajectory tracking tasks without using observers. The approach is naturally extended, using Active Disturbance Rejection Control ideas to the control of significantly perturbed SISO nonlinear systems, affected by unknown endogenous nonlinearities in the presence of exogenous disturbances and unmodeled dynamics.

REFERENCES

[1] J. Preminger, J. Rootenberg, Some considerations relating to control systems employing the invariance principle, IEEE Transactions on Automatic Control 9 (3) (1964) 209–215.

[2] G. Schipanov, Theory and methods of designing automatic regulators, Automatika i Telemekhanika 4 (1939) 49–66.

[3] A.S. Poznyak, V.I. Utkin, A.S. Vostrikov, Russia [control education], IEEE Control Systems 16 (2) (1996) 37–40.

[4] V.S. Kulebakin, The use of the principle of invariance in physically realizable systems, Dokl. Akad. Nauk. SSSR 60 (1948) 231–234.

[5] B.N. Petrov, The application of conditions of invariance, in: Transactions of the First All-Union Conference on the Theory of Automatic Regulation, Izvest. Akad. Nauk. SSSR, ML, 1955.

[6] L. Finkelstein, State and advances of general principles of measurement and instrumentation science, Measurement 3 (2) (1985) 2–6.

[7] F.G. Shinskey, Feedforward control applied, ISA Journal (1963) 61–65.

[8] K. Narendra, L. McBride, Multiparameter self-optimizing systems using correlation techniques, IEEE Transactions on Automatic Control 9 (1) (1964) 31–38.

[9] C.D. Johnson, Accommodation of external disturbances in linear regulator and servomechanism problems, IEEE Transactions on Automatic Control 16 (6) (1971) 635–644.

[10] C.D. Johnson, Discrete-time disturbance-accommodating control theory with applications to missile digital control, Journal of Guidance, Control, and Dynamics 4 (2) (1981) 116–125.

[11] P.N. Nikiforuk, K. Tamura, Design of a disturbance accommodating adaptive control system and its application to a DC-servo motor system with Coulomb friction, Journal of Dynamic Systems, Measurement, and Control 110 (4) (1988) 343–349.

[12] Jingqing Han, A class of extended state observers for uncertain systems, Control and Decision 10 (1) (1995) 85–88.

[13] Zhiqiang Gao, Yi Huang, Jingqing Han, An alternative paradigm for control system design, in: Decision and Control, 2001, Proceedings of the 40th IEEE Conference on, vol. 5, IEEE, 2001, pp. 4578–4585.

[14] Yi Huang, Jingqing Han, Analysis and design for the second order nonlinear continuous extended states observer, Chinese Science Bulletin 45 (21) (2000) 1938–1944.

[15] Yi Huang, Kekang Xu, Jingqing Han, James Lam, Flight control design using extended state observer and non-smooth feedback, in: Proceedings of the IEEE Conference on Decision and Control, IEEE, 2001; The Journal's web site is located at http://www.ieeecss.org.

[16] Huanpao Huang, Liqiang Wu, Jingqing Han, Gao Feng, Yongjun Lin, A new synthesis method for unit coordinated control system in thermal power plant-ADRC control scheme, in: 2004 International Conference on Power System Technology, vol. 1, PowerCon 2004, IEEE, 2004, pp. 133–138.

[17] J. Han, From PID to active disturbance rejection control, IEEE Transactions on Industrial Electronics 56 (3) (2009) 900–906.

[18] Yi Hou, Zhiqiang Gao, Fangjun Jiang, Brian T. Boulter, Active disturbance rejection control for web tension regulation, in: Decision and Control, 2001, Proceedings of the 40th IEEE Conference on, vol. 5, IEEE, 2001, pp. 4974–4979.

[19] Robert Miklosovic, Zhiqiang Gao, A robust two-degree-of-freedom control design technique and its practical application, in: Industry Applications Conference, 2004, 39th IAS Annual Meeting, Conference Record of the 2004 IEEE, vol. 3, IEEE, 2004, pp. 1495–1502.

[20] Zhiqiang Gao, Scaling and bandwidth-parameterization based controller tuning, in: Proceedings of the American Control Conference, vol. 6, 2006, pp. 4989–4996.

[21] Gang Tian, Zhiqiang Gao, From Poncelet's invariance principle to active disturbance rejection, in: 2009 American Control Conference, IEEE, 2009, pp. 2451–2457.

[22] Qing Zheng, Zhiqiang Gao, On practical applications of active disturbance rejection control, in: Proceedings of the 29th Chinese Control Conference, IEEE, 2010, pp. 6095–6100.

[23] Xing Chen, Donghai Li, Zhiqiang Gao, Chuanfeng Wang, Tuning method for second-order active disturbance rejection control, in: Control Conference (CCC), 2011 30th Chinese, IEEE, 2011, pp. 6322–6327.

[24] Zhiqiang Gao, Gang Tian, Extended active disturbance rejection controller, US Patent 8,180,464, 2012.

[25] Li Dazi, Pan Ding, Zhiqiang Gao, Fractional active disturbance rejection control, ISA Transactions 62 (2016) 109–119.

[26] Xiaohui Qi, Jie Li, Yuanqing Xia, Zhiqiang Gao, On the robust stability of active disturbance rejection control for SISO systems, Circuits, Systems, and Signal Processing 36 (1) (2017) 65–81.

[27] R. Madoński, M. Kordasz, P. Sauer, Application of a disturbance-rejection controller for robotic-enhanced limb rehabilitation trainings, ISA Transactions 53 (4) (2014) 899–908.

[28] Bao-Zhu Guo, Zhi-Liang Zhao, On convergence of the nonlinear active disturbance rejection control for MIMO systems, SIAM Journal on Control and Optimization 51 (2) (2013) 1727–1757.

[29] Bao-Zhu Guo, Zhi-liang Zhao, On the convergence of an extended state observer for nonlinear systems with uncertainty, Systems & Control Letters 60 (6) (2011) 420–430.

[30] Qing Zheng, L. Gao, Zhiqiang Gao, On stability analysis of active disturbance rejection control for nonlinear time-varying plants with unknown dynamics, in: Proceedings of the IEEE Conference on Decision and Control, New Orleans, LA, Dec 2007, pp. 12–14.

[31] Bao-Zhu Guo, Zhiliang Zhao, Active Disturbance Rejection Control for Nonlinear Systems: An Introduction, John Wiley & Sons, 2016.

[32] Hongyinping Feng, Bao-Zhu Guo, A new active disturbance rejection control to output feedback stabilization for a one-dimensional anti-stable wave equation with disturbance, IEEE Transactions on Automatic Control (2017), http://dx.doi.org/10.1109/TAC.2016.2636571.

[33] Bao-Zhu Guo, Ze-Hao Wu, Output tracking for a class of nonlinear systems with mismatched uncertainties by active disturbance rejection control, Systems & Control Letters 100 (2017) 21–31.

[34] Hebertt Sira-Ramirez, Vicente Feliu Batlle, Robust $\Sigma-\Delta$ modulation-based sliding mode observers for linear systems subject to time polynomial inputs, International Journal of Systems Science 42 (4) (2011) 621–631.

[35] Alberto Luviano-Juarez, John Cortes-Romero, Hebertt Sira-Ramirez, Synchronization of chaotic oscillators by means of generalized proportional integral observers, International Journal of Bifurcation and Chaos 20 (05) (2010) 1509–1517.

[36] E.W. Zurita-Bustamante, J. Linares-Flores, E. Guzmán-Ramírez, H. Sira-Ramírez, A comparison between the GPI and PID controllers for the stabilization of a DC–DC Buck converter: a field programmable gate array implementation, IEEE Transactions on Industrial Electronics 58 (11) (2011) 5251–5262.

[37] Marco A. Arteaga-Pérez, Alejandro Gutiérrez-Giles, On the GPI approach with unknown inertia matrix in robot manipulators, International Journal of Control 87 (4) (2014) 844–860.
[38] M. Ramírez-Neria, H. Sira-Ramírez, R. Garrido-Moctezuma, A. Luviano-Juárez, Active disturbance rejection control of singular differentially flat systems, in: Society of Instrument and Control Engineers of Japan (SICE), 2015 54th Annual Conference of the, IEEE, 2015, pp. 554–559.
[39] Lidia M. Belmonte, Rafael Morales, Antonio Fernández-Caballero, José A. Somolinos, A tandem active disturbance rejection control for a laboratory helicopter with variable-speed rotors, IEEE Transactions on Industrial Electronics 63 (10) (2016) 6395–6406.
[40] H. Sira-Ramírez, C. López-Uribe, M. Velasco-Villa, Linear observer-based active disturbance rejection control of the omnidirectional mobile robot, Asian Journal of Control 15 (1) (2013) 51–63.
[41] H. Sira-Ramírez, F. Gonzalez-Montanez, J. Cortés-Romero, A. Luviano-Juárez, A robust linear field-oriented voltage control for the induction motor: experimental results, IEEE Transactions on Industrial Electronics 60 (8) (2013) 3025–3033.
[42] H. Sira-Ramirez, J. Linares-Flores, C. Garcia-Rodriguez, M.A. Contreras-Ordaz, On the control of the permanent magnet synchronous motor: an active disturbance rejection control approach, IEEE Transactions on Control Systems Technology (ISSN 1063-6536) 22 (5) (2014) 2056–2063, http://dx.doi.org/10.1109/TCST.2014.2298238.
[43] Mario Ramirez-Neria, Hebertt Sira-Ramírez, R. Garrido-Moctezuma, Alberto Luviano-Juarez, On the linear active disturbance rejection control of the Furuta pendulum, in: 2014 American Control Conference, IEEE, 2014, pp. 317–322.
[44] M. Ramirez-Neria, H. Sira-Ramirez, R. Garrido-Moctezuma, Alberto Luviano-Juarez, On the linear active disturbance rejection control of the inertia wheel pendulum, in: 2015 American Control Conference (ACC), IEEE, 2015, pp. 3398–3403.
[45] Mario Ramirez-Neria, J.L. Garcia-Antonio, Hebertt Sira-Ramírez, Martin Velasco-Villa, Rafael Castro-Linares, On the linear active rejection control of Thomson's jumping ring, in: 2013 American Control Conference, IEEE, 2013, pp. 6643–6648.
[46] John Cortés-Romero, Harvey Rojas-Cubides, Horacio Coral-Enriquez, Hebertt Sira-Ramírez, Alberto Luviano-Juárez, Active disturbance rejection approach for robust fault-tolerant control via observer assisted sliding mode control, Mathematical Problems in Engineering 2013 (2013).
[47] R. Madoński, P. Herman, Survey on methods of increasing the efficiency of extended state disturbance observers, ISA Transactions 56 (2015) 18–27.
[48] Michel Fliess, Cédric Join, Hebertt Sira-Ramirez, Complex continuous nonlinear systems: their black box identification and their control, IFAC Proceedings Volumes 39 (1) (2006) 416–421.
[49] M. Fliess, C. Join, H. Sira-Ramirez, Non-linear estimation is easy, International Journal of Modelling, Identification and Control 4 (1) (2008) 12–27.
[50] H. Sira-Ramírez, C. García-Rodríguez, J. Cortés-Romero, A. Luviano-Juárez, Algebraic Identification and Estimation Methods in Feedback Control Systems, John Wiley & Sons, 2014.
[51] Michel Fliess, Hebertt Sira-Ramirez, Control via state estimations of some nonlinear systems, in: IFAC Symposium on Nonlinear Control Systems (NOLCOS 2004), 2004.
[52] Michel Fliess, Cédric Join, Intelligent PID controllers, in: 16th Mediterranean Conference on Control and Automation, 2008.
[53] Michel Fliess, Cédric Join, Commande sans modèle et commande à modèle restreint, e-STA Sciences et Technologies de l'Automatique 5 (4) (2008) 1–23.
[54] M. Fliess, C. Join, Model-free control, International Journal of Control 86 (12) (2013) 2228–2252.
[55] H. Sira-Ramírez, J. Linares-Flores, Alberto Luviano-Juarez, J. Cortés-Romero, Ultramodelos Globales y el Control por Rechazo Activo de Perturbaciones en Sistemas No lineales Diferencialmente Planos, Revista Iberoamericana de Automática e Informática Industrial RIAI 12 (2) (2015) 133–144.

[56] Michel Fliess, Model-free control and intelligent PID controllers: towards a possible trivialization of nonlinear control?, IFAC Proceedings Volumes 42 (10) (2009) 1531–1550.

[57] Michel Fliess, Cédric Join, Mamadou Mboup, Hebertt Sira-Ramirez, Vers une commande multivariable sans modèle, in: Actes Conf. Francoph. Automat. (CIFA 2006), Bordeaux, 2006.

[58] Mamadou Mboup, Cédric Join, Michel Fliess, Numerical differentiation with annihilators in noisy environment, Numerical Algorithms 50 (4) (2009) 439–467.

[59] Da-Yan Liu, Olivier Gibaru, Wilfrid Perruquetti, Error analysis of Jacobi derivative estimators for noisy signals, Numerical Algorithms 58 (1) (2011) 53–83.

[60] Da-yan Liu, Olivier Gibaru, Wilfrid Perruquetti, Differentiation by integration with Jacobi polynomials, Journal of Computational and Applied Mathematics 235 (9) (2011) 3015–3032.

[61] Hassane Abouaïssa, Michel Fliess, Cédric Join, On ramp metering: towards a better understanding of ALINEA via model-free control, International Journal of Control (2017), http://dx.doi.org/10.1080/00207179.2016.1193223.

[62] Lghani Menhour, Brigitte D'Andréa-Novel, Michel Fliess, Dominique Gruyer, Hugues Mounier, A new model-free design for vehicle control and its validation through an advanced simulation platform, in: Control Conference (ECC), 2015 European, IEEE, 2015, pp. 2114–2119.

[63] Brigitte D'Andréa-Novel, Clément Boussard, Michel Fliess, Oussama El Hamzaoui, Hugues Mounier, Bruno Steux, Commande sans modèle de la vitesse longitudinale d'un véhicule électrique, in: Sixième Conférence Internationale Francophone d'Automatique (CIFA 2010), 2010.

[64] Sungwoo Choi, Brigitte D'Andréa-Novel, Michel Fliess, Hugues Mounier, Jorge Villagra, Model-free control of automotive engine and brake for stop-and-go scenarios, in: Control Conference (ECC), 2009 European, IEEE, 2009, pp. 3622–3627.

[65] Jorge Villagrá, Vicente Milanés, Joshué Pérez, Teresa de Pedro, Control basado en PID inteligentes: Aplicación al control de crucero de un vehículo a bajas velocidades, Revista Iberoamericana de Automática e Informática Industrial RIAI 7 (4) (2010) 44–52.

[66] Cédric Join, John Masse, Michel Fliess, Etude préliminaire d'une commande sans modèle pour papillon de moteur (A model-free control for an engine throttle: a preliminary study), Journal Européen des Systèmes Automatisés (JESA) 42 (2–3) (2008) 337–354.

[67] Taghreed Mohammad Ridha, Claude Moog, Emmanuel Delaleau, Michel Fliess, Cédric Join, A variable reference trajectory for model-free glycemia regulation, in: SIAM Conference on Control & Its Applications (SIAM CT15), 2015.

[68] Frédéric Lafont, Jean-François Balmat, Nathalie Pessel, Michel Fliess, A model-free control strategy for an experimental greenhouse with an application to fault accommodation, Computers and Electronics in Agriculture 110 (2015) 139–149.

[69] Jérôme De Miras, Cédric Join, Michel Fliess, Samer Riachy, Stéphane Bonnet, Active magnetic bearing: a new step for model-free control, in: 52nd IEEE Conference on Decision and Control (CDC 2013), 2013, pp. 7449–7454.

[70] Jérôme De Miras, Samer Riachy, Michel Fliess, Cédric Join, Stéphane Bonnet, Vers une commande sans modèle d'un palier magnétique, in: 7e Conf. Internat. Francoph. Automatique, Grenoble, 2012; Available at http://hal.archives-ouvertes.fr/hal-00682762/en.

[71] Loïc Michel, Cédric Join, Michel Fliess, Pierre Sicard, Ahmed Chériti, Model-free control of dc/dc converters, in: 2010 IEEE 12th Workshop on Control and Modeling for Power Electronics (COMPEL), IEEE, 2010, pp. 1–8.

[72] Cédric Join, Gérard Robert, Michel Fliess, Model-free based water level control for hydroelectric power plants, in: 1st IFAC Conference on Control Methodologies and Technology for Energy Efficiency, IFAC Proceedings Volumes 43 (1) (2010) 134–139.

[73] Loïc Michel, Pilotage optimal des IGBT et commande sans-modèle des convertisseurs de puissance, Ph.D. thesis, Université du Québec à Trois-Rivières, 2012.

[74] Pierre-Antoine Gédouin, Emmanuel Delaleau, Jean-Matthieu Bourgeot, Cédric Join, Shab-nam Arbab Chirani, Sylvain Calloch, Experimental comparison of classical PID and model-free control: position control of a shape memory alloy active spring, Control Engineering Practice 19 (5) (2011) 433–441.

[75] Maxime Doublet, Cédric Join, Frédéric Hamelin, Model-free control for unknown delayed systems, in: Control and Fault-Tolerant Systems (SysTol), 2016 3rd Conference on, IEEE, 2016, pp. 630–635.

[76] Jorge Villagra, Carlos Balaguer, A model-free approach for accurate joint motion control in humanoid locomotion, International Journal of Humanoid Robotics 8 (01) (2011) 27–46.

[77] Jing Wang, Analyse et commande sans modèle de quadrotors avec comparaisons, Ph.D. thesis, Paris 11, 2013.

[78] Romain Bourdais, Michel Fliess, Wilfrid Perruquetti, Towards a model-free output tracking of switched nonlinear systems, IFAC Proceedings Volumes 40 (12) (2007) 504–509.

[79] Samer Riachy, Michel Fliess, High-order sliding modes and intelligent PID controllers: first steps toward a practical comparison, IFAC Proceedings Volumes 44 (1) (2011) 10982–10987.

[80] Richard H. Middleton, G.C. Goodwin, Digital Control and Estimation: A Unified Approach, Prentice Hall Professional Technical Reference, ISBN 0132116650, 1990.

[81] A. Feuer, G.C. Goodwin, Sampling in Digital Signal Processing and Control, Springer Science & Business Media, 1996.

Chapter 2

Generalities of ADRC

Chapter Points

- This chapter introduces the framework in which Active Disturbance Rejection control is to be developed for regulation or output reference trajectory tracking in nonlinear single- and multivariable perturbed systems.
- At the heart of the ADRC methodology lies the need for a flatness property or of a partially linearizable structure with stable zero dynamics (minimum phase systems). The exploitation of the flatness property leads to a vast simplification in the controller design efforts. Flatness is natural in the ADRC option and offers no obstacle to either observer-based designs or direct disturbance cancellation possibilities. A linear observer-based disturbance cancelling controller easily extends to a rather large class of nonlinear differentially flat systems, including systems with input delays, fractional-order controlled dynamic systems, discrete-time controlled systems, just to mention a few.
- This chapter places special attention to direct disturbance cancelling controllers not using disturbance observers. The surprising fact resides in the linearity of such controllers and their intimate relation to classical compensation networks. An important extension resides in the immersion of the technique in the sliding-mode control of nonlinear systems.
- The conceptual justification of ADRC lies in the realm of ultralocal disturbance models in the context of additive unknown but bounded disturbances. However, a technical justification is also possible in the context of Lyapunov stability theory-based observer and controller designs. Disturbance annihilation constitutes an alternative.

2.1 INTRODUCTION

Active Disturbance Rejection Control (ADRC) is fundamentally based on the possibility of online estimating unknown disturbance inputs affecting the plant behavior by means of suitable observers and proceed to cancelling them via an appropriate feedback control law using the gathered disturbance estimate. The feedback control law usually requires full knowledge of the state, but the lack of verification of this assumption is not removed from reality. The challenge then is to design a single-state observer that simultaneously estimates the unmeasured state variables of the system and the unknown disturbances. Naturally, the task at hand is quite ambitious if one strives for full generality of the plant

Active Disturbance Rejection Control of Dynamic Systems. http://dx.doi.org/10.1016/B978-0-12-849868-2.00002-2
Copyright © 2017 Elsevier Inc. All rights reserved.

description and the possibly assumed nonlinear nature of the state-dependent output maps. Moreover, the entire problem may be cast in a stochastic description of uncertainties, or more dramatically, the plant mathematical model may be subject to the presence of various types of time delays. A solution to the most general disturbance rejection problem is far from being solvable within the realm of present-day rigorous mathematical control theory approaches. Some compromises have to be naturally enforced, especially from the engineering viewpoint, which makes the problem tractable from a theoretical viewpoint and implementable from a practical standpoint.

We will address nonlinear SISO systems or MIMO systems, expressed in input–output form. In particular, we are interested in the class of SISO nonlinear systems that are feedback linearizable systems via static feedback whose linearizing output is available for measurement. Alternatively, we shall also consider MIMO systems that are feedback linearizable by means of static or dynamic feedback with measurable linearizing outputs. In one word, we are interested in the class of nonlinear differentially flat systems with measurable flat outputs. In the SISO case, the plant model is addressed as an input-to-flat output additively perturbed model. The disturbances are a lumped aggregation of exogenous (i.e., purely time-dependent) and endogenous (state-dependent) terms of unknown nature. In the MIMO case, the plant models will be usually simplified to a set of pure decoupled integration subsystems with additive lumped endogenous and exogenous disturbances. The nature of the disturbances will be additive, state-dependent, and totally unknown. Stochastic aspects will be usually avoided from the analysis, but the efforts will be geared toward actual laboratory implementation in prototype plants where the noise influences are unavoidable.

Differentially flat systems are rather common in engineering problems, and their theoretical treatment can be approach from either a highly sophisticated differential algebra approach of infinite-dimensional differential geometry of diffieties, or, alternatively, their study can be approached from a rather intuitive stand point. We adopt the latter in the interest of exploiting concrete application examples with relative ease and for the benefit of an engineering oriented audience.

We relegate to Chapter 4 the exploration of more complex descriptions of the given plant, such as the presence of input delays and other infinite-dimensional effects. In particular, we devote some attention to systems with the presence of fractional derivatives in the system dynamics.

In this chapter, we introduce ADRC in a tutorial fashion starting from the scratch by dealing with perturbed one-dimensional pure integration systems. At this stage, already a straightforward Lyapunov stability approach allows for a clear perspective on how to proceed in higher-dimensional systems. The treatment is elementary and with the intention of making the developments look simple enough. This is precisely a remarkable advantage of flatness, which will be systematically exploited later on. Slowly, we increase the complexity of the plant regarding the order of the system showing that we can envision the inherent simplicity of the ADRC approach throughout.

2.2 MAIN THEORETICAL ISSUES

2.2.1 Extended State Observers: A Tutorial Introduction

We introduce observer-based ADRC by examining, in a tutorial fashion, a simple example that will also serve as a basis for a desired generalization. Consider the perturbed system

$$\dot{y} = u + \xi(t) \tag{2.1}$$

where $\xi(t)$ is a uniformly absolutely bounded disturbance with, likewise, uniformly absolutely bounded first-order time derivative $\dot{\xi}(t)$. We, therefore, hypothesize the existence of finite constants K_0 and K_1 such that

$$\sup_t |\xi(t)| \le K_0, \quad \sup_t |\dot{\xi}(t)| \le K_1, \tag{2.2}$$

Consider, for $\lambda > 0$, the following disturbance observer for the given system:

$$\dot{\hat{y}} = u + z + \lambda(y - \hat{y}) \tag{2.3}$$

where z is an auxiliary variable that needs to be determined on the basis of the stability of the (redundant) output estimation error and of the disturbance estimation error. The disturbance estimation error is defined as $\eta = \xi(t) - z$. The output estimation error $e = y - \hat{y}$ then evolves according to

$$\dot{e} = -\lambda e + (\xi(t) - z) = -\lambda e + \eta \tag{2.4}$$

Let γ be a strictly positive constant parameter. Consider the following positive definite Lyapunov function candidate in the (e, η) space:

$$V(e, \eta) = \frac{1}{2}e^2 + \frac{1}{2\gamma}\eta^2 \tag{2.5}$$

The time derivative of the function $V(e, \eta)$ along the solutions of the error dynamics (2.4) has the form

$$\dot{V} = e\dot{e} + \frac{1}{\gamma}\eta\dot{\eta}$$

$$= -\lambda e^2 + \eta e + \frac{1}{\gamma}\eta(\dot{\xi}(t) - \dot{z})$$

$$= -\lambda e^2 + \eta\left[e + \frac{1}{\gamma}(\dot{\xi}(t) - \dot{z})\right] \tag{2.6}$$

It is clear that the (forbidden) choice

$$\dot{z} = \dot{\xi}(t) + \gamma e \tag{2.7}$$

leads to the negative semidefinite time derivative of the function V. Indeed, $\dot{V} = -\lambda e^2 \leq 0$. This means that V is bounded, or, expressed in a more customary fashion, $V < \infty$. This implies that the following integral is also finite:

$$-\lambda \lim_{t\to\infty} \int_0^t e^2(\sigma)d\sigma < \infty \tag{2.8}$$

In other words, e is square integrable and, necessarily, $e \to 0$, i.e., $\hat{y} \to y$. If such is the case, $\dot{z} \to \dot{\xi}(t)$ and $\dot{V} \to 0$. The set $\{(e, \eta) \mid \dot{V} = 0\}$ is given by $e = 0$. The observer trajectories asymptotically converge to $e = 0$ and to $\eta = z - \xi(t) = 0$.

Under these circumstances, the (injected) observation error dynamics satisfies the asymptotically exponentially stable dynamics:

$$\ddot{e} = -\lambda\dot{e} - \gamma e \tag{2.9}$$

for $\lambda, \gamma > 0$, as required above.

Unfortunately, $\xi(t)$ is unknown, and the previously chosen dynamics for z, is unfeasible.

The above development, nevertheless, motivates a modified proposal for the dynamics of z. We feasibly set

$$\dot{z} = \gamma e = \gamma(y - \hat{y}) \tag{2.10}$$

We then have the following suggested observer:

$$\dot{\hat{y}} = u + z + \lambda(y - \hat{y})$$
$$\dot{z} = \gamma(y - \hat{y}) \tag{2.11}$$

for the perturbed system. This observer constitutes what is known in the literature of disturbance observers as an Extended State Observer (ESO). The estimation error e system satisfies the following injected dynamics:

$$\dot{e} = \xi(t) - z - \lambda e$$
$$\dot{z} = \gamma e \tag{2.12}$$

i.e.,

$$\ddot{e} + \lambda \dot{e} + \gamma e = \dot{\xi}(t) \tag{2.13}$$

The injected estimation error dynamics is a perturbed linear dynamic system whose homogeneous (i.e., unperturbed) corresponding dynamics is globally asymptotically exponentially stable whenever the design parameters γ and λ are chosen to be strictly positive. Let ϵ be a small, strictly positive parameter intended to be used in a singular perturbation approach for the analysis of the stability of the high-gain perturbed observation error system. Set λ and γ to be of the form

$$\lambda = \frac{2\zeta \omega_n}{\epsilon}, \quad \gamma = \frac{\omega_n^2}{\epsilon^2} \tag{2.14}$$

where ζ and ω_n are strictly positive parameters denoting (when ϵ is set to be 1), respectively, a damping ratio and a natural frequency. This choice of the observer parameters in terms of ϵ makes the eigenvalues of the homogeneous linear estimation error system to be located on the following rays, containing the origin, in the left half of the complex plane:

$$s_{1,2} = -\frac{\zeta \omega_n}{\epsilon} \pm j\frac{\omega_n}{\epsilon}\sqrt{1 - \zeta^2}, \quad \epsilon > 0 \tag{2.15}$$

The above choice of λ and γ sets the injected dynamics to be:

$$\ddot{e} + \left(\frac{2\zeta \omega_n}{\epsilon}\right)\dot{e} + \left(\frac{\omega_n^2}{\epsilon^2}\right)e = \dot{\xi}(t) \tag{2.16}$$

Multiplying throughout by ϵ^2 and setting $\epsilon = 0$ yield the degenerate reduced dynamics as the equilibrium point $e = 0$. On the other hand, under the time scale transformation $d\tau = dt/\epsilon$, the formal setting of ϵ to zero yields the boundary layer dynamics:

$$\frac{d^2 e}{d\tau^2} + 2\zeta \omega_n \frac{de}{d\tau} + \omega_n^2 e = 0 \tag{2.17}$$

which is also globally asymptotically exponentially stable. According to a variant of Tikhonov's theorem, the perturbed system trajectories will be eventually found after a finite time t_ϵ, dwelling uniformly in an arbitrary small neighborhood of the origin of the $(e, \dot{e})$-space. This neighborhood, for a given small desired bound on $|\eta|$, becomes smaller as ϵ is chosen to be smaller.

In terms of a Lyapunov stability analysis, used previously, we prescribe a Lyapunov function candidate $V(e, \eta)$ as defined in (2.5). We obtain the following time derivative for V along the injected dynamics, with the new feasible choice for the z dynamics:

$$\dot{V} = -\lambda e^2 + \frac{(\xi(t) - z)\dot{\xi}(t)}{\gamma} \leq -\lambda e^2 + \frac{|(\xi - z)||\dot{\xi}(t)|}{\gamma}$$

$$\leq -\lambda e^2 + \frac{|(\xi - z)|K_1}{\gamma} \qquad (2.18)$$

Let $\delta_2 > 0$ be a positive scalar quantity. Consider the set $|\eta| \leq \delta_2$. Outside the set,

$$B = \{(e, \eta) \mid e^2 = \frac{\delta_2 K_1}{\gamma \lambda}, \;\; |\eta| \leq \delta_2\} \qquad (2.19)$$

the time derivative of V is strictly negative, whereas the sign of $\dot{V}$ is not definite inside the set B. We have the following result: Let ζ and ω_n be strictly positive constants. Given arbitrarily small positive constants $\delta_1 > 0$ and $\delta_2 > 0$, there exist $\epsilon > 0$ and t_h such that, for all $t > t_h$, the trajectories of the perturbed injected dynamics

$$\ddot{e} + \left(\frac{2\zeta \omega_n}{\epsilon}\right)\dot{e} + \left(\frac{\omega_n^2}{\epsilon^2}\right)e = \dot{\xi}(t) \qquad (2.20)$$

uniformly satisfies, for all $t > t_\epsilon$, $|e(t)| < \delta_1$ and $|\eta(t)| = |\xi(t) - z(t)| < \delta_2$, provided that

$$\epsilon < \sqrt[3]{\frac{2\delta_1^2 \zeta \omega_n^3}{\delta_2 K_1}} \qquad (2.21)$$

2.2.2 A Second Order Extension of the Disturbance Observer: GPI Observers

There are several ways to extend the above result. One of them is to further extend the dimension of the disturbance observer in order to possibly include an approximate estimate option for the first-order time derivative of the disturbance input signal $\xi(t)$.

Consider again the perturbed system (2.1) along with an additional uniform absolute boundedness assumption regarding the second-order time derivative of the perturbation input, i.e., suppose

$$\sup_t |\xi^{(i)}(t)| = K_i, \;\; i = 0, 1, 2. \qquad (2.22)$$

Then, consider the following extended observer for the perturbed first-order system:

$$\dot{y}_0 = u + y_1 + \lambda_2(y - y_0)$$

$$\dot{y}_1 = y_2 + \lambda_1(y - y_0)$$
$$\dot{y}_2 = \lambda_0(y - y_0) \tag{2.23}$$

The observation error system is given by

$$\dot{e}_0 = \xi(t) - y_1 - \lambda_2 e_0$$
$$\dot{y}_1 = y_2 + \lambda_1 e_0$$
$$\dot{y}_2 = \lambda_0 e_0 \tag{2.24}$$

where e_0 stands for the redundant observation error of the output, i.e., $e_0 = y - y_0$. The variable e_0 satisfies the following perturbed linear dynamics:

$$e_0^{(3)} + \lambda_2 \ddot{e}_0 + \lambda_1 \dot{e}_0 + \lambda_0 e_0 = \ddot{\xi}(t) \tag{2.25}$$

The choice of the observer design parameter set $\{\lambda_0, \lambda_1, \lambda_2\}$ may be settled in accordance with the characteristic polynomial of a desired asymptotically exponentially stable, unperturbed injected observation error dynamics. For suitable strictly positive parameters ζ, ω_n, and p, this polynomial may be of the form

$$p(s) = \left(s^2 + \left(\frac{2\zeta\omega_n}{\epsilon}\right)s + \left(\frac{\omega_n^2}{\epsilon^2}\right)\right)\left(s + \frac{p}{\epsilon}\right)$$
$$= s^3 + \left(\frac{2\zeta\omega_n + p}{\epsilon}\right)s^2 + \left(\frac{2\zeta\omega_n p + \omega_n^2}{\epsilon^2}\right)s + \left(\frac{\omega_n^2 p}{\epsilon^3}\right) \tag{2.26}$$

Hence, we choose

$$\lambda_2 = \frac{2\zeta\omega_n + p}{\epsilon}, \quad \lambda_1 = \frac{2\zeta\omega_n p + \omega_n^2}{\epsilon^2}, \quad \lambda_0 = \frac{\omega_n^2 p}{\epsilon^3} \tag{2.27}$$

Similarly to the previous case, given an arbitrarily small ultimate bound for the supremum of the absolute value of the estimation error phase vector $(e_0, \dot{e}_0, \ddot{e}_0)$, and for a given desirable ultimate bound on the disturbance estimation errors $\eta_1 = \xi(t) - y_1$, $\eta_2 = \dot{\xi}(t) - y_2$, there exist a sufficiently small value of ϵ and a finite time $t_\epsilon > 0$ after which, the perturbed response of the injected observation error dynamics (2.25) remains absolutely uniformly bounded, in an asymptotic exponentially dominated manner, by the given small neighborhood around the origin of the estimation error phase space. This neighborhood is determined by ϵ and the a priori bounds K_i, $i = 0, 1, 2$, on the corresponding time derivatives of the disturbance signal.

Notice that the observer (2.23) can be rewritten as

$$\dot{y}_0 = u + \lambda_2 e_0 + \lambda_1 \int_0^t e_0(\sigma_1)d\sigma_1 + \lambda_0 \int_0^t \int_0^{\sigma_1} e_0(\sigma_2)d\sigma_2 d\sigma_1 \tag{2.28}$$

and the corresponding estimation error dynamics for e_0, obtained by subtracting the perturbed plant dynamics from the observer expression (2.28), is given by

$$\dot{e}_0 = -\lambda_2 e_0 - \lambda_1 \int_0^t e_0(\sigma_1)d\sigma_1 - \lambda_0 \int_0^t \int_0^{\sigma_1} e_0(\sigma_2)d\sigma_2 d\sigma_1 + \xi(t) \quad (2.29)$$

Independently of the system order, the possibility of dynamically extending the traditional observer by means of a suitable linear combination of nested output error integrations results in the possibility of estimating a certain number of time derivatives of the disturbance signal. In linear control theory, controllers involving suitable linear combinations of nested integrations of output tracking, or output stabilization errors, receive the name of Generalized Proportional Integral (GPI) controllers. Injected error dynamics containing similar linear combinations of multiple integrals of the estimation error will be, accordingly, addressed as GPI observers (see Appendix B).

2.2.3 The mth-Order GPI Observer for an Elementary Perturbed Plant

Let $(\int^{(j)} \phi)$ denote the following iterated integral:

$$\left(\int^{(j)} \phi\right) = \int_0^t \int_0^{\sigma_1} \cdots \int_0^{\sigma_{j-1}} \phi(\sigma_j) d\sigma_j \cdots d\sigma_1 \quad (2.30)$$

with $(\int^{(0)} \phi) = \phi$.

For the elementary linear perturbed plant $\dot{y} = u + \xi(t)$, an mth-order extended observer of the form

$$\dot{y}_0 = u + \sum_{i=0}^{m} \lambda_i \left(\int^{(m-i)} (y - y_0)\right) \quad (2.31)$$

yields an estimation error dynamics ($e_0 = y - y_0$) given by the perturbed integro-differential equation

$$\dot{e}_0 = -\sum_{i=0}^{m} \lambda_i \left(\int^{(m-i)} (y - y_0)\right) + \xi(t) \quad (2.32)$$

It is not difficult to show that this equation is equivalent to the perturbed estimation error linear dynamics

$$\dot{e}_0 = \xi(t) - y_1 - \lambda_m (y - y_0)$$
$$\dot{y}_1 = y_2 + \lambda_{m-1}(y - y_0)$$

$$\vdots$$

$$\dot{y}_{m-1} = y_m + \lambda_1(y - y_0)$$
$$\dot{y}_m = \lambda_0(y - y_0) \tag{2.33}$$

or, in an input–output form,

$$e_0^{(m+1)} + \lambda_m e_0^{(m)} + \cdots + \lambda_1 \dot{e}_0 + \lambda_0 e_0 = \xi^{(m)}(t) \tag{2.34}$$

As the order m of the observer extension is increased, to guarantee a uniformly asymptotically exponentially dominated decay of the vector norm of estimation error phase variables, we must assume that the perturbation input signal $\xi(t)$ exhibits uniformly absolutely bounded higher-order time derivatives. These must exhibit an order compatible with the order of the observer extension. The higher the extension order, the higher the number of time derivatives of the disturbance signal that may be estimated. Clearly, this may be achieved for sufficiently high observer gains guaranteeing injected observer eigenvalues located sufficiently far into the left half of the complex plane.

A polynomial in the complex variable s with real coefficients $p(s) = s^{m+1} + \sum_{j=0}^{m} \lambda_j s^j$ is said to be Hurwitz if the roots of the polynomial are all located in the open left half of the complex plane. In such a case, one usually states that the set of coefficients $\{\lambda_0, \lambda_1, \ldots, \lambda_m\}$ is a set of Hurwitz coefficients.

Let $\{\lambda_0, \lambda_1, \cdots, \lambda_m\}$ be a set of Hurwitz coefficients. We may prescribe the following injected dynamics:

$$e_0^{(m+1)} + \frac{\lambda_m}{\epsilon} e_0^{(m)} + \cdots + \frac{\lambda_1}{\epsilon^m} \dot{e}_0 + \frac{\lambda_0}{\epsilon^{m+1}} e_0 = \xi^{(m)}(t) \tag{2.35}$$

The boundedness assumption on $\sup_t |\xi^{(m-1)}(t)|$ guarantees, via a singular perturbation analysis, the uniform asymptotic exponentially dominated convergence of the trajectories of the vector of estimation error phase variables $(e_0(t), \dot{e}_0(t), \ldots, e^{(m)}(t))$ toward the interior of a small (as small as desired) ϵ-dependent neighborhood centered around the origin of the estimation error phase space. This, in turn, guarantees sufficiently close estimates of the time derivatives of the disturbance input signal $\xi(t)$.

2.2.4 Simultaneously Estimating the Phase Variables, the Disturbance Input Signal, and Its Successive Time Derivatives

Continuing with our tutorial approach to disturbance observer-based ADRC, let us now consider the following nth-order pure integration perturbed system:

$$y^{(n)} = u + \xi(t) \tag{2.36}$$

The canonical form of the perturbed linear plant is simply expressed as

$$\frac{d}{dt} \begin{bmatrix} y_0 \\ y_1 \\ \vdots \\ y_{n-1} \end{bmatrix} = A \begin{bmatrix} y_0 \\ y_1 \\ \vdots \\ y_{n-1} \end{bmatrix} + b(u + \xi(t)), \quad y = c^T \begin{bmatrix} y_0 \\ y_1 \\ \vdots \\ y_{n-1} \end{bmatrix} \quad (2.37)$$

with A, b, and c^T given by

$$A = \begin{bmatrix} 0 & 1 & 0 & \cdots & 0 \\ 0 & 0 & 1 & \cdots & 0 \\ \vdots & \vdots & \vdots & \ddots & \vdots \\ 0 & 0 & 0 & \cdots & 1 \\ 0 & 0 & 0 & \cdots & 0 \end{bmatrix}, \quad b = \begin{bmatrix} 0 \\ 0 \\ \vdots \\ 0 \\ 1 \end{bmatrix}, \quad c^T = [1 \; 0 \; \cdots \; 0 \; 0] \quad (2.38)$$

We prescribe the following mth-order extended state observer (GPI observer):

$$\frac{d}{dt} \hat{y}_0 = \hat{y}_1 + \lambda_{m+n-1}(y_0 - \hat{y}_0)$$

$$\frac{d}{dt} \hat{y}_1 = \hat{y}_2 + \lambda_{m+n-2}(y_0 - \hat{y}_0)$$

$$\vdots$$

$$\frac{d}{dt} \hat{y}_{n-1} = u + z_1 + \lambda_m(y_0 - \hat{y}_0)$$

$$\dot{z}_1 = z_2 + \lambda_{m-1}(y_0 - \hat{y}_0)$$

$$\dot{z}_2 = z_3 + \lambda_{m-2}(y_0 - \hat{y}_0)$$

$$\vdots$$

$$\dot{z}_{m-1} = z_m + \lambda_1(y_0 - \hat{y}_0)$$

$$\dot{z}_m = \lambda_0(y_0 - \hat{y}_0) \quad (2.39)$$

The estimation error injected dynamics is given by

$$\frac{d}{dt} e_0 = e_1 - \lambda_{m+n-1} e_0$$

$$\frac{d}{dt} e_1 = e_2 - \lambda_{m+n-2} e_0$$

$$\vdots$$

$$\frac{d}{dt}e_{n-1} = \xi(t) - z_1 - \lambda_m e_0$$

$$\dot{z}_1 = z_2 + \lambda_{m-1} e_0$$

$$\dot{z}_2 = z_3 + \lambda_{m-2} e_0$$

$$\vdots$$

$$\dot{z}_{m-1} = z_m + \lambda_1 e_0$$

$$\dot{z}_m = \lambda_0 e_0 \tag{2.40}$$

Clearly, the dynamics of the estimation error $e_0 = y - \hat{y}_0$ satisfies the linear perturbed differential equation

$$e_0^{(m+n)} + \lambda_{m+n-1} e_0^{(m+n-1)} + \cdots + \lambda_1 \dot{e}_0 + \lambda_0 e_0 = \xi^{(m)} \tag{2.41}$$

Since for a uniformly absolutely bounded mth-order time derivative of the disturbance signal, the trajectories of the observation error and its associated phase variables $(e_0(t), \ldots e_{n-1}(t))$ can be made to ultimately converge, in a finite time, to the interior of an arbitrarily small neighborhood centered around the origin of the observation error phase space, from the equation

$$\frac{d}{dt}e_{n-1} = \xi(t) - z_1 - \lambda_m e_0 \tag{2.42}$$

it follows that the trajectory of $z_1(t)$ will represent a sufficiently close estimate of the disturbance signal trajectory $\xi(t)$. Hence, it follows that $z_2(t) \approx \dot{z}_1(t)$, i.e., $z_2(t)$ will approximate the first-order time derivative of $\xi(t)$ and, in general, $z_i(t)$, $i = 1, 2, \ldots, m-1$, will represent, respectively, close online estimates of the trajectories of $\xi^{(i-1)}(t)$.

The possibility of estimating time derivatives of the unknown disturbance signal $\xi(t)$ will be especially useful in disturbance prediction schemes. These are naturally associated with the ADRC of systems exhibiting fixed known input delays (see Chapter 4).

Example 1. Consider the following exogenously perturbed actuated pendulum system:

$$\ddot{\theta} = -\frac{g}{L}\sin\theta + \zeta(t) + u =: \xi(t) + u \tag{2.43}$$

where $\zeta(t)$ and its first-order time derivative are uniformly absolutely bounded, and the nonlinear state-dependent gravitational term $-(g/L)\sin\theta$ is considered to be a certain function of time, $-(g/L)\sin\theta(t)$, which is, evidently, uniformly absolutely bounded. Let $\theta = \theta_1$ and $\dot{\theta} = \theta_2$, and let their corresponding estimates be denoted, respectively, as $\hat{\theta}_1$ and $\hat{\theta}_2$. We also define $\xi(t) =$

$-(g/L)\sin\theta(t) + \zeta(t)$ and address the term $-(g/L)\sin\theta(t)$ as the endogenous part of the disturbance and $\zeta(t)$ as the exogenous part of the disturbance.

Consider the following ESO:

$$\frac{d}{dt}\hat{\theta}_1 = \hat{\theta}_2 + \lambda_2(\theta - \hat{\theta}_1)$$

$$\frac{d}{dt}\hat{\theta}_2 = u + z + \lambda_1(\theta - \hat{\theta}_1)$$

$$\frac{d}{dt}z = \lambda_0(\theta - \hat{\theta}_1) \tag{2.44}$$

Here, the variable z is supposed to estimate the total disturbance input to the system, constituted by the sum of the endogenous perturbation input $-(g/L)\sin\theta(t)$ and the exogenous perturbation input $\zeta(t)$.

The estimation error $e = \theta - \hat{\theta}$ evolves according to the linear perturbed dynamics:

$$e^{(3)} + \lambda_2\ddot{e} + \lambda_1\dot{e} + \lambda_0 e = \dot{\xi}(t) \tag{2.45}$$

Set λ_2, λ_1, and λ_0 to adopt the following strictly positive values:

$$\lambda_2 = \frac{1}{\epsilon}(p_o + 2\zeta_o\omega_{no}), \quad \lambda_1 = \frac{1}{\epsilon^2}(2p_o\zeta_o\omega_{no} + \omega_{no}^2), \quad \lambda_0 = \frac{1}{\epsilon^3}\omega_{no}^2 p_o \tag{2.46}$$

with $\zeta_o, \omega_{no}, p_o > 0$ and ϵ being a small parameter bestowing a high-gain character to the ESO design.

We used the following simulation data:

$$\zeta(t) = 0.5\left(1 + e^{-\sin(3t)\sin(t)}\right)\cos(0.5t), \quad m = 1.0 \text{ [kg]},$$

$$g = 9.8 \text{ [m/s}^2], \quad L = 0.7 \text{ [m]},$$

$$\zeta_o = 1, \quad \omega_{no} = 2, \quad p_o = 2$$

Fig. 2.1 depicts the performance of the ESO under open-loop control conditions ($u = 0$ for $t < 2$, $u = 1$ for $t > 2$) for various values of the design parameter ϵ. The redundant estimate of the angular position and the estimate of the angular velocity are indistinguishable for the three used values of $\epsilon = 0.01, 0.005, 0.001$. The bottom of the figure shows the value of the joint integral square error ($ISE(t)$) of the total disturbance estimate, the redundant angular position estimate, and the angular velocity estimate:

$$ISE(t) = \int_0^t \left[(\theta(\sigma) - \hat{\theta}_1(\sigma))^2 + (\dot{\theta}(\sigma) - \hat{\theta}_2(\sigma))^2 + (z(\sigma) - \xi(\sigma))^2\right]d\sigma \tag{2.47}$$

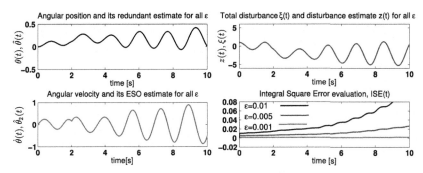

FIGURE 2.1 Evaluation of high-gain Extended State Observer performance via Integral Square Estimation Errors.[1]

The $ISE(t)$ performance index rapidly grows for higher values of ϵ (for $\epsilon = 0.01$ and $\epsilon = 0.005$), and it is nearly constant for $\epsilon = 0.001$ (indicating the existence of a nearly constant estimation error, very close to zero, with negligible growth).

Example 2. A second simulation example depicts a totally different nature of the total disturbance. Consider the following nonlinear controlled spring damper system, with horizontal displacement denoted by x, exhibiting significant position and velocity dependent nonlinearities:

$$\ddot{x} = -x^2(\dot{x})^3 - x^7 + u \tag{2.48}$$

We regard the endogenous terms $-x^2(t)(\dot{x})^3(t) - x^7(t)$ as the total disturbance affecting the system. The simplified system to be considered is the following:

$$\ddot{x} = \xi(t) + u \tag{2.49}$$

Define $x_1 = x$ and $x_2 = \dot{x}$. The ESO observer is given, in this case, by

$$\dot{\hat{x}}_1 = \hat{x}_2 + \lambda_2(x_1 - \hat{x}_1)$$
$$\dot{\hat{x}}_2 = z + u + \lambda_1(x_1 - \hat{x}_1)$$
$$\dot{z} = \lambda_0(x_1 - \hat{x}_1)$$

Using the same set of design parameters for the ESO, as in the previous example, we obtain the following performance of the ESO observer under the same open-loop conditions. The $ISE(t)$ performance index behaves in a remarkable analogous manner as in the previous example. See Fig. 2.2.

---

1. For interpretation of the references to color in this and the following figures, the reader is referred to the web version of this chapter.

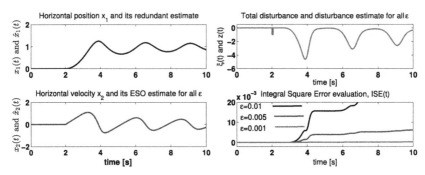

FIGURE 2.2 Evaluation of high-gain Extended State Observer performance via Integral Square Estimation Errors.

The above simulation examples depict a crucial fact on which observer-based ADRC control schemes work: State-dependent, unknown, nonlinearities, and exogenous disturbances (and unmodeled dynamics effects) can all be treated in the same footing as unstructured disturbances. Estimation of the effects of such disturbances can be efficiently carried out via a high-gain Extended State Observer on the basis of a substantially simplified (pure integration) system.

2.2.5 The ADRC Scheme for Pure Integration Systems

In reference to the system

$$y^{(n)} = u + \xi(t), \tag{2.50}$$

the perturbed nth-order pure integration system admits a high-gain GPI observer, which arbitrarily closely estimates the phase variables associated with the output $(y, \dot{y}, \ldots, y^{(n-1)})$ and a certain number of time derivatives of the disturbance signal $\xi(t)$, including the disturbance signal itself. The next step simply consists in prescribing a feedback control law that cancels the disturbance on the basis of its online estimate and imposes a (perturbed) closed-loop dynamics, which approximately solves a stabilization or an output reference trajectory tracking task. We assume that an output reference trajectory tracking task is at hand for a given smooth reference trajectory $y^*(t)$.

The feedback controller, for a smooth output reference trajectory tracking task in which $y \to y^*(t)$ represents the desired objective, is given by

$$u = [y^*(t)]^{(n)} - z_1 - \sum_{j=0}^{n-1} \kappa_j(\hat{y}_j - [y^*(t)]^{(j)}) \tag{2.51}$$

where

$$\frac{d}{dt}\hat{y}_0 = \hat{y}_1 + \lambda_{m+n-1}(y_0 - \hat{y}_0)$$

$$\frac{d}{dt}\hat{y}_1 = \hat{y}_2 + \lambda_{m+n-2}(y_0 - \hat{y}_0)$$

$$\vdots$$

$$\frac{d}{dt}\hat{y}_{n-1} = u + z_1 + \lambda_m(y_0 - \hat{y}_0)$$

$$\dot{z}_1 = z_2 + \lambda_{m-1}(y_0 - \hat{y}_0)$$

$$\dot{z}_2 = z_3 + \lambda_{m-2}(y_0 - \hat{y}_0)$$

$$\vdots$$

$$\dot{z}_{m-1} = z_m + \lambda_1(y_0 - \hat{y}_0)$$

$$\dot{z}_m = \lambda_0(y_0 - \hat{y}_0) \tag{2.52}$$

Define the output tracking error as $e_y = y - y^*(t)$. Define, this time, the estimation error by, $e_o = y - \hat{y}_0$. We use $e_o^{(j)} = y^{(j)} - \hat{y}_j$.

Clearly, the estimation error dynamics for e_o is given, as before, by

$$e_o^{(m+n)} + \lambda_{m+n-1}e_o^{(m+n-1)} + \cdots + \lambda_1\dot{e}_o + \lambda_0 e_o = \xi^{(m)} \tag{2.53}$$

After substituting $\hat{y}_j$ into (2.51) by $y^{(j)} - e_o^{(j)}$, the tracking error $e_y = y - y^*(t)$ evolves satisfying the following closed-loop dynamics:

$$e_y^{(n)} + \kappa_{n-1}e_y^{(n-1)} + \cdots + \kappa_0 e_y = \xi(t) - z_1 + \sum_{j=0}^{n-1} \kappa_j e_o^{(j)} \tag{2.54}$$

The set of coefficients $\{\kappa_0, \kappa_1, \cdots, \kappa_{n-1}\}$ is chosen to be a Hurwitz set. The disturbance term $\xi - z_1$ is arbitrarily small, and so are the terms $e_o^{(j)}$ for all j. Then, the right-hand side of (2.54) is a small perturbation term affecting an exponentially asymptotically stable dynamics. The net result is that there is no need to choose high gains for the linear feedback controller (2.51). We state that such gains need only be moderately high.

2.2.6 Generalized Proportional Integral Control

In a completely dual manner to the extended observer-based ADRC scheme, we now pose the observer-less approach for robust control of linear perturbed systems.

Consider, in the spirit of a tutorial introduction, the elementary linear perturbed scalar system

$$\dot{y} = u + \xi(t) \tag{2.55}$$

with $\xi(t)$ an unknown but uniformly absolutely bounded function of time with uniformly absolutely bounded first-order time derivative $\dot{\xi}(t)$. As before, we hypothesize that $\sup_t |\xi(t)| \le K_0$ and $\sup_t |\dot{\xi}(t)| \le K_1$. Suppose that we desire to track a given smooth reference signal $y^*(t)$. Let z be an auxiliary variable destined to online compensate for the unknown disturbance on the basis of satisfying some dynamics yet to be determined. We propose

$$u = \dot{y}^*(t) - k_1(y - y^*(t)) - z \tag{2.56}$$

The partially closed-loop system then reads as

$$\dot{e}_y = -k_1 e_y + (\xi(t) - z) \tag{2.57}$$

with $e_y = y - y^*(t)$ being the reference trajectory tracking error and k_1 a strictly positive scalar gain.

A Lyapunov function candidate, defined in the (e_y, η) space, with $\eta = \xi(t) - z$ is given by

$$V(e_y, \eta) = \frac{1}{2} e_y^2 + \frac{1}{\gamma} \eta^2 \tag{2.58}$$

The time derivative of V is obtained as

$$\begin{aligned} \dot{V}(e_y, \eta) &= e_y(-k_1 e_y + \eta) + \frac{1}{\gamma} \eta(\dot{\xi}(t) - \dot{z}) \\ &= -k_1 e_y^2 + \eta(e_y + \frac{1}{\gamma}(\dot{\xi}(t) - \dot{z})) \end{aligned} \tag{2.59}$$

Clearly, the choice $\dot{z} = \dot{\xi}(t) + \gamma e_y$ yields $\dot{V} = -k_1 e_y^2 \le 0$, thus implying that V is bounded and that e_y is square integrable and thus inducing the asymptotic convergence of e_y to zero, i.e., $e_y \to 0$. Then $\dot{z} \to \dot{\xi}(t)$. Letting $\gamma = k_0$, the proposed feedback controller adopts the form

$$u = \dot{y}^*(t) - k_1(y - y^*(t)) - k_0 \int_0^t (y - y^*(\sigma))d\sigma - \xi(t) \tag{2.60}$$

i.e., the controller of the perturbed first-order plant is none other than a classical proportional integral controller with an exact cancellation of the unknown disturbance.

However, due to the lack of information about $\xi(t)$, the variable z can not be chosen to obey

$$\dot{z} = \dot{\xi}(t) - k_0 e_y \qquad (2.61)$$

and, at most, we may propose the following closely related controller:

$$\dot{z} = -k_0 e_y, \quad \text{i.e.} \quad z = -k_0 \int_0^t e_y(\sigma) d\sigma \qquad (2.62)$$

This new controller choice yields the following closed-loop system:

$$\dot{e}_y = -k_1 e_y - k_0 \int_0^t e_y(\sigma) d\sigma + \xi(t) \qquad (2.63)$$

which is equivalent to the system

$$\ddot{e}_y + k_1 \dot{e}_y + k_0 e_y = \dot{\xi}(t) \qquad (2.64)$$

The closed-loop system is constituted by a perturbed asymptotically exponentially stable dynamics. The situation is identical to that of the disturbance observer exposed in the previous subsection. The choice of the controller gains k_0 and k_1 as high gains of the form

$$k_1 = \frac{2\zeta \omega_n}{\epsilon}, \quad k_0 = \frac{\omega_n^2}{\epsilon^2} \qquad (2.65)$$

with ϵ a strictly positive, sufficiently small parameter, and $\zeta, \omega_n > 0$, yields a tracking error closed-loop response of the perturbed system whose trajectory can be made to stay, ultimately, arbitrarily close to zero, uniformly in time, after some finite time t_ϵ.

2.2.7 An Iterated Tracking Error Integral Compensation

Consider again the perturbed first-order system

$$\dot{y} = u + \xi(t) \qquad (2.66)$$

with the following feedback controller:

$$u = \dot{y}^*(t) - k_2(y - y^*(t)) - k_1 \int_0^t (y - y^*(\sigma_1)) d\sigma_1$$

$$- k_0 \int_0^t \int_0^{\sigma_1} (y - y^*(\sigma_2)) d\sigma_2 d\sigma_1 \qquad (2.67)$$

The closed-loop output tracking error $e_y = y - y^*(t)$ is given by

$$\dot{e}_y = -k_2 e_y - \rho_1 + \xi(t)$$
$$\dot{\rho}_1 = \rho_2 + k_1 e_y$$
$$\dot{\rho}_2 = k_0 e_y \qquad (2.68)$$

or by

$$e_y^{(3)} + k_2 \ddot{e}_y + k_1 \dot{e}_y + k_0 e_y = \ddot{\xi}(t) \qquad (2.69)$$

Here, we are precisely in the same situation depicted by equation (2.25) for the observation error of the extended GPI disturbance observer case. For a disturbance signal $\xi(t)$, exhibiting a uniformly absolutely bounded second-order time derivative, we will arrive exactly at the same result but for the tracking error e_y, instead of the observation error e_0.

For uniformly absolutely bounded disturbances with similarly bounded time derivatives, a high-gain controller guarantees the uniform exponentially dominated asymptotic convergence of the tracking error phase trajectories to a small neighborhood of the tracking error phase space. The farther to the left of the complex axis the roots of the characteristic polynomial are located, the smaller the neighborhood around the origin. The result is termed to be dual to the observer based ADRC, but, as in the previous integral controller case, it evades the need for using disturbance observers.

The above controller is called a GPI controller, and its robustness with respect to unstructured disturbances follows from the same results obtained for the GPI observer. In this elementary case, we can similarly use an mth-order iterated tracking error integrals conformed in a suitable linear combination.

2.2.8 An mth-Order Iterated Tracking Error Integral Compensation

For the first-order perturbed system, the controller

$$u = \dot{y}^*(t) - \sum_{j=0}^{m} k_{m-j} \left(\int^{(j)} (y - y^*(t)) \right) \qquad (2.70)$$

yields the closed-loop system

$$\dot{e}_y = -k_m e_y - \rho_1 + \xi(t)$$
$$\dot{\rho}_1 = \rho_2 + k_{m-1} e_y$$
$$\vdots$$
$$\dot{\rho}_{m-2} = \rho_{m-1} + k_2 e_y$$

$$\dot{\rho}_{m-1} = \rho_m + k_1 e_y$$
$$\dot{\rho}_m = k_0 e_y \tag{2.71}$$

which is equivalent to

$$e_y^{(m+1)} + k_m e_y^{(m)} + \cdots + k_2 \ddot{e}_y + k_1 \dot{e}_y + k_0 e_y = \xi^{(m)}(t) \tag{2.72}$$

The choice of suitable high-gain design coefficients drives the output reference tracking error e_y and its time derivatives to a small (as small as desired) neighborhood of the origin of the tracking error phase space, provided that the mth-order time derivative of the disturbance input $\xi(t)$ is uniformly absolutely bounded. Notice that this is precisely the case of equation (2.34) for the mth-order GPI extended observer in the dual disturbance estimation problem for the first-order case.

2.3 THE NEED FOR FLATNESS-BASED ADRC DESIGNS

Flatness, or differential flatness, is a property that generalizes the concept of linear controllability to the nonlinear systems case. In the linear time-invariant and time-varying cases, it entirely coincides with the controllability concept. In the nonlinear single-input system case, it is equivalent to feedback linearizability by means of state and input coordinate transformations. In the nonlinear multivariable case, flatness identifies those systems that are either statically or dynamically feedback linearizable. The concept of flatness is extendable to some interesting classes of infinite-dimensional systems, such as systems with finite input and state delays, systems exhibiting fractional derivatives, and systems described by linear partial differential equations controlled from the boundary.

Flatness trivializes the system structure, under state transformation and partial feedback, to that of a pure integration system.

Example 3. Consider the following inertia-rotational spring system with externally applied torque u:

$$J_1 \ddot{\theta}_1 = -k(\theta_1 - \theta_2) + u$$
$$J_2 \ddot{\theta}_2 = -k(\theta_2 - \theta_1) \tag{2.73}$$

The system is flat, with flat output given by the angular displacement of the second inertia θ_2. Indeed, all system variables are parameterizable in terms of this special output and its time derivatives, up to finite order:

$$\theta_1 = \frac{J_2}{k} \ddot{\theta}_2 + \theta_2$$

$$u = \frac{J_1 J_2}{k} \theta_2^{(4)} + (1 + J_2) \ddot{\theta}_2 \tag{2.74}$$

Notice that we can in principle bring θ_1 to a complete rest by means of a simple enough ESO-based ADRC stabilizing controller, such as

$$u = -z_1 - J_1 \left(2\zeta\omega_n \hat{\dot{\theta}}_1 + \omega_n^2 \theta_1 \right)$$

$$\frac{d}{dt}\hat{\theta}_1 = \hat{\dot{\theta}}_1 + \lambda_2(\theta_1 - \hat{\theta}_1)$$

$$\frac{d}{dt}\hat{\dot{\theta}}_1 = u + z_1 + \lambda_1(\theta_1 - \hat{\theta}_1)$$

$$\frac{d}{dt}z_1 = \lambda_0(\theta_1 - \hat{\theta}_1) \tag{2.75}$$

The scheme leaves uncontrolled an oscillatory zero dynamics governed by

$$\ddot{\theta}_2 = -\left(\frac{k}{J_2} \right)\theta_2 \tag{2.76}$$

which depends on the values of the initial conditions.

On the other hand, the input-to-flat output model has the same order of the plant, and hence the flat output does not have any zero dynamics.

2.3.1 The Nonlinear Flat Case

The main motivation for introducing either extended observer-based ADRC control schemes or direct GPI controllers for pure integration systems of any order rests in the relevance of such a simple plant dynamics in connection with nth-order differentially flat nonlinear systems. For SISO flat systems with measurable flat output, the system is known to be equivalent, after a state and input coordinate transformation, to an nth-order linear controllable system whose input–output canonical description generally allows one to solve (modulo input gain singularities associated to the nonlinear plant dynamics) for the highest time derivative of the flat output (i.e., $y^{(n)}$) in terms of a linear combination of the first $n-1$ time derivatives of the flat output (flat output phase variables) and of the control input multiplied by some nonzero constant gain.

As customary in ADRC, the phase variable-dependent terms of the linear input–output description of the feedback linearized system can be regarded as unknown endogenous disturbances. The scaling of the presumably known constant input gain allows one to deal with the control of the nonlinear system in terms of the simplified dynamics

$$y^{(n)} = u + \xi(t)$$

This is one of the major simplifications of the flatness-based ADRC control scheme and one that has been proven to efficiently work in a large variety of physical situations, as it is demonstrated in subsequent chapters of this book and in the many bibliographical references that contain closely related expositions and engineering application case studies.

2.3.2 The nth-Order Integration Plant

Pure integration plants of second and higher orders require structural phase variable reconstruction based on suitable iterated integrations of linear combinations of the control input and of the output. This procedure is pursued, in all detail, in Appendix B of this book, in a tutorial fashion. The net result is the close connection of GPI control with classical compensation networks using extra integration. We summarize the results of the Appendix without further analysis.

For the perturbed plant

$$y^{(n)} = u + \xi(t) \tag{2.77}$$

the GPI controller for trajectory tracking tasks is written in a mixture of complex variable notation and time domain notation as follows:

$$u = u^*(t) - \left[\frac{\kappa_{m+n-1} s^{m+n-1} + \kappa_{m+n-2} s^{m+n-2} + \cdots + \kappa_1 s + \kappa_0}{s^m (s^{n-1} + \kappa_{2n+m-2} s^{n-2} + \cdots + \kappa_{n+m+1} s + \kappa_{n+m})} \right] (y - y^*(t)) \tag{2.78}$$

where the nominal control input is computed as $u^*(t) = [y^*(t)]^{(n)}$, and where $y^*(t)$ is the given smooth output reference trajectory.

The closed-loop system evolves governed by

$$e_y^{(2n+m-1)} + \kappa_{2n+m-2} e_y^{(2n+m)} + \cdots + \kappa_1 \dot{e}_y + \kappa_0 e_y$$
$$= \xi^{(m+n-1)}(t) + \kappa_{2n+m-2} \xi^{(m+n)}(t) + \cdots + \kappa_{n+m} \xi(t) \tag{2.79}$$

The set of design coefficients $\{\kappa_{2n+m-2}, \kappa_{2n+m-3}, \cdots, \kappa_1, \kappa_0\}$ is chosen as a Hurwitz set, which makes the closed-loop tracking error system into a low-pass filter processing the disturbance input $\xi(t)$. In a transfer function form, the tracking error is obtained as

$$e_y(t) = \left[\frac{s^m (s^{n-1} + \kappa_{2n+m-2} s^{n-2} + \cdots + \kappa_{n+m+1} s + \kappa_{n+m})}{s^{2n+m-1} + \kappa_{2n+m-2} s^{2n+m-2} + \cdots + \kappa_1 s + \kappa_0} \right] \xi(t) \tag{2.80}$$

At high frequencies, the filter is approximated by the attenuation effects of a multiple integrator of the form

$$e_y(s) \approx \frac{\xi(s)}{s^n} \tag{2.81}$$

with an attenuation slope of $20\,n$ [db/dec].

At low frequencies, the filter is approximated by the pure differentiator

$$e_y(s) \approx \left[\frac{\kappa_{n+m} s^m}{\kappa_0} \right] \xi(s) \qquad (2.82)$$

with a magnifying slope of $20\,m$ db/dec. The design guideline indicates that the amount of extra integrations m should be kept as small as possible, with $m = 1$ being the lowest possible choice. Incidentally, this option is dual to the original one-dimensional extension of the traditional Extended State Observer-based method for ADRC, proposed in early works, such as in that of J. Han [1]. A compromise is therefore established between the number of additional integrations and the prevailing filter behavior in the bandwidth interval.

Notice that a high-gain design, characterized by the use of an ϵ-dependent description of the characteristic polynomial roots in the stable portion of the complex plane, implies that $\kappa_0 \gg \kappa_{n+m}$, introducing yet an additional attenuation factor for the low-frequency behavior of the closed-loop system, thus counteracting the amplifying effects introduced by the low frequency differentiator.

Example 4. Consider the perturbed third-order integration system

$$y^{(3)} = u + \xi(t) \qquad (2.83)$$

where $\xi(t)$ is an unknown but uniformly absolutely bounded signal with uniform absolutely bounded first-order time derivative. Define $e_y = y - y^*(t)$, $e_u = u - u^*(t)$ where $u^*(t)$ is nominally computed as $[y^*(t)]^{(3)}$.

Consider the following controller, written in the complex frequency domain via the standard operator s:

$$u = [y^*(t)]^{(3)} - \left[\frac{\kappa_3 s^3 + \kappa_2 s^2 + \kappa_1 s + \kappa_0}{s(s^2 + \kappa_5 s + \kappa_4)} \right] (y - y^*(t)) \qquad (2.84)$$

The closed-loop perturbed system evolves in accordance with

$$\begin{aligned} e_y(s) &= \left[\frac{s(s^2 + \kappa_5 s + \kappa_4)}{s^6 + \kappa_5 s^5 + \kappa_4 s^4 + \kappa_3 s^3 + \kappa_2 s^2 + \kappa_1 s + \kappa_0} \right] \xi(s) \\ &= \left[\frac{n(s)}{d(s)} \right] \xi(s) \end{aligned} \qquad (2.85)$$

Suppose that the set of coefficients $\{\kappa_5, \kappa_4, \cdots, \kappa_1, \kappa_0\}$ is chosen so that the denominator polynomial $d(s) = s^6 + \kappa_5 s^5 + \kappa_4 s^4 + \cdots + \kappa_1 s + \kappa_0$ is strictly Hurwitz, i.e., all its roots are in the open left half of the complex plane.

We could locate the roots of the closed-loop characteristic polynomial via the auxiliary polynomial:

$$(s^2 + 2\zeta\frac{\omega_n}{\epsilon}s + \frac{\omega_n^2}{\epsilon^2})^3 = s^6 + \kappa_5 s^5 + \kappa_4 s^4 + \cdots + \kappa_1 s + \kappa_0 \tag{2.86}$$

Here,

$$\kappa_0 = \frac{\omega_n^6}{\epsilon^6}, \quad \kappa_4 = \frac{12\zeta^2\omega_n^2 + 3\omega_n^2}{\epsilon^2} \tag{2.87}$$

The quotient κ_4/κ_0 affecting as a factor, the equivalent closed-loop differentiator for low frequencies is of the form

$$\frac{\kappa_4}{\kappa_0} = \epsilon^4(\frac{12\zeta^2 + 3}{\omega_n^4}) \tag{2.88}$$

For high frequencies, the closed-loop system attenuates the disturbance input $\xi(t)$ via three equivalent integrations i.e., an attenuation of 60 [db/dec].

Example 5. As an illustration of the performance of the previously derived controller for a third-order plant, consider the following nonlinear model of a synchronous machine (see Singh and Titli [2]):

$$\dot{y}_1 = y_2$$
$$\dot{y}_2 = B_1 - A_1 y_2 - A_2 y_3 \sin y_1 - \frac{B_1}{2}\sin 2y_1$$
$$\dot{y}_3 = u - C_1 y_3 + C_2 \cos y_1 \tag{2.89}$$

The control input u represents the voltage applied to the machine and the variable to be controlled, y_1 represents an angular position deviation variable that should be controlled toward a desired nonzero equilibrium point. All constant coefficients of the system are assumed to be known. The objective is to stabilize the plant toward the equilibrium point:

$$(\bar{y}_1, \bar{y}_2, \bar{y}_3) = (0.7461, 0, 7.7438), \quad \bar{u} = 1.1$$

The system is differentially flat, with flat output given by $y = y_1$. The simplified input-to-flat output dynamics is then given by

$$y^{(3)} = v + \xi(t) \tag{2.90}$$

where ξ represents an endogenous nonlinear additive term depending on the phase variables y_1, y_2, y_3, which is here treated as an unstructured total disturbance. The auxiliary input v satisfies $v = -A(\sin y)u$. Hence, a singularity is

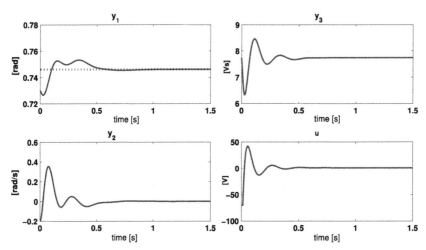

FIGURE 2.3 Closed-loop response of third order nonlinear system, controlled via a GPI linear controller.

present in $y = 0$ that leaves u undefined. This is the reason why the system is not controlled toward an equilibrium value where $y = 0$.

Fig. 2.3 depicts the closed-loop performance of the linearly controlled nonlinear system. In order to examine the robustness with respect to control input saturations, a saturation limit of 70 [V] was enforced on the input signal $u(t)$. The obtained equilibrium up to a fifth decimal was:

$$\bar{y}_1 = 0.74609 \text{ [rad]}, \quad \bar{y}_2 = -0.00004 \text{ [rad/s]},$$
$$\bar{y}_3 = 7.74371 \text{ [V-s]}, \quad \bar{u} = 1.10091 \text{ [V]}$$

The parameter values were set to be

$$A_1 = 0.2703, \quad A_2 = 12.012, \quad B_1 = 39.1892, \quad B_2 = -48.048,$$
$$C_1 = 0.3222, \quad C_2 = 1.9$$

The design parameters for the GPI controller in the closed-loop characteristic polynomial $p(s)$ were set to be

$$p(s) = (s^2 + 2\zeta \frac{\omega_n}{\epsilon} s + \frac{\omega_n^2}{\epsilon^2})^3, \quad \zeta = 0.707, \quad \omega_n = 2, \quad \epsilon = 0.1$$

Fig. 2.4 depicts, for different values of ϵ, the evolution of the Integral Square Error (ISE) associated with the output reference stabilization problem. Recall that,

$$ISE(t) = \int_0^t (y(\sigma) - \bar{y})^2 d\sigma \qquad (2.91)$$

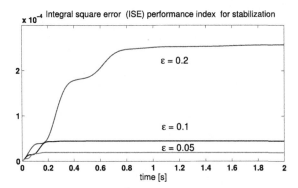

FIGURE 2.4   Integral square error evolution for the output stabilization error, controlled via a GPI linear controller.

Generically speaking, the higher the gain (the smaller ϵ), the smaller the final constant value of the ISE performance index. The seemingly constant steady-state value of the performance index indicates that the stabilization error is arbitrarily close to zero, as expected, with negligible growth. Naturally, this implication is not valid when the control input magnitude is restricted and saturations are present on the closed-loop performance of the system.

The corresponding steady-state values of the ISE performance index, as a function of the parameter ϵ, are given by

$$ISE(0.2) = 2.5713 \times 10^{-4}, \quad ISE(0.1) = 4.4753 \times 10^{-5},$$
$$ISE(0.05) = 1.958 \times 10^{-5}$$

Imposing the saturation constraint to the input of 70 [V] in absolute value, a minimum for the performance index is found for, approximately, $\epsilon = 0.1$ with a value of $ISE(0.1) = 3.059 \times 10^{-5}$ smaller than for the unsaturated controller case.

2.4   ULTRALOCAL DISTURBANCE MODELS

We begin the treatment of global ultramodels in terms of a particularization of a well-known concept of trajectory-equivalent systems, developed by Coleman [3].

Definition 1. Two systems

$$\dot{y} = f(t, y), \quad y(t_0) = y_0, \ y \in \mathbb{R}^n,$$
$$\dot{z} = g(t, z), \quad z(t_0) = z_0, \ z \in \mathbb{R}^n \tag{2.92}$$

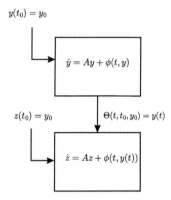

$$y(t_0) = y_0$$

$$\dot{y} = Ay + \phi(t, y)$$

$$z(t_0) = y_0 \qquad \Theta(t, t_0, y_0) = y(t)$$

$$\dot{z} = Az + \phi(t, y(t))$$

FIGURE 2.5   Trajectory-equivalent systems.

are said to be *trajectory-equivalent* whenever $y(t) = z(t)$ for all $t \geq 0$ (in particular, $y_0 = z_0$). They are *asymptotically trajectory-equivalent* if for $y_0 \neq z_0$, $e = y(t) - z(t) \rightarrow 0$. They are said to be *asymptotically exponentially trajectory-equivalent* if the convergence of the error $e = y(t) - z(t)$ to zero is exponential.

Trajectory-equivalent systems do not necessarily look alike, nor they have to be of the same nature, as the following example illustrates.

Example 6. The linear system $\dot{y} = A(t)y$, $y(t_0) = y_0$, $y \in R^n$, is trajectory-equivalent to the pure integration system $\dot{z} = A(t)\Phi_A(t, t_0)y_0$, $z(t_0) = y_0$, $z \in R^n$, where $\Phi_A(t, t_0)$ is the fundamental matrix associated with $A(t)$. Naturally, $\dot{y} = A(t)y$, $y(t_0) = y_0$ and $\dot{z} = A(t)y(t)$, $z(t_0) = y_0$ are trajectory-equivalent.

In reference to Fig. 2.5, consider the following system of time-varying nonlinear differential equations:

$$\dot{y} = Ay + \phi(t, y), \quad y(t_0) = y_0, \quad y \in R^n \qquad (2.93)$$

where A is an $n \times n$ Hurwitz matrix, and $\phi(t, y)$ is a vector of nonlinearities possibly including exogenous signals. We denote by $y(t) = \Theta(t, t_0, y_0)$ the **solution trajectory** of the nonlinear differential equation (2.93). Let z be an n-dimensional vector and consider the following linear system with an exogenous time-varying injection:

$$\dot{z} = Az + \phi(t, \Theta(t, t_0, y_0)) = Az + \phi(t, y(t))$$
$$z(t_0) = y_0 + b \qquad (2.94)$$

Proposition 1. *For $b \neq 0$, systems (2.93) and (2.94) are asymptotically exponentially trajectory-equivalent. If $b = 0$, then the systems are trajectory-equivalent.*

Proof. Certainly, letting $e = y(t) - z(t)$, we have:

$$\dot{e} = Ae + \phi(t, y(t)) - \phi(t, \Theta(t, t_0, y_0)) = Ae \qquad (2.95)$$

with $e(t_0) = b$. The result follows. $\qquad \qquad \square$

We emphasize that in the above demonstration we have used the following two facts: a) The time varying vector $y(t) = \Theta(t, t_0, y_0)$ trivially satisfies the identity (see Pontryagin [4], Ch. 1, p. 19):

$$\dot{\Theta}(t, t_0, y_0) = A\Theta(t, t_0, y_0) + \phi(t, \Theta(t, t_0, y_0))$$
$$\Theta(t_0, t_0, y_0) = y_0 \qquad (2.96)$$

i.e.,

$$\dot{y}(t) = Ay(t) + \phi(t, y(t)), \quad y(t_0) = y_0 \qquad (2.97)$$

b) The linear differential equation for z includes a copy of the term $\phi(t, y)$ particularized for the solution $y = y(t) = \Theta(t, t_0, y_0)$ of the nonlinear differential equation. As such, for every fixed initial condition y_0, at a given instant t_0, the term, $\phi(t, \Theta(t, t_0, y_0))$ is a time function expressed as $\xi(t) = \phi(t, \Theta(t, t_0, y_0)) = \phi(t, y(t))$.

Example 7. Consider the following nonlinear time-varying system:

$$\dot{x}_1 = x_2 + \phi_1(t, x_1, x_2)$$
$$\dot{x}_2 = -8x_1 - 16x_2 + \phi_2(t, x_1, x_2) \qquad (2.98)$$

with $x_1(0) = 1$, $x_2(0) = -1$,

$$\phi_1(t, x_1, x_2) = \sin(0.1t)\cos(t)e^{-x_1^2}\cos(0.1x_2)$$
$$\phi_2(t, x_1, x_2) = \cos(0.1t)\sin(t)e^{-x_2^2}\sin(0.1x_1) \qquad (2.99)$$

and the linear trajectory equivalent system

$$\dot{z}_1 = z_2 + \phi_1(t, x_1(t), x_2(t))$$
$$\dot{z}_2 = -8z_1 - 16z_2 + \phi_2(t, x_1(t), x_2(t)) \qquad (2.100)$$

with $z_1(0) = 0$, $z_2(0) = 0$,

Fig. 2.6 depicts the evolution of the nonlinear state variables x_1, x_2 and the convergent evolution of the linear state variables z_1, z_2, along with the corresponding tracking errors $e_1 = x_1 - z_1$ and $e_2 = x_2 - z_2$.

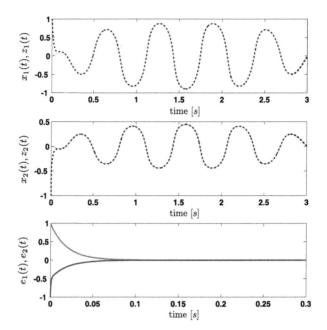

FIGURE 2.6 States evolution of asymptotically exponentially trajectory-equivalent systems.

Theorem 1. *The nonlinear system (2.93) is trajectory-equivalent to the forced linear system* $\dot{z} = Az + \xi(t)$, $z(t_0) = y_0$, *if and only if*

$$\xi(t) = \phi(t, y(t)) \ \forall \, t \geq 0 \tag{2.101}$$

Proof. Suppose that the two systems are trajectory-equivalent. Then $e(t) = y(t) - z$ is identically zero for all t. Moreover, $\dot{e}$ is also identically zero. Therefore,

$$\dot{e} = Ae + \left[\phi(t, y(t)) - \xi(t)\right] = \phi(t, y(t)) - \xi(t) = 0 \tag{2.102}$$

which implies that $\phi(t, y(t)) = \xi(t)$ for all t.

Consider now the systems $\dot{y} = Ay + \phi(t, y)$, $y(0) = y_0$, and $\dot{z} = Az + \phi(t, y(t))$, $z(t_0) = y_0$. The error $e = y(t) - z(t)$ evolves governed by

$$\dot{e} = Ae, \quad e(t_0) = 0 \tag{2.103}$$

Therefore, $e(t)$ is identically zero for all t, and both systems are trajectory-equivalent. □

The consequence of this simple fact is that we may view the nonlinear system (2.93) as a linear system with an exogenous time-varying injection term $\xi(t) =$

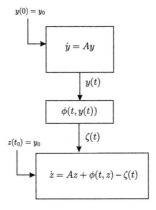

FIGURE 2.7   Exogenous cancelation of state-dependent nonlinearities.

$\phi(t, y(t))$,

$$\dot{z} = Az + \xi(t), \quad z(t_0) = y_0. \tag{2.104}$$

These two systems, (2.93) and (2.104), are *identical* in the precise sense that their trajectories are the same over any time interval. Consequently, the distinction between z and y becomes irrelevant. Any pertinent consideration on the system (2.93) may be examined on the linear trajectory equivalent system (2.103), viewed now without any ambiguity as

$$\dot{y} = Ay + \xi(t), \quad y(0) = y_0 \tag{2.105}$$

which is devoid of the state-dependent nonlinear structure. We address the linear perturbed system (2.105) as the *ultralocal model* of the nonlinear system (2.93). In practise, the time-valued nonlinearities, lumped in $\xi(t)$, constitute an unknown term, thus acting as a proper disturbance to the system.

2.4.1 Exogenous Cancelation of Nonlinearities

Let $A \in \mathbb{R}^{n \times n}$ be a Hurwitz matrix. Consider the systems

$$\dot{y} = Ay, \quad y(t_0) = y_0,$$
$$\dot{z} = Az + \phi(t, z) - \zeta(t), \quad z(t_0) = y_0 \tag{2.106}$$

with (see Fig. 2.7)

$$\zeta(t) = \phi(t, y(t)) = \phi(t, \Phi_A(t, t_0)y_0) \tag{2.107}$$

The two previous systems are trajectory-equivalent.

Proof. For all t, the error $e = y - z(t)$ evolves according to

$$\dot{e} = Ae + \phi(t, y(t)) - \phi(t, z(t)) = Ae + \phi(t, z(t) + e) - \phi(t, z(t)) \quad (2.108)$$

with $e(t_0) = 0$. It follows that $\dot{e}(t_0) = 0$, and this in turn implies that $\ddot{e}(t_0) = 0$, and thus all time derivatives of $e(t)$ are zero at $t = t_0$. Therefore, $e(t) \equiv 0$ for all $t \geq t_0$. $\qquad\square$

Example 8. Consider the following linear system:

$$\dot{x}_1 = x_2$$
$$\dot{x}_2 = -8x_1 - 16x_2 \quad (2.109)$$

with $x_1(0) = 1$, $x_2(0) = -1$ and the nonlinear systems

$$\dot{z}_1 = z_2 + \phi_1(t, z_1, z_2) - u\phi_1(t, x_1(t), x_2(t))$$
$$\dot{z}_2 = -8z_1 - 16z_2 + \phi_2(t, z_1, z_2) - u\phi_2(t, x_1(t), x_2(t)) \quad (2.110)$$

with $z_1(0) = 1$, $z_2(0) = -1$, and $u \in \{0, 1\}$ and

$$\phi_1(t, z_1, z_2) = \sin(0.1t)\cos(t)e^{-z_1^2}\cos(0.1z_2)$$
$$\phi_2(t, z_1, z_2) = \cos(0.1t)\sin(t)e^{-z_2^2}\sin(0.1z_1) \quad (2.111)$$

The two systems are trajectory-equivalent whenever $u = 1$ for all t.

Fig. 2.8 illustrates that when the online exogenous cancelation is not activated (and $u = 0$ for all t), the solution trajectories of the state variables of the two systems are substantially different; but they entirely coincide when the exogenous cancelation is activated (and u is set to the value of 1 for all times).

This result implies that the effect of additive state-dependent nonlinearities may be exactly canceled from the nonlinear system behavior via the injection of a precise exogenously generated time-varying signal vector.

2.4.2   The Case of Nonlinear Controlled Systems

We are particularly interested in those cases where the additive nonlinearities ($\phi(t, y)$) are uncertain due to: a) the lack of knowledge of certain parameters, or b) due to the presence of unmodeled state-dependent nonlinearities, or c) the combination of these two previous cases with the presence of uncertain exogenous time-varying vector signals. From the previous results we consider all such terms as a single lumped, unstructured, and time-varying disturbance term $\xi(t)$. When the disturbance term $\xi(t)$ cannot be exactly on line reconstructed and only an approximate on line knowledge of its time behavior is available (say, via the

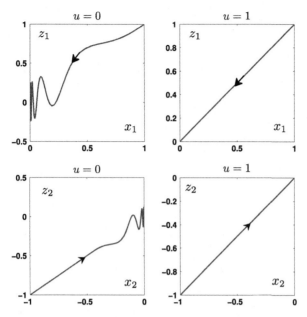

FIGURE 2.8   Exogenous cancelation of state-dependent nonlinearities.

use of a suitable asymptotic observer), approximate cancelation may still be effectively achieved. In controlled systems, the effects of this approximate cancelation can be further counteracted to achieve arbitrarily close output trajectory tracking features to differentially flat systems. The underlying problem, though, is clearly linear.

Consider the nonlinear scalar, differentially flat system

$$y^{(n)} = \psi(t, y)u + \phi(t, y, \dot{y}, \ldots, y^{(n-1)}) \tag{2.112}$$

with the following set of initial conditions:

$$Y_0 = \{y(t_0), \dot{y}(t_0), \ldots, y^{(n-1)}(t_0)\} \tag{2.113}$$

We refer to the function $\psi(t, y)$ as to the *control input gain* of the system. The term $\phi(t, y, \dot{y}, \ldots, y^{(n-1)})$ will be addressed as the *drift function*.

Proposition 2. *For a given smooth control input function $u(t)$, let $y(t) = \Theta(t, t_0, Y_0, u(t))$ denote the solution trajectory of system (2.112) from the set of initial conditions Y_0. The system*

$$z^{(n)} = \psi(t, z)u + \phi(t, y(t), \dot{y}(t), \ldots, y^{(n-1)}(t)) \tag{2.114}$$

is trajectory-equivalent to system (2.112), provided that the initial conditions for z and its time derivatives $Z_0 = \{z(t_0), \dots, z^{(n-1)}(t_0)\}$ coincide with Y_0, i.e., $z^{(j)}(t_0) = y^{(j)}(t_0)$, $j = 0, 1, \dots, n-1$.

Proof. The proof is based on the fact that the error signal $e = y(t) - z(t)$ satisfies

$$e^{(n)} = \left[\psi(t, y(t)) - \psi(t, z(t)) \right] u(t) \tag{2.115}$$

with $e^{(j)}(t_0) = 0$ for all $j = 0, 1, \dots, n-1$. Clearly, $e^{(n)}(t_0) = 0$. It is not difficult to establish that $e^{(n+k)}(t_0) = 0$ for all integers k. Thus, $e(t) \equiv 0$ for all $t \geq t_0$, and the systems are trajectory-equivalent for the given control input function $u(t)$.

As before, we denote by the time function $\xi(t)$ the additive drift function $\phi(t, y(t), \dot{y}(t), \dots, y^{(n-1)}(t))$, without regard for any particular internal structure. We refer to the trajectory-equivalent system

$$y^{(n)} = \psi(t, y)u + \xi(t) \tag{2.116}$$

as the *ultralocal model* associated with system (2.116). $\qquad\square$

2.4.3 Ultralocal Models and Flatness

It is desired to drive the flat output y of the system

$$y^{(n)} = \psi(t, y)u + \phi(t, y, \dot{y}, \dots, y^{(n-1)}) \tag{2.117}$$

to track a given smooth reference trajectory $y^*(t)$, regardless of the unknown but uniformly bounded nature of the drift function $\phi(t, y, \dot{y}, \dots, y^{(n-1)})$.

Regarding the controlled system (2.117), we make the following assumptions:

1. The drift function $\phi(t, y, \dot{y}, \dots, y^{(n-1)})$ is completely unknown, whereas the control input gain $\psi(t, y)$ is perfectly known. Let ϵ be a strictly positive real number. The control input gain $\psi(t, y(t))$ is assumed to be uniformly bounded away from zero, i.e., $\inf_t |\psi(t, y(t))| \geq \epsilon > 0$ for any solution $y(t)$ of the controlled system. In particular, it is bounded away from zero for the given output reference trajectory $y^*(t)$.

2. It is assumed that a solution $y(t)$ exists, uniformly in t, for every given set of initial conditions: Y_0, specified at time $t = t_0$, and a given sufficiently smooth control input function $u(t)$. Given a desired flat output reference trajectory $y^*(t)$, the flatness of the system and the previous assumption allow for a straightforward calculation of the corresponding (unique) open-loop control input $u^*(t)$.

3. Let m be a given integer. As a time function, the mth time derivative of $\xi(t) = \phi(t, y(t), \dot{y}(t), \ldots, y^{(n-1)}(t))$ is uniformly absolutely bounded (in an "almost everywhere" sense where needed). In other words, there exists a constant K such that[2]

$$\sup_t |\xi^{(m)}(t)| = \sup_t |\phi^{(m)}(t, y(t), \dot{y}(t), \ldots, y^{(n-1)}(t))| \leq K \qquad (2.118)$$

2.4.4 A High-Gain Observer-Based Approach

Setting $y_1 = y$, $y_2 = \dot{y}, \ldots, y_n = y^{(n-1)}$, a state space model for the ultralocal model of such an uncertain system is given by

$$\dot{y}_j = y_{j+1}, \quad j = 1, \ldots, n-1$$
$$\dot{y}_n = \psi(t, y_1)u + \xi(t) \qquad (2.119)$$

We propose the following observer (called Generalized Proportional Integral observer or, simply, GPI observer) for the phase variables $\{y_1, y_2, \ldots, y_n\}$ associated with the flat output y, characterized by the states $\hat{y}_1, \ldots, \hat{y}_n$ and complemented by m output estimation error iterated integral injections, characterized by the variable z_1. We have

$$\dot{\hat{y}}_j = \hat{y}_{j+1} + \lambda_{n+m-j}(y_1 - \hat{y}_1), \quad j = 1, \ldots, n-1$$
$$\dot{\hat{y}}_n = \psi(t, y_1)u + z_1 + \lambda_m(y_1 - \hat{y}_1)$$
$$\dot{z}_i = z_{i+1} + \lambda_{m-i}(y_1 - \hat{y}_1), \quad i = 1, \ldots, m-1$$
$$\dot{z}_m = \lambda_0(y_1 - \hat{y}_1) \qquad (2.120)$$

Let the estimation error e_y be defined as $e_y = e_1 = y_1 - \hat{y}_1 = y - \hat{y}_1$ with $e_2 = y_2 - \hat{y}_2$, etc.,

$$\dot{e}_j = e_{j+1} - \lambda_{n+m-j}e_1, \quad j = 1, \ldots, n-1$$
$$\dot{e}_n = \xi(t) - z_1 - \lambda_m e_1$$
$$\dot{z}_i = z_{i+1} + \lambda_{m-i}e_1, \quad i = 1, \ldots, m-1$$
$$\dot{z}_m = \lambda_0 e_1 \qquad (2.121)$$

---

2. This assumption cannot be verified a priori when $\phi(t, y, \dot{y}, \ldots, y^{(n-1)})$ is completely unknown. However, in cases where the nonlinearity is known except for some of its parameters, its validity can be assessed with some work. Also, if $\xi^{(m)}(t)$ is not uniformly absolutely bounded almost everywhere, then solutions $y(t)$ for (2.117) do not exist for any finite $u(t)$ (see Gliklikh [5]).

The estimation error $e_y = e_1$ satisfies, after elimination of all variables z, the following $(n+m)$th order perturbed linear differential equation:

$$e_y^{(n+m)} + \lambda_{n+m-1}e_y^{(n+m-1)} + \cdots + \lambda_1\dot{e}_y + \lambda_0 e_y = \xi^{(m)}(t) \qquad (2.122)$$

Theorem 2. *Suppose that all previous assumptions are valid. Let the coefficients λ_j, $j = 0, 1, \ldots, n + m - 1$, of the characteristic polynomial in the complex variable s,*

$$p_o(s) = s^{n+m} + \lambda_{n+m-1}s^{n+m-1} + \cdots + \lambda_1 s + \lambda_0 \qquad (2.123)$$

be chosen so that, for a sufficiently large real number $N > 0$, the polynomial $p_o(s)$ exhibits all its roots to the left of the line $\{s \in \mathbb{C} \,|\, \text{Re}(s) \leq -N\}$ in the complex plane $\mathbb{C}$. Then, the trajectories of the estimation error $e_y(t)$ and its time derivatives $e^{(j)}(t)$, $j = 1, \ldots, n + m - 1$, globally converge toward a small as desired sphere of radius ρ, denoted by $S(0, \rho)$, centered at the origin of the estimation error phase space $\{e_y, \dot{e}_y, \ldots, e_y^{(n+m-1)}\}$, where they remain ultimately bounded. The larger the value of N, the smaller the radius of the sphere $S(0, \rho)$. Similarly, the variable z_1 and its time derivatives z_j, $j = 1, \ldots, m$, track arbitrarily closely to the unknown time function $\xi(t)$ and its time derivatives $\xi^{(j)}(t)$, $j = 1, \ldots, m$.

Proof. Let $x = (e_1, \ldots, e_{n+m})^T$ denote the phase variables of (2.122). The perturbed linear system (2.122) is of the form $\dot{x} = Ax + b\xi^{(m)}(t)$, with A being a Hurwitz matrix written in companion form and b being a vector of zeroes except for the last component equal to 1. Let $|\text{Re}(\sigma_{max}A)|$ denote the absolute value of the negative real part of the largest eigenvalue of the matrix A (i.e., the closest to the imaginary axis). Clearly, $|\text{Re}(\sigma_{max}A)| \geq N$. The Hurwitzian character of A implies that, for every constant, $(n + m) \times (n + m)$, symmetric, positive definite matrix $Q = Q^T > 0$, there exists a symmetric, positive definite $(n + m) \times (n + m)$ matrix $P = P^T > 0$ such that $A^T P + PA = -Q$. The Lyapunov function candidate $V(x) = \frac{1}{2}x^T Px$ exhibits a time derivative along the solutions of the closed-loop system given by $\dot{V}(x,t) = \frac{1}{2}x^T(A^T P + PA)x + x^T Pb\xi^{(m)}(t)$. This function satisfies $\dot{V}(x,t) \leq -\|x\|^2\|P\||\text{Re}(\sigma_{max}(A))| + K\|x\|\|P\| = -\|x\|\|P\|(|\text{Re}(\sigma_{max}(A))|\|x\| - K)$. This function is strictly negative everywhere outside the sphere $S(0, \rho) = \{x \in R^{m+n} \mid \|x\| \leq \rho = K/|\text{Re}(\sigma_{max}(A))| \leq K/N\}$. Hence, all trajectories $x(t)$ starting outside this sphere converge toward its interior, and all the trajectories starting inside $S(0, \rho)$ will never abandon it. The more negative the real part of the dominant eigenvalue of A, the larger the quantity $|\text{Re}(\sigma_{max}(A))|$, and, hence, the smaller the radius of the ultimate bounding sphere $S(0, \rho)$ in the x space.

From (2.121) it follows that

$$z_1 = \phi(t, y_1(t), y_2(t), \dots, y_n(t)) - \lambda_m e_1 - \dot{e}_n$$
$$= \xi(t) - \lambda_m e_1 - \dot{e}_n \qquad (2.124)$$

Hence, as e_1 and $\dot{e}_n$ evolve toward the small bounding sphere in the estimation error phase space, the trajectory of z_1 tracks, arbitrarily closely, the unknown disturbance function $\xi(t) = \phi(t, y_1(t), y_2(t), \dots, y_n(t))$. In general, z_i, $i = 1, \dots, m$, converges toward a small as desired vicinity of $\xi^{(i-1)}(t)$, $i = 1, \dots, m$. Since $z_1 - \xi(t)$ is ultimately bounded, the approximate estimate z_1 of the ultralocal model disturbance term $\xi(t)$ is thus self-updating. From the definition of the estimation errors for y and its time derivatives, it follows that $\hat{y}_j$, $j = 1, \dots, n$, reconstruct, in an arbitrarily close fashion, the time derivatives of y. $\qquad \square$

2.4.5 The Disturbance Canceling Linear Controller

The trajectory tracking controller is given by

$$u = \frac{1}{\psi(t, y)} \left[[y^*(t)]^{(n)} - \sum_{k=0}^{n-1} \gamma_k \left(\hat{y}_{k+1} - [y^*(t)]^{(k)} \right) - z_1 \right] \qquad (2.125)$$

with the set of coefficients $\{\gamma_0, \dots, \gamma_{n-1}\}$ chosen so that $p_c(s) = s^n + \gamma_{n-1} s^{n-1} + \dots + \gamma_0$ exhibits all its roots in the left half of $\mathbb{C}$. The closed-loop flat output tracking error $e = y - y^*(t)$ is governed by

$$e^{(n)} + \gamma_{n-1} e^{(n-1)} + \dots + \gamma_0 e = (\xi(t) - z_1) + \sum_{k=1}^{n-1} \gamma_k e_y^{(k)} \qquad (2.126)$$

Theorem 3. *The disturbance rejection output feedback controller (2.125) drives the trajectory of the controlled system flat output $y(t)$ toward a small as desired vicinity of the origin of the output tracking error phase space $(e, \dot{e}, \dots, e^{(n-1)})$, provided that the set of coefficients $\{\gamma_0, \dots, \gamma_{n-1}\}$ is chosen so that $p_c(s)$ is a Hurwitz polynomial with roots sufficiently far from the imaginary axis in $\mathbb{C}$.*

Proof. According to the previous theorem, the term $\xi(t) - z_1$ and the terms $e_y^{(k)}$, $k = 1, 2, \dots, n - 1$, evolve in a small as desired neighborhood of the origin. It follows that the right-hand side of the linear system (2.126) evolves, in a uniformly ultimately bounded fashion, within a sufficiently small neighborhood of the origin of the output tracking error phase space. Using the same arguments as in the proof of the previous theorem, it follows that the tracking error e and its time derivatives converge toward a small as desired vicinity of the tracking

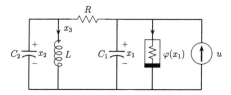

FIGURE 2.9 Controlled Chua's circuit.

error phase space coordinates $z = (e, \dot{e}, \cdots, e^{(n-1)})$, provided that the roots of $p_c(s)$ are located sufficiently to the left of the imaginary axis in $\mathbb{C}$. □

Remark 1. Implementation of the GPI observer-based ADR algorithm requires low-pass filtering along with "clutching" of the observer output signals. The integer m, depicting the order of approximation of the lumped disturbance signal $\xi(t)$, is typically chosen as $m = 3$ or, at most, $m = 5$ (see [6] for experimental details)[3].

2.4.6 An Illustrative Experimental Example

In reference to Fig. 2.9, let $\varphi(x_1)$ represent the so-called Chua's diode, given by [7], [8]

$$\varphi(x_1) = m_0 x_1 + \frac{(m_1 - m_0)}{2}(|x_1 + B_p| - |x_1 - B_p|) \tag{2.127}$$

The inductor current $y = x_3$ is the flat output. The input–flat output model is given by

$$RLC_1C_2 y^{(3)} + L(C_1 + C_2)\ddot{y} + RC_1\dot{y} + y + \theta(y, \dot{y}, \ddot{y}) = u \tag{2.128}$$

with $\theta(y, \dot{y}, \ddot{y}) = \varphi(x_1)$ and $x_1 = RLC_2\ddot{y} + L\dot{y} + Ry$. The constant-gain linear global ultralocal model is given by

$$y^{(3)} = (1/RLC_1C_2) u + \xi(t) \tag{2.129}$$

It is desired that the output $y(t)$ of (2.129) tracks the reference signal $\zeta(t) = y^*(t)$ of the following unrelated Lorenz system:

$$\dot{\zeta}_1 = \sigma(\zeta_2 - \zeta_1), \ \dot{\zeta}_2 = r\zeta_1 - \zeta_2 - \zeta_1\zeta_3, \ \dot{\zeta}_3 = \zeta_1\zeta_2 - b\zeta_3 \tag{2.130}$$

---

3. As a justification for such a low degree, recall John von Neumann's statement: "With four parameters, I can fit an elephant, and with five, I can make him wiggle his trunk!"

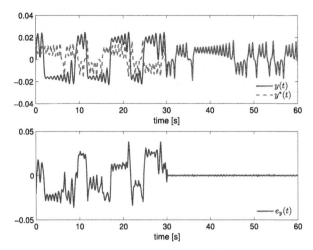

FIGURE 2.10 Performance of the observer-based ADR controller for Chua's circuit ($e_y = y - y^*(t)$).

with $\zeta(t) = \epsilon \zeta_1(t)$. We used the following linear observer-based disturbance canceling control:

$$
\begin{aligned}
u &= RLC_1C_2\big[[y^*(t)]^{(3)} - \gamma_2(y_3 - \ddot{y}^*(t)) - \gamma_1(y_2 - \dot{y}^*(t)) \\
&\quad - \gamma_0(y_1 - y^*(t)) - z_1\big] \\
\dot{y}_1 &= y_2 + \lambda_5(y - y_1), \quad \dot{y}_2 = y_3 + \lambda_4(y - y_1) \\
\dot{y}_3 &= (1/RLC_1C_2)u + z_1 + \lambda_3(y - y_1) \\
\dot{z}_1 &= z_2 + \lambda_2(y - y_1), \quad \dot{z}_2 = z_3 + \lambda_1(y - y_1) \\
\dot{z}_3 &= \lambda_0(y - y_1)
\end{aligned}
\tag{2.131}
$$

The characteristic polynomial associated with the output estimation error is given by $p_o(s) = s^6 + \lambda_5 s^5 + \ldots + \lambda_1 s + \lambda_0$, whereas that associated with the closed-loop system is $p_c(s) = s^3 + \gamma_2 s^2 + \gamma_1 s + \gamma_0$. The ADR control law was implemented by means of a National Instruments Data Acquisition Card PCI-6259. The control output was synthesized through the analog output, and the control law was devised using Matlab Simulink software in a real-time implementation environment using the xPc target hardware in the loop simulation with a sampling period of 0.1 [ms]. The circuit parameters were set to be $C_1 = 23.5$ [μF], $C_2 = 235$ [μF], $R = 1550$ [Ω], and the inductance $L = 42.3$ [H]. The parameters of Chua's diode were set to be $m_0 = -0.409$ [ms], $m_1 = -0.758$ [ms], $B_p = 1.8$ [V]. The simulated Lorenz system parameters were set as $\sigma = 10$, $r = 28$, $b = 8$, with $\epsilon = 0.75 \times 10^{-4}$. We let Chua's circuit evolve without any control action for a period of 30 [s], and then we switched

FIGURE 2.11 Picture of oscilloscope showing controlled circuit performance.

"ON" the observer-based ADR controller. Figs. 2.10 and 2.11 depict the performance of the ADR controller in the trajectory tracking task.

REFERENCES

[1] J. Han, From PID to active disturbance rejection control, IEEE Transactions on Industrial Electronics 56 (3) (2009) 900–906.

[2] M.G. Singh, A. Titli, Systems: Decomposition, Optimisation, and Control, Pergamon, 1978.

[3] C. Coleman, Local trajectory equivalence of differential systems, Proceedings of the American Mathematical Society 16 (1966) 890–892.

[4] M.G. Singh, Ordinary Differential Equations, Addison–Wesley, 1962.

[5] Y.E. Gliklikh, Necessary and sufficient conditions for global-in-time existence of solutions of ordinary, stochastic and parabolic differential equations, Abstract and Applied Analysis 2006 (2006) 1–17.

[6] H. Sira-Ramírez, M. Ramírez-Neria, A. Rodríguez-Ángeles, On the linear control of nonlinear mechanical systems, in: 49th IEEE Conference on Decision and Control, Atlanta, Georgia, USA, 2010, pp. 1999–2004.

[7] L. Fortuna, M. Frasca, M.G. Xibilia, Chua's circuit implementations: yesterday, today and tomorrow, Nonlinear Science, Series A 65 (2009).

[8] H. Sira-Ramírez, A. Luviano-Juárez, J. Cortés-Romero, Flatness-based linear output feedback control for disturbance rejection and tracking tasks on a Chua's circuit, International Journal of Control 85 (5) (2012) 594–602.

Chapter 3

Merging Flatness, GPI Observation, and GPI Control with ADRC

Chapter Points

- Differential flatness is shown to be of invaluable help in ADRC control of nonlinear systems, independently of the chosen feedback scheme: Extended State Observer-based ADRC, GPI observe-based ADRC, or flat filtering techniques. Flatness trivializes the trajectory planning issues and the feedback controller design aspects.
- GPI observers are natural generalizations of single-dimensional extended state observers. Their advantages relate to the possibilities of estimating time derivatives of additive disturbances, a valuable asset in controlling input-delayed systems.
- Flat filtering control of nonlinear simplified flat systems constitutes a dual alternative to GPI observation-based ADRC. Flat filtering is a robustified version of GPI control as developed for linear time-invariant systems and can be classified as a linear robust feedback scheme for nonlinear systems. This chapter provides illustrative applications of flat filtering to text book examples and to real laboratory prototypes.

3.1 INTRODUCTION

The differential flatness property in control systems allows for the trivialization of the controller design task, reducing the feedback control problem to that corresponding to a linear controllable time-invariant system ([1], [2], and Appendix A for background on flatness). For an approximate, yet effective, disturbance observer design, the input-to-flat output relation is simplified to a linear perturbed system. The perturbation input lumps both external disturbance inputs and state-dependent nonlinear terms into a single uniformly absolutely bounded disturbance function. This viewpoint transforms the original control task into one of robustly controlling, via a linear output feedback controller, an externally perturbed chain of integrators. The additive disturbance input can be effectively, though approximately, estimated via a linear high-gain observer. This information is used in approximately cancelling (at the controller stage) its effects on the

Active Disturbance Rejection Control of Dynamic Systems. http://dx.doi.org/10.1016/B978-0-12-849868-2.00003-4
Copyright © 2017 Elsevier Inc. All rights reserved.
51

trajectory tracking performance quality. The effects of the unknown disturbance input on the output reconstruction error dynamics (at the observer stage) may be attenuated via a suitable linear combination of iterated integral injections of the output estimation error. This is precisely the dual procedure to that characterizing disturbance input attenuation in Generalized Proportional Integral (GPI) Control, and, hence, the observers we advocate may be properly called GPI observers. Both in GPI control and GPI observer design, the need for appropriate linear combinations of iterated integral errors is completely equivalent to the hypothesis of a self-updating internal model of the perturbation input (local ultramodel) as a time polynomial approximation.

GPI observer-based output feedback control of nonlinear uncertain systems is considered as an Active Disturbance Rejection Control, which deals with the problem of canceling, from the controller actions, endogenous and exogenous unknown additive disturbance inputs affecting the system. Perturbation effects are made available via suitable linear or nonlinear estimation efforts. GPI observers are also intimately related to a radically new viewpoint in nonlinear state estimation, based on differential algebra, developed by [3].

From a practical perspective, the use of simple controllers with the least possible information of a complex system is a challenging control problem. Robust strategies that involve the use of observers for unknown dynamics and external disturbance have been the basis of the Active Disturbance Rejection Control (ADRC) approach [4], [5], which relies on the use of Extended State Observers of linear or nonlinear nature, which use input–output information to reconstruct the lumped disturbances for their further cancellation. The effectiveness of this scheme has been shown through many academic and industrial applications (see [6], [7], [8], [9], [10], [11], and references therein). ADRC does not require a detailed mathematical description of the system, thanks to the online estimation and rejection of unmodeled elements [12]. The original idea of Han was to establish a direct connection between the classical PID control and other modern control theory techniques. In this sense, an alternative approach is the use of Classical Compensation Networks by means of the Generalized Proportional Integral Control (GPI) [13], [14], [15].

Classical Compensation Networks controlling linear systems constitute dynamical systems that are naturally flat [16], [14], [1]. Although flatness here is not to be taken in a strict control-oriented sense, but in a new filtering sense (the input of the compensation network is actually constituted by the plant output signal, and the compensation network output is actually the system input). All variables in the compensator or filter are indeed expressible in terms of the filtered output signal and a finite number of its time derivatives, a general defining property of flat systems [17].

In this chapter, the concept of flat filtering, as a disturbance rejection tool for output feedback control design in controllable linear systems, is analyzed in the context of GPI control and ADRC. Flat filtering is based on the fact that a GPI controller is viewed as a dynamical linear system that exhibits as a natural flat output a filtered version of the plant output signal. This property is particularly helpful in the design of efficient output feedback stabilization schemes and in solving output reference trajectory tracking tasks. The approach is naturally extended applying Active Disturbance Rejection Control ideas to the control of significantly perturbed differentially flat SISO nonlinear systems, affected by unknown endogenous nonlinearities, in the presence of exogenous disturbances and unmodeled dynamics.

3.2 GPI EXTENDED STATE OBSERVER DESIGN

Here, we propose a natural dual extension of Generalized Proportional Integral control to the state observer design problem for linear systems. First, as a parallel to what we did for the GPI controller design case, we undertake the case of estimating via GPI observers unstable perturbation input signals of polynomial type. Suitable dynamic output estimation error injections to the traditional Luenberger observer results in a robust behavior capable of producing asymptotic exponentially stable estimations of the system state variables and of the unstable inputs.

In particular, for the most common case of constant unknown perturbation inputs to the plant, the proposed observers produce asymptotically convergent state estimates, a feature not shared by traditional observers of Luenberger type. The proposed robust observers with dynamic injections may be regarded as average observer designs when the additional restriction is imposed of output error injections taking values on a discrete set of the form $\{-1, 1\}$, as in sliding-mode observers of Luenberger type. This restriction is natural for remotely observed systems with output estimation error values being transmitted in the form of a binary-valued sequence.

The results here obtained are shown to be rather pertinent in unknown unstable perturbation estimation. An application of our proposed estimation scheme for the identification of temporarily growing coupling torques and Coulomb friction terms in closed-loop controlled flexible appendages undergoing unforeseen collisions is also here presented.

3.2.1 An Introductory Example

We begin with an illustrative design example of elementary nature that exhibits the lack of robustness of traditional observers of Luenberger type even to bounded perturbation inputs. Consider the following linear perturbed system:

$$\dot{x}_1 = x_2, \quad \dot{x}_2 = \xi(t), \quad y = x_1 \qquad (3.1)$$

where $\xi(t)$ is a perturbation input signal. Assume, momentarily, that $\xi(t)$ is bounded, i.e., for some positive constant A, $\sup_t |\xi(t)| = A$.

A traditional observer of Luenberger type, designed on the basis of the unperturbed plant model, is customarily designed as follows:

$$\dot{\hat{x}}_1 = \hat{x}_2 + \lambda_2 (y - \hat{y}), \quad \dot{\hat{x}}_2 = \lambda_1 (y - \hat{y})$$
$$\hat{y} = \hat{x}_1, \quad \lambda_1, \lambda_2 > 0 \qquad (3.2)$$

The perturbed reconstruction error state: $(e_1, e_2) = (y - \hat{y}, x_2 - \hat{x}_2)$, evolves according to the perturbed dynamics

$$\dot{e}_1 = e_2 - \lambda_2 e_1$$
$$\dot{e}_2 = -\lambda_1 e_1 + \xi(t)$$
$$e_1 = x_1 - \hat{x}_1 = y - \hat{y} \qquad (3.3)$$

The injected dynamics of e_1 evolves according to

$$\ddot{e}_1 + \lambda_2 \dot{e}_1 + \lambda_1 e_1 = \xi(t) \qquad (3.4)$$

Clearly, only for finite time fading perturbation inputs or perturbations with compact support, we can guarantee the eventual convergence to zero of the estimation error e_1, provided that λ_1 and λ_2 are properly chosen.

Rigorously speaking, the Luenberger observer is nonrobust with respect to any practical system input perturbation. Nevertheless, the approximate estimation of x_2 is still possible for bounded perturbations using a suitably large design parameter λ_1. This last property has been extensively used in the literature of sliding-mode observer applications (see, e.g., McCann et al. [18]).

To complete the example, we assume that $\xi(t)$ is a time-polynomial perturbation of the quadratic type given by $\xi(t) = p_0 + p_1 t + p_2 t^2$ with $\{p_0, p_1, p_2\}$ being constant, but completely unknown, coefficients, thus yielding a family of polynomial perturbation inputs of second degree.

Consider the following observer with dynamic output reconstruction error injection and zero initial conditions for the variables z[1]:

$$\dot{\hat{x}}_1 = \hat{x}_2 + \lambda_4(y(t) - \hat{y}(t))$$
$$\dot{\hat{x}}_2 = \lambda_3(y(t) - \hat{y}(t)) + z_1$$

---

1. Note that we take three extra state variables z in the extended observer in order to match the order of the model of the unknown family of polynomial perturbation inputs, namely, $d^3\xi(t)/dt^3 = 0$.

$$\dot{z}_1 = z_2 + \lambda_2 \left(y(t) - \hat{y}(t) \right)$$
$$\dot{z}_2 = z_3 + \lambda_1 \left(y(t) - \hat{y}(t) \right)$$
$$\dot{z}_3 = \lambda_0 \left(y(t) - \hat{y}(t) \right)$$
$$\hat{y} = \hat{x}_1 \qquad (3.5)$$

As before, letting $e_1 = x_1(t) - \hat{x}_1(t) = y(t) - \hat{y}(t)$ and $e_2 = x_2(t) - \hat{x}_2(t)$, the estimation error dynamics evolves now according to

$$\dot{e}_1 = e_2 - \lambda_4 e_1$$
$$\dot{e}_2 = \xi(t) - \lambda_3 e_1 - z_1$$
$$\dot{z}_1 = z_2 + \lambda_2 e_1$$
$$\dot{z}_2 = z_3 + \lambda_1 e_1$$
$$\dot{z}_3 = \lambda_0 e_1 \qquad (3.6)$$

The perturbed integro-differential equation satisfied by the output error e_1 is just obtained to be[2]

$$\ddot{e}_1 + \lambda_4 \dot{e}_1 + \lambda_3 e_1 + \lambda_2 \left(\int e_1 \right) + \lambda_1 \left(\int^{(2)} e_1 \right) + \lambda_0 \left(\int^{(3)} e_1 \right) = \xi(t) \qquad (3.7)$$

Then, clearly, for the assumed second-degree time-polynomial perturbation input $\xi(t)$, the characteristic polynomial of the estimation error for the proposed observer is given, in this case, by

$$p(s) = s^5 + \lambda_4 s^4 + \lambda_3 s^3 + \lambda_2 s^2 + \lambda_1 s + \lambda_0 \qquad (3.8)$$

The proper choice of the set of design coefficients $\{\lambda_4, \ldots, \lambda_0\}$ as those of a corresponding degree Hurwitz polynomial results in an asymptotically exponentially stable output reconstruction error dynamics. The exponential stability of the origin of the reconstruction error space is thus independent of the parameters in the signal $\xi(t)$, and, hence, the observer is robust with respect to such a perturbation input.

Note that since $e_2 = \dot{e}_1 + \lambda_4 e_1$, if e_1 is asymptotically exponentially driven to zero, so is e_2. Also, we have that z_1 asymptotically tracks the unknown unstable signal $\xi(t)$, i.e., $\lim_{t \to \infty} z_1(t) = \xi(t)$. Similarly, $z_2 \to \dot{\xi}(t)$ and $z_3 \to \ddot{\xi}(t)$.

---

2. Here, $(\int^{(n)} \eta)$ stands for the expression

$$\int_0^t \int_0^{\sigma_1} \cdots \int_0^{\sigma_{n-1}} \eta(\sigma_n) d\sigma_n \cdots d\sigma_1$$

with

$$\left(\int^{(1)} \eta \right) = \left(\int \eta \right) = \int_0^t \eta(\sigma_1) d\sigma_1$$

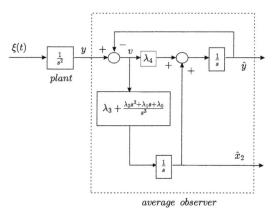

FIGURE 3.1   Robust observer with dynamic output error injection.

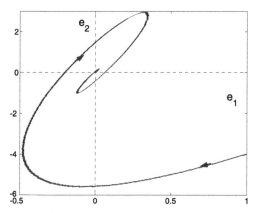

FIGURE 3.2   Robust state estimation error trajectories for a linear plant undergoing unstable polynomial perturbations.

Fig. 3.1 depicts the observer scheme exhibiting the proposed dynamic output error injection.

Fig. 3.2 depicts the trajectories of the estimation error variables e_1, e_2 of the observer design based on suitable dynamic output error injections. The estimation error state variables e_1, e_2 are seen to converge toward zero in an asymptotically exponential fashion.

3.2.2   Problem Formulation and Main Results

Given a linear perturbed SISO observable system

$$y(s) = \frac{1}{s^n + a_{n-1}s^{n-1} + \cdots + a_1 s + a_0}\xi(s) \tag{3.9}$$

where all the system coefficients

$$a_i, i = 0, 1, \ldots, n-1 \tag{3.10}$$

are assumed to be perfectly known, devise an asymptotic observer that is robust with respect to polynomial perturbation inputs of the form

$$\xi(t) = \sum_{i=0}^{r-1} p_i t^i \tag{3.11}$$

where all the coefficients p_i are completely unknown.

The given system has the following state representation in observable canonical form:

$$\dot{x} = Ax + b\xi(t)$$
$$y = \begin{bmatrix} 1 & 0 & 0 & \cdots & 0 \end{bmatrix} x \tag{3.12}$$

with

$$A = \begin{bmatrix}
-a_{n-1} & 1 & 0 & \cdots & 0 & 0 & \cdots & 0 \\
-a_{n-2} & 0 & 1 & \cdots & 0 & 0 & \cdots & 0 \\
\vdots & \vdots & \vdots & \ddots & \vdots & \vdots & \vdots & \vdots \\
-a_m & 0 & 0 & \cdots & 1 & 0 & \cdots & 0 \\
-a_{m-1} & 0 & 0 & \cdots & 0 & 1 & \cdots & 0 \\
\vdots & \vdots & \vdots & \ddots & \vdots & \vdots & \ddots & \vdots \\
-a_1 & 0 & 0 & \cdots & 0 & 0 & 0 & 1 \\
-a_0 & 0 & 0 & \cdots & 0 & 0 & 0 & 0
\end{bmatrix}$$

$$b = \begin{bmatrix} 0 & 0 & \cdots & 0 & 0 & \cdots & 0 & 1 \end{bmatrix}^T \tag{3.13}$$

An observer, capable of overcoming the effects of the time polynomial perturbation input to the plant, is given by

$$\dot{\hat{x}} = \begin{bmatrix}
-a_{n-1} & 1 & 0 & \cdots & 0 & 0 & \cdots & 0 \\
-a_{n-2} & 0 & 1 & \cdots & 0 & 0 & \cdots & 0 \\
\vdots & \vdots & \vdots & \ddots & \vdots & \vdots & \vdots & \vdots \\
-a_m & 0 & 0 & \cdots & 1 & 0 & \cdots & 0 \\
-a_{m-1} & 0 & 0 & \cdots & 0 & 1 & \cdots & 0 \\
\vdots & \vdots & \vdots & \ddots & \vdots & \vdots & \ddots & \vdots \\
-a_1 & 0 & 0 & \cdots & 0 & 0 & 0 & 1 \\
-a_0 & 0 & 0 & \cdots & 0 & 0 & 0 & 0
\end{bmatrix} \hat{x}$$

$$
+ \begin{bmatrix} \lambda_{n+r-1} \\ \lambda_{n+r-2} \\ \vdots \\ \lambda_{r+m} \\ \lambda_{r+m-1} \\ \vdots \\ \lambda_{r+1} \\ \lambda_r \end{bmatrix} (y - \hat{y}) + \begin{bmatrix} 0 \\ 0 \\ \vdots \\ 0 \\ 0 \\ \vdots \\ 0 \\ 1 \end{bmatrix} z_1 \tag{3.14}
$$

$$
\hat{y} = \begin{bmatrix} 1 & 0 & 0 \cdots & 0 \end{bmatrix} \hat{x}
$$

$$
\dot{z}_1 = z_2 + \lambda_{r-1}(y - \hat{y})
$$
$$
\dot{z}_2 = z_3 + \lambda_{r-2}(y - \hat{y})
$$
$$
\vdots
$$
$$
\dot{z}_{r-1} = z_r + \lambda_1(y - \hat{y})
$$
$$
\dot{z}_r = \lambda_0(y - \hat{y}) \tag{3.15}
$$

The vector of estimation errors $e = x - \hat{x}$ satisfies the following perturbed dynamics:

$$
\dot{e} = \begin{bmatrix}
-a_{n-1} - \lambda_{n+r-1} & 1 & 0 & \cdots & 0 & 0 & \cdots & 0 \\
-a_{n-2} - \lambda_{n+r-2} & 0 & 1 & \cdots & 0 & 0 & \cdots & 0 \\
\vdots & \vdots & \vdots & \ddots & \vdots & \vdots & \vdots & \vdots \\
-a_m - \lambda_{r+m} & 0 & 0 & \cdots & 1 & 0 & \cdots & 0 \\
-a_{m-1} - \lambda_{r+m-1} & 0 & 0 & \cdots & 0 & 1 & \cdots & 0 \\
\vdots & \vdots & \vdots & \ddots & \vdots & \vdots & \ddots & \vdots \\
-a_1 - \lambda_{r+1} & 0 & 0 & \cdots & 0 & 0 & 0 & 1 \\
-a_0 - \lambda_r & 0 & 0 & \cdots & 0 & 0 & 0 & 0
\end{bmatrix} e
$$

$$
+ \begin{bmatrix} 0 \\ 0 \\ \vdots \\ 0 \\ 0 \\ \vdots \\ 0 \\ 1 \end{bmatrix} \xi(t) - \begin{bmatrix} 0 \\ 0 \\ \vdots \\ 0 \\ 0 \\ \vdots \\ 0 \\ 1 \end{bmatrix} z_1 \tag{3.16}
$$

$$\dot{z}_1 = z_2 + \lambda_{r-1}(y - \hat{y})$$
$$\dot{z}_2 = z_3 + \lambda_{r-2}(y - \hat{y})$$
$$\vdots$$
$$\dot{z}_{r-1} = z_r + \lambda_1(y - \hat{y})$$
$$\dot{z}_r = \lambda_0(y - \hat{y}) \tag{3.17}$$

From this expression it is not difficult to see that the output observation error $e_1 = y - \hat{y}$ satisfies the following perturbed integro-differential equation:

$$e_1^{(n)} + (a_{n-1} + \lambda_{n+r-1})e_1^{(n-1)} + \cdots + (a_0 + \lambda_r)e_1$$
$$+ \sum_{i=1}^{r} \lambda_{i-1} \left(\int^{(r-i+1)} e_1 \right) = \xi(t) \tag{3.18}$$

The dynamics of the output reconstruction error, given that $\xi(t)$ is a time polynomial of order $r - 1$, exhibits the following characteristic equation[3]:

$$e_1^{(n+r)} + (a_{n-1} + \lambda_{n+r-1})e_1^{(n+r-1)} + \cdots$$
$$+ (a_0 + \lambda_r)e_1^{(r)} + \lambda_{r-1}e_1^{(r-1)} + \cdots + \lambda_0 e_1 = 0 \tag{3.19}$$

Clearly, the coefficients of the characteristic polynomial can be adjusted by means of a suitable specification of the design gains

$$\{\lambda_{n+r-1}, \ldots, \lambda_1, \lambda_0\} \tag{3.20}$$

so that the output estimation error e_1 exponentially asymptotically converges to zero. This may be easily achieved from the relations

$$\lambda_{i+r} = \kappa_{i+r} - a_i, \quad i = 0, 1, \ldots, n - 1,$$
$$\lambda_j = \kappa_j, \quad j = 0, 1, \ldots, r - 1 \tag{3.21}$$

where the coefficients κ are those of a desired $(n + r)$th-degree characteristic polynomial of the Hurwitz type with all its roots in the open left half portion of the complex plane,

$$p(s) = s^{n+r} + \kappa_{n+r-1}s^{n+r-1} + \cdots + \kappa_1 s + \kappa_0 \tag{3.22}$$

---

3. According to classical usage, we term the homogeneous closed loop differential equation of the reconstruction error: the *characteristic equation*. The *characteristic polynomial* refers to the polynomial factor, in the complex variable s of the Laplace transform, accompanying the reconstruction error signal.

From the differential parameterization resulting from the designed average observer

$$e_2 = \dot{e}_1 + \kappa_{n+r-1} e_1$$
$$e_3 = \ddot{e}_1 + \kappa_{n+r-1} \dot{e}_1 + \kappa_{n+r-2} e_1$$
$$\vdots$$
$$e_n = e_1^{(n-1)} + \kappa_{n+r-1} e_1^{(n-2)} + \cdots + \kappa_{r+1} e_1 \qquad (3.23)$$

it follows that the exponential asymptotic convergence of e_1 to zero induces the same type of convergence to zero on the state estimation errors $e_2, e_3, \ldots, e_n$.

Also, from zero initial conditions for all z, then z_1 tends asymptotically exponentially to the unknown perturbation input $\xi(t)$, and, hence, $z_j(t) \rightarrow \xi^{(j-1)}(t)$, $j = 1, 2, \ldots r$. Indeed, the state variables z of the observer are also differentially parameterizable in terms of e_1 and $\xi(t)$ as follows:

$$z_1 = -\left[e_1^{(n)} + \kappa_{n+r-1} e_1^{(n-1)} + \cdots + \kappa_{r+1} \dot{e}_1 + \kappa_r e_1 \right] + \xi(t)$$
$$z_2 = \dot{z}_1 - \kappa_{r-1} e_1$$
$$\quad = -\left[e_1^{(n+1)} + \kappa_{n+r-1} e_1^{(n)} + \cdots + \kappa_{r+1} \ddot{e}_1 + \kappa_r \dot{e}_1 + \kappa_{r-1} e_1 \right] + \dot{\xi}(t)$$
$$\vdots$$
$$z_r = -\left[e_1^{(n+r-1)} + \kappa_{n+r-1} e_1^{(n+r-2)} + \cdots + \kappa_1 e_1 \right] + \xi^{(r-1)}(t) \qquad (3.24)$$

We have proven the following result:

Given the nth-order linear observable SISO system considered above subject to the unstable perturbation input of the polynomial type, then, the average $(n + r)$th-order observer defined by

$$\dot{\hat{x}} = A\hat{x} + \lambda_a(y - \hat{y}) + b\omega, \quad \dot{z} = Fz + \lambda_b(y - \hat{y})$$
$$\omega = [1, 0, \ldots, 0]z,$$
$$\hat{y} = [1, 0, \ldots, 0]\hat{x}$$

$$A = \begin{bmatrix} -a_{n-1} & 1 & 0 & \cdots & 0 & 0 \\ -a_{n-2} & 0 & 1 & \cdots & 0 & 0 \\ \vdots & \vdots & \ddots & \vdots & \vdots \\ -a_1 & 0 & 0 & \cdots & 0 & 1 \\ -a_0 & 0 & 0 & \cdots & 0 & 0 \end{bmatrix}, \quad b = \begin{bmatrix} 0 \\ 0 \\ \vdots \\ 0 \\ 1 \end{bmatrix}$$

$$\lambda_a = \begin{bmatrix} \lambda_{n+r-1} \\ \lambda_{n+r-2} \\ \vdots \\ \lambda_{r+1} \\ \lambda_r \end{bmatrix}, \quad F = \begin{bmatrix} 0 & 1 & 0 & \cdots & 0 \\ 0 & 0 & 1 & \cdots & 0 \\ \vdots & \vdots & \vdots & \ddots & \vdots \\ 0 & 0 & 0 & \cdots & 1 \\ 0 & 0 & 0 & \cdots & 0 \end{bmatrix}$$

$$\lambda_b = \begin{bmatrix} \lambda_{r-1} & \lambda_{r-2} & \cdots & \lambda_1 & \lambda_0 \end{bmatrix}^T \tag{3.25}$$

exponentially asymptotically estimates the n system states with an output esti-
mation error $e_1 = y - \hat{y}$ governed by an $(n+r)$th-order characteristic equation
whose design coefficients may be entirely chosen at will in order to guarantee
location of the roots of the corresponding $(n+r)$th-degree characteristic poly-
nomial in the left half of the complex plane. Moreover, the r states in the vector
z of the observer, exponentially asymptotically estimate the unknown time poly-
nomial perturbation input $\xi(t)$ and its first $r-1$ time derivatives.

Note that the observer design demands the knowledge of the polynomial
input degree r. In those cases in which this degree is unknown, the excess of
integrations in the average observer scheme does not preclude the accurate estimation of the lower-order perturbation input.

Fig. 3.3 depicts the block diagram structure of the proposed robust Luenberger observer with dynamic output error injections for the studied observable
linear plant subject to unstable perturbation inputs of the polynomial type.

3.2.3   An Application Example

Consider the lumped mass model of a horizontally moving single-link flexible
arm with a point mass concentrated at the tip of the appendage provided with a
reduction gear of ratio $N{:}1$. See Fig. 3.4.

The dynamics of the motor shaft angle is given by

$$J_m \ddot{\theta}_m + v \dot{\theta}_m + \frac{1}{N^2} g(\dot{\theta}_m) + \frac{1}{N^2} \Gamma = \frac{K}{N} u \tag{3.26}$$

where u is the voltage signal applied for controlling the motor armature circuit,
J_m is the combined inertia of the motor and the gear, v is the viscous friction
coefficient, $g(\dot{\theta}_m)$ is the Coulomb friction torque assumed to be of the form
$g(\dot{\theta}_m) = G \operatorname{sign}(\dot{\theta}_m)$ with G being completely unknown, and Γ is the coupling
torque between the motor and the link assumed to be of the form

$$\Gamma = c(\theta_m - \theta_t) \tag{3.27}$$

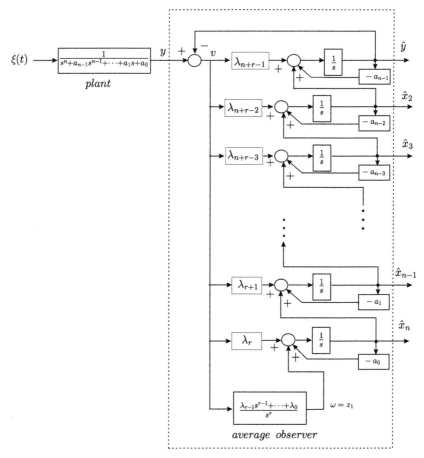

FIGURE 3.3 Robust observer with dynamic injections for estimation of the plant states subject to time polynomial inputs.

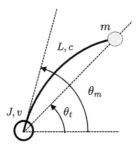

FIGURE 3.4 Single-link flexible arm.

with θ_t denoting the angular position of the tip of the arm, as illustrated in the figure. The constant c is found to be given by $c = mL^2\omega_0^2$ with ω_0 being the natural vibration frequency of the arm clamped at its base.

We assume that the motor is being controlled so that a constant positive angular velocity ω_r is achieved. It is known that in steady-state free motion, the coupling torque Γ may be made negligible if tip vibrations are prevented. They can be efficiently avoided by using prefilters that shape the command reference for the motor in such a way that the arm vibration frequency is removed.

When the flexible arm collides with a fixed obstacle, often the tip remains in contact, most of the time, with the obstacle surface. This is in contradistinction to standard rigid arms, where rebounds happen as a consequence of the interaction between the control and the mechanical constraint. After collision, provided that a good servo-controller is available for the motor, the angle of the motor θ_m keeps growing at a constant rate while the tip angle remains at a fixed value θ_0.

We deduce that the coupling torque grows linearly after the collision instant. Hence, a ramp perturbation input is expected to affect the closed-loop motor dynamics. We thus placed a second-order integration dynamic injection into our observer.

In our simulation model, during the free motion phase, the perturbation $\xi(t)$ is related to the unknown Coulomb friction term alone (i.e., $\xi(t) = -G/(J_m N^2)$ with G constant for positive values of $\dot{\theta}$). After the collision, the perturbation $\xi(t)$ is given by the sum of the Coulomb friction term and the linear growing torque scaled by the factor $J_m N^2$.

Then, for a collision time occurring at time $t = t_i$, under the assumption of $\dot{\theta} > 0$, we have the following model of the flexible arm-motor system:

$$\ddot{\theta}_m + \frac{v}{J_m}\dot{\theta}_m = \frac{K}{J_m N}u + \xi(t)$$

$$\xi(t) = \begin{cases} -G/(J_m N^2) & \text{for } t \le t_i \\ -1/(J_m N^2)\left[G + mL^2\omega_0^2(\theta - \theta_0)\right] & \text{for } t > t_i \end{cases} \qquad (3.28)$$

with t_i being the collision instant approximately estimated as $t_0 + \omega_0/\omega_r$ with ω_r being the constant angular reference velocity tracked by the motor shaft and t_0 the initial time.

We propose the following robust GPI observer:

$$\dot{\theta}_{me} = \omega_{me} + \lambda_3(y - \hat{y})$$

$$\dot{\omega}_{me} = -\frac{v}{J_m}\omega_{me} + \lambda_2(y - \hat{y}) + z_1 + \frac{K}{N J_m}u$$

$$\dot{z}_1 = z_2 + \lambda_1(y - \hat{y})$$

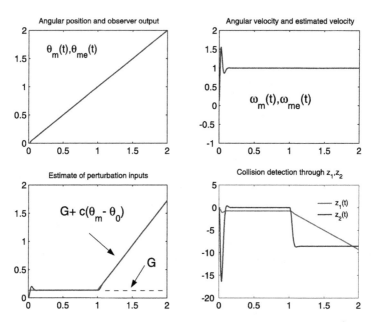

FIGURE 3.5 Collision detection in a controlled flexible link via robust GPI observer.[4]

$$\dot{z}_2 = \lambda_0(y - \hat{y})$$
$$\hat{y} = \theta_{me} \qquad\qquad (3.29)$$

where the set of design coefficients λ_i, $i = 0, 1, 2, 3$, are chosen so that the characteristic polynomial of the estimation error system dynamics given by

$$p(s) = s^4 + (\lambda_3 + \frac{\nu}{J_m})s^3 + (\lambda_2 + \frac{\nu}{J_m}\lambda_3)s^2 + \lambda_1 s + \lambda_0 \qquad\qquad (3.30)$$

has all its roots in the strict left half of the complex plane, i.e., it is made into a Hurwitz polynomial. We have chosen the λ coefficients so that the characteristic polynomial matches those of the desired polynomial $p_d(t) = (s^2 + 2\zeta_o\omega_o s + \omega_o^2)$ with $\zeta_o, \omega_o > 0$.

Fig. 3.5 depicts the performance of the proposed robust observer in the estimation of the state variables of the motor. The figure also shows the estimation of the unstable input signal represented by the sum of the Coulomb friction torque and the temporarily growing coupling torque. Note that while the collision has not taken place, the excess of integrators does not preclude the estimation of the unknown constant perturbation input represented by the Coulomb friction term (which would have only required one integration in the

---

4. For interpretation of the references to color in this and the following figures, the reader is referred to the web version of this chapter.

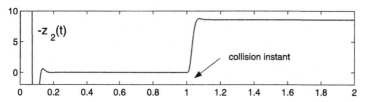

FIGURE 3.6 Collision detection in a controlled flexible link via robust GPI observer.

output error injection). The monitoring of the unstable variable z_1 or of its time derivative z_2, also available from the observer, yields, therefore, a quite efficient collision detection scheme.

Fig. 3.6 shows that for the estimation of the variable z_2, the observer needs less than 0.1 s to converge toward the real value of the perturbation. Such a time delay for detecting a collision is deemed quite acceptable for flexible robots.

It should be stressed that designing algorithms to rapidly detect collisions in robots is of the utmost interest since they allow one to switch from position control to force control in real time when contact is produced. This results in the avoidance of possible damages either in the robot or in the surrounding environment.

3.3 AN OBSERVER-BASED APPROACH TO GPI PERTURBATION REJECTION

Consider the perturbed system

$$y^{(n)} = u + \varphi(t) \tag{3.31}$$

where $\varphi(t)$ is a sufficiently differentiable bounded signal. We first establish the properties of an observer similar to that developed before in the presence of a nonpolynomial perturbation input. The presence of the control input does not introduce any change into the previous or present developments.

We propose the following observer:

$$
\begin{aligned}
\hat{y}^{(n)} &= u + \lambda_{n+r-1}(y - \hat{y})^{(n-1)} + \ldots + \lambda_r(y - \hat{y}) + z_1 \\
\dot{z}_1 &= z_2 + \lambda_{r-1}(y - \hat{y}) \\
\dot{z}_2 &= z_3 + \lambda_{r-2}(y - \hat{y}) \\
&\ \vdots \\
\dot{z}_{r-1} &= z_r + \lambda_1(y - \hat{y}) \\
\dot{z}_r &= \lambda_0(y - \hat{y})
\end{aligned}
\tag{3.32}
$$

The output estimation error $e_1 = (y - \hat{y})$ satisfies the perturbed injected dynamics

$$e_1^{(n+r)} + \lambda_{(n+r-1)} e_1^{(n+r-1)} + \cdots + \lambda_r e_1^{(r)} + \lambda_{r-1} e_1^{(r-1)} + \cdots + \lambda_0 e_1 = \varphi^{(r)}(t) \tag{3.33}$$

If $\varphi^{(r)}(t)$ is bounded by a small ϵ-neighborhood of zero and the roots of the polynomial

$$p(s) = s^{n+r} + \lambda_{n+r-1} s^{n+r-1} + \cdots + \lambda_1 s + \lambda_0 \tag{3.34}$$

are chosen, by means of a suitable specification of the set of coefficients $\{\lambda_{n+r-1}, \cdots, \lambda_0\}$ to be sufficiently deep into the left half of the complex plane, then the estimation error e_1 and its time derivatives will be sufficiently small to be bounded by a small ϵ-dependent radius ball centered around the origin of the error phase space.

As before, the quantity z_1 provides an ϵ-approximate online estimate of $\varphi(t)$.

3.3.1   An Example

Consider the system

$$\ddot{y} = au + \varphi(t) \tag{3.35}$$

We use the observer-based controller

$$u = \frac{1}{a} \left\{ \ddot{y}^*(t) - \left[\frac{k_2 s^2 + k_1 s + k_0}{s(s + k_3)} \right] (y - y^*(t)) - \hat{\varphi}(t) \right\} \tag{3.36}$$

where $\hat{\varphi}(t)$ is generated via the following observer:

$$\begin{aligned}
\hat{\varphi}(t) &= z_1 \\
\dot{\hat{x}}_1 &= \hat{x}_2 + \lambda_4(y - \hat{y}) \\
\dot{\hat{x}}_2 &= au + z_1 + \lambda_3(y - \hat{y}) \\
\dot{z}_1 &= z_2 + \lambda_2(y - \hat{y}) \\
\dot{z}_2 &= z_3 + \lambda_1(y - \hat{y}) \\
\dot{z}_3 &= \lambda_0(y - \hat{y}) \\
y &= x_1, \quad \hat{y} = \hat{x}_1
\end{aligned} \tag{3.37}$$

We used the following system for simulation purposes:

$$a = 1, \quad \varphi(t) = e^{-\sin^3(t)} \sin(5t) \tag{3.38}$$

and the following observer:

$$
\begin{aligned}
\dot{\hat{x}}_1 &= \hat{x}_2 + \lambda_4(y - \hat{x}_1) \\
\dot{\hat{x}}_2 &= u + z_1 + \lambda_3(y - \hat{x}_1) \\
\dot{z}_1 &= z_2 + \lambda_2(y - \hat{x}_1) \\
\dot{z}_2 &= z_3 + \lambda_1(y - \hat{x}_1) \\
\dot{z}_3 &= \lambda_0(y - \hat{x}_1)
\end{aligned} \tag{3.39}
$$

We used the following observer-based controller with amplitude saturation on the control input:

$$
\begin{aligned}
u_{calculated} &= -\hat{\varphi}(t) - 2\zeta_c \omega_c \hat{x}_2 - \omega_c^2 y \\
u &= u_{max} \operatorname{sat}(u_c) \\
u_{max} &= 5
\end{aligned} \tag{3.40}
$$

with $\zeta_c = 1$ and $\omega_c = 2.5$.

To block the "peak" involved at the beginning of the transient response, we use the following expression for the perturbation estimate:

$$
\begin{aligned}
\hat{\varphi}(t) &= \text{if } t < \delta, \text{ then } 0 \text{ else } z_1 \\
\hat{y} &= \text{if } t < \delta, \text{ then } 0 \text{ else } \hat{x}_2
\end{aligned}
$$

with $\delta = 0.1$. We used the following design parameter values for the observer:

$$
\begin{aligned}
\lambda_4 &= (p + 4\zeta\omega) \\
\lambda_3 &= (2\omega^2 + 4\zeta^2\omega^2 + 4\zeta\omega p) \\
\lambda_2 &= (4\zeta\omega^3 + 2\omega^2 p + 4\zeta^2\omega^2 p) \\
\lambda_1 &= (4\zeta\omega^3 p + \omega^4) \\
\lambda_0 &= \omega^4 p
\end{aligned}
$$

with $\zeta = 20$, $\omega = 100$, and $p = 100$.

Fig. 3.7 illustrates the stabilization in spite of the lumped disturbance input φ which is reconstructed after a small transient time (see Fig. 3.8). The time derivative estimation convergence is illustrated in Fig. 3.9.

3.4 THE BUCK CONVERTER

Fig. 3.10 shows the electrical circuit of the dc/dc buck converter. Using the Kirchhoff laws, we get the following average model of the buck converter:

The system of differential equations describing the dynamics of the Buck converter is obtained through the application of Kirchoff's current and voltage

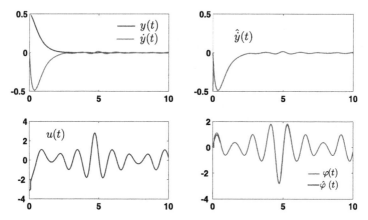

FIGURE 3.7 Performance of GPI observer-based stabilization controller for perturbed system.

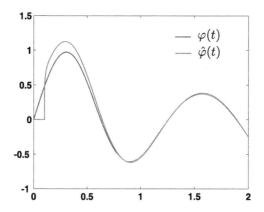

FIGURE 3.8 Inset showing the perturbation estimation via a GPI-observer.

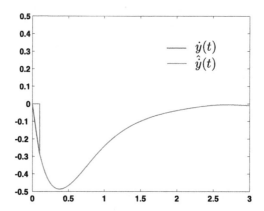

FIGURE 3.9 Inset showing the state estimation via a GPI-observer.

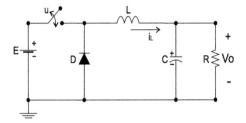

FIGURE 3.10 The electrical circuit of Buck converter.

laws and is given by

$$L\frac{di_L}{dt} = -v_o + Eu \tag{3.41}$$

$$C\frac{dv_o}{dt} = i_L - (\frac{1}{R})v_o \tag{3.42}$$

where i_L is the converter input current, and v_o is the converter output voltage. The control input is represented by the variable u, which models a switch position function taking values on the discrete set $\{0, 1\}$. If it is considered an average value in the entry of control in each period of switching of the interruptor (Mosfet Transistor), this average control input u_{av} is bounded to interval $[0, 1]$. From (3.41) we observed that the buck converter system is a second-order linear system.

The Kalman controllability and the observability matrix of the system is given by

$$C = \begin{bmatrix} \frac{E}{L} & 0 \\ 0 & \frac{E}{LC} \end{bmatrix} \tag{3.43}$$

$$O = \begin{bmatrix} 0 & 1 \\ \frac{1}{C} & -\frac{1}{RC} \end{bmatrix} \tag{3.44}$$

The determinant of the controllability matrix is

$$\frac{E^2}{L^2C} \neq 0$$

Therefore the system is controllable and flat [1]. Moreover, the flat output of the system is the buck converter output voltage variable

$$F = v_o \tag{3.45}$$

The flatness differential of the system implies that all variables of state of the system and the variable of input control are parameterizable in terms of $F = v_o$

and a finite number of its derivatives with respect to time:

$$F = v_o$$

$$i_L = C\dot{F} + \frac{1}{R}F \tag{3.46}$$

and the average control input is

$$u = \frac{LC}{E}\ddot{F} + \frac{L}{RE}\dot{F} + \frac{1}{E}F \tag{3.47}$$

Moreover, the average model given in [1] we can see that the system is observable when v_o is taken as the output variable, i.e., the Kalman observability matrix given as

$$\mathcal{O} = \begin{bmatrix} C \\ CA \end{bmatrix} = \begin{bmatrix} 0 & 1 \\ \frac{1}{C} & -\frac{1}{RC} \end{bmatrix} \tag{3.48}$$

complies with the property to be of full range. This implies that all variables of state of the system that are not the flat output can be estimated through integral reconstructors involving the input and output.

3.4.1 Problem Formulation

Given a dc/dc buck converter, it is desired to have the output voltage of the converter track a given smooth reference voltage profile in the presence of sudden static load changes in parallel with the load resistance of the Buck converter.

3.4.2 A GPI Observer-Based ADRC Approach

To increase the robustness of the output voltage of Buck converter, we propose an Active Disturbance Rejection Control with respect to the average input control law presented in (3.47). From (3.47) we have

$$u = \frac{LC}{E}\ddot{F} + \xi(t) \tag{3.49}$$

where

$$\xi(t) = \frac{L}{RE}\dot{F} + \frac{1}{E}F \tag{3.50}$$

Here, $\xi(t)$ represents the included unmodeled dynamics and external disturbance inputs. A GPI disturbance observer can be proposed as follows:

$$\dot{\hat{F}}_0 = F_1 + \lambda_3(F - F_0)$$

$$\dot{\hat{F}}_1 = \frac{E}{LC}u + z_1 + \lambda_2(F - F_0)$$
$$\dot{\hat{z}}_1 = z_2 + \lambda_1(F - F_0)$$
$$\dot{\hat{z}}_2 = \lambda_0(F - F_0) \tag{3.51}$$

where F_0 represents the redundant estimation of F. The estimation error $e = F - F_0$ satisfies the following linear perturbed differential equation:

$$e^{(4)} + \lambda_3 e^{(3)} + \lambda_2 e^{(2)} + \lambda_1 \dot{e} + \lambda_0 e = 0 \tag{3.52}$$

The coefficients $(\lambda_3, \lambda_2, \lambda_1, \lambda_0)$ are chosen sufficiently high so as to have the poles corresponding with characteristic polynomial $p(s) = s^4 + k_3 s^3 + k_2 s^2 + k_1 s + k_0$ far in the left half of the complex plane.

An Active Disturbance Rejection Controller (ADRC) is synthesized cancelling the disturbance term $(\xi(t))$ in terms of $\hat{z}_1$ is proposed as

$$u = \frac{LC}{E}\left[[F^*(t)]^{(2)} - k_1\left(F_1 - \dot{F}^*(t)\right) - k_0\left(F - F^*(t)\right) \right] \tag{3.53}$$

The coefficients k_1 and k_0 correspond with a Hurwitz polynomial of second order of the form $p(s) = s^2 + 2\zeta \omega_n s + \omega_n^2$ to guarantee that the poles are in the left half of the complex plane with $\zeta > 0$ and $\omega_n > 0$.

3.4.3 A Flat Filtering Approach

Consider the second-order unperturbed system

$$\ddot{F} = u \tag{3.54}$$

Ii is desired to asymptotically track a given smooth output reference signal $F^*(t)$. The nominal control input is

$$u^*(t) \doteq \ddot{F}^*(t) \tag{3.55}$$

The output tracking error dynamics is given by

$$e_F = F - F^*(t), \quad e_u = u - u^*(t) \tag{3.56}$$

The GPI Control eliminates the need for asymptotic state observers through integral reconstructors that involve the input and output [14]. For this case, an integral reconstructor of $\dot{e}_F$ is obtained as

$$\hat{\dot{e}}_F = \int_0^t e_u(\sigma)d\sigma, \quad \dot{e}_F = \hat{\dot{e}}_F + \hat{\dot{e}}_F(0) \tag{3.57}$$

The incorporation of the estimator of $\hat{e}_F$ in a linear feedback controller implies that this law may be affected by constant values. To correct this destabilizing effect caused by the mistakes of estimate and by external perturbations, we used the compensation through integrations of error of the flat output. The following linear feedback controller is proposed:

$$e_u = -k_3 \int_0^t e_u(\sigma)d\sigma - k_2 e_F - k_1 \int_0^t e_F(\sigma)d\sigma - k_0 \int_0^t \int_0^\tau e_F(\sigma)d\tau d\sigma$$

(3.58)

where the coefficients $(k_3, \ldots, k_0)$ are chosen to guarantee the stability of the tracking error e_F.

Defining $\gamma = \int_0^t e_F(\sigma)d\sigma - (k3/k0)\dot{e}_F(0)$, the closed-loop tracking error may be represented as

$$e_u = \ddot{e}_F = -k_3 \int_0^t e_u(\sigma)d\sigma - k_2 e_F - k_1 \gamma - k_0 \eta$$

(3.59)

$$\dot{\gamma} = e_F$$

(3.60)

$$\dot{\eta} = \gamma$$

(3.61)

According to (3.58), the following implicit expression for the linear feedback controller is obtained:

$$e_u = -k_2 e_F - k_1 \int_0^t (-k_1 e_F(\sigma)d\sigma - k_3 e_u(\sigma)d\sigma) - k_0 \int_0^t \int_0^\tau e_F(\sigma)d\tau d\sigma$$

(3.62)

The values of the design parameters $\{k_3, k_2, k_1, k_0\}$ are chosen so that the closed-loop characteristic polynomial

$$p(s) = s^4 + k_3 s^3 + k_2 s^2 + k_1 s + k_0$$

(3.63)

has all its roots in the left half of the complex plane.

The controller parameters were chosen according to the following desired closed loop characteristic polynomial

$$p(s) = (s^2 + 2\zeta \omega_n s + \omega_n{}^2)^2$$

(3.64)

The Laplace transform of (3.62) allows us to obtain an expression for the linear feedback output tracking controller; this expression is a lead compensator network, and Fig. 3.11 shows us the block diagram representation of the second-order plant with GPI controller system

$$e_u = -\left[\frac{k_2 s^2 + k_1 s + k_0}{s(s + k_3)}\right] e_F$$

(3.65)

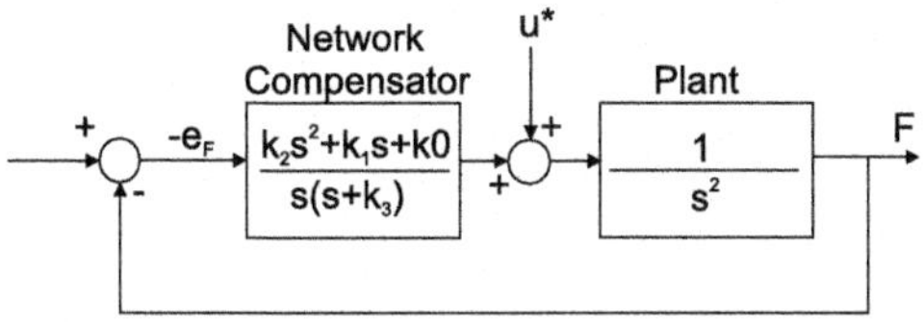

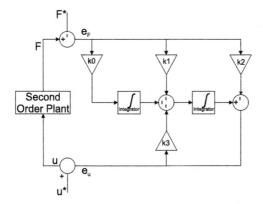

FIGURE 3.11 Lead network realization with GPI controller for a second-order plant.

FIGURE 3.12 GPI control of a second-order plant via integral reconstructor.

The expression in (3.65) can be represented in state variables. Defining $x = e_F/(s(s + k_3))$, x can be interpreted like a filter, and the representation in state variables is

$$x = \frac{e_F}{s(s + k_3)}$$
$$x = x_1$$
$$\dot{x}_1 = x_2$$
$$\dot{x}_2 = -k_3 x_2 + e_F \qquad (3.66)$$

The feedback controller (3.66) is synthesized in the time domain as

$$u = \ddot{F}^*(t)(k_2 k_3 - k_1)\phi - k_0\varphi - k_2 e_F$$
$$\dot{\varphi} = e_F = F(t) - F^*(t)$$
$$\dot{\phi} = \varphi \qquad (3.67)$$

The last expression is known as the flat-filtered based controller [17]. In Fig. 3.12 we can show the flat-filtered control for a second-order plant via integral reconstructor.

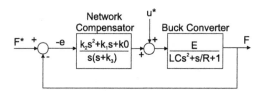

FIGURE 3.13 Block diagram for a Buck converter with flat filtering control.

3.4.3.1 The Buck Converter Case

According to (3.41) and (3.42), the model of the Buck converter is a second-order plant. Therefore it is easy to implement a Flat-filtered controller (Fig. 3.13). An input–output transfer function for the Buck converter can be expressed in the form

$$F(s) = \frac{Eu(s)}{LCs^2 + \frac{1}{R}s + 1} \tag{3.68}$$

Substituting the flat-filtered control u into the transfer function (3.68), we get a resulting characteristic polynomial with the lead network compensator in the form

$$m(s) = s^4 + (k_3 + \frac{1}{RLC})s^3 + \frac{1}{LC}(\frac{k_3}{R} + Ek_2 + 1)s^2$$
$$+ \frac{1}{LC}(k_3 + Ek_1)s + \frac{Ek_0}{LC} \tag{3.69}$$

To ensure asymptotic stability of the tracking error e_F, all roots in the polynomial $m(s)$ need to be in the left side of the complex plane. A fourth-order polynomial Hurwitz is of the form

$$H(s) = (s^2 + 2\xi\omega_n s + \omega_n^2)^2$$

where

$$k_3 = 4\xi\omega_n - \frac{1}{RLC} > 0$$

$$k_2 = 2LC\omega_n^2(2\xi^2 + 1) - \frac{4\xi\omega_n}{R} + \frac{1}{R^2LC} - 1 > 0$$

$$k_1 = 4\xi\omega_n(\omega_n^2 LC - 1) + \frac{1}{RLC} > 0$$

$$k_0 = LC\omega_n^4 > 0$$

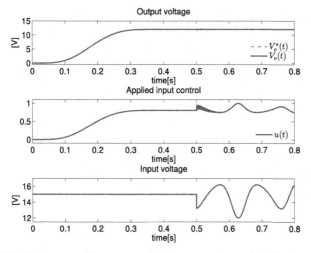

FIGURE 3.14 Trajectory tracking response for the output voltage of the Buck converter with flat-filtering approach and variable input voltage.

3.4.4 Simulation Results

For simulation results, a buck converter was designed with the following specifications: $L = 3$ [mH], $C = 470$ [μF], $R = 150$ [Ω], $E = 15$ [V]. Due to the linearity of the system and to corroborate the robustness of the control, the input of the Buck converter was implemented as

$$E(t) = \begin{cases} 15 & \text{if } t \in [0, 0.5) \\ 15 + 2e^{-\sin^2(20t)}\cos^2(50t) & \text{if } t \in [0.5, \infty) \end{cases} \tag{3.70}$$

The gains values for the Flat-filtering approach are

$$\omega_n = 2500$$
$$\zeta = 1$$

In Fig. 3.14 a rest-to-rest trajectory tracking is shown for the output voltage in the Buck circuit. The initial voltage is set to 0 [V], and the final voltage was set to 12 [V] in a time lapse of 0.4 [s]. The input voltage was set to 15 [V], and after 0.5 [s], this value is variable.

3.4.5 Experimental Results

A buck converter was designed with the following specifications: $L = 3$ [mH], $C = 470$ [μF], $R = 150$ [Ω], $E = 15$ [V], $f = 5$ [kHz].

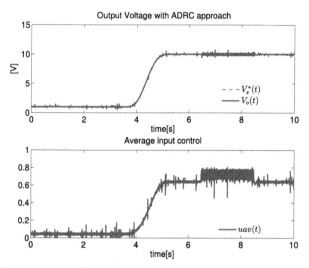

FIGURE 3.15 Trajectory tracking response for the output voltage of the Buck converter with ADRC approach and variable input average voltage with static load change.

A rest-to-rest output voltage reference trajectory tracking task from 3.5 [s] to 5.5 [s] was implemented for the output voltage of the Buck converter for three different control laws. The first control is an ADR controller and the second control is a flat-filtering approach. These controls were implemented on an STM32F4 Discovery target. The only variable measured was the output voltage of the converter. The control law, Bezier polynomial, and PWM were implemented on the target with a PWM frequency of 5 [kHz].

In Fig. 3.15 an ADR control response for the desired trajectory tracking is shown. A static load variation was activated at 6.5 [s]. This change corresponds with a resistor in parallel with the resistor of the Buck converter. The average input control is shown in the same figure. The gains values for the ADRC approach were

$$\omega_n = 10$$
$$\zeta = 1$$

However, the implementation for the Flat-filtering approach presented a better performance than the ADRC implementation. In Fig. 3.16, we present the response of the output voltage from the converter. The average input control has less noise than the output voltage. The gains values for the Flat-filtering approach were

$$\omega_n = 450$$
$$\zeta = 1.5$$

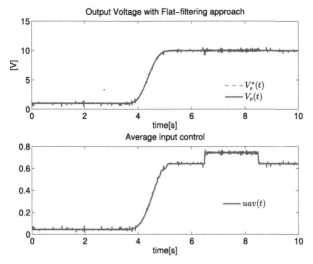

FIGURE 3.16 Trajectory tracking response for the output voltage of the Buck converter with Flat-filtering approach and variable input average voltage with static load change.

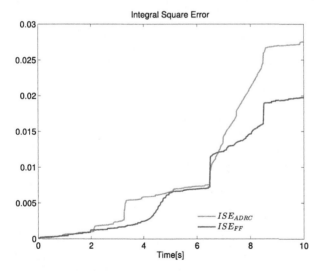

FIGURE 3.17 Integral Square Error for the trajectory tracking response with static load change.

Both control laws present similar performance, but if we compare the Integral Square Error criterion (ISE), which gives us a better comparison between the different control schemes, the Flat-filtering presents the best performance in this experiment. In Fig. 3.17 we show the ISE for different control laws.

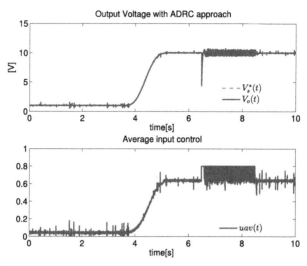

FIGURE 3.18 Trajectory tracking response for the output voltage of the Buck converter with ADRC approach and variable input average voltage with dynamic load change.

A second experiment for the Buck converter was performed for the comparison of the different control laws. In this case a dynamic load, in the form of a DC motor, was connected in cascade with the Buck converter.

For the ADRC approach in the second experiment, in Fig. 3.18, we present the reference voltage and the average input control. The average control presents a saturation, but the control maintains the output voltage.

Finally, in Fig. 3.19 we present implementation of the Flat-filtering. It presents a saturation in the control input, but the response presents less noise than the ADRC and PID implementations.

For comparison purposes, as the static load, the ISE criterion is presented in Fig. 3.20. The ADRC presents a better performance than the Flat-Filtering.

3.5 EXAMPLE: A PENDULUM SYSTEM

Consider a pendulum system, consisting of a bar of length L, mass M, and an extra mass m_e attached to its tip (see Fig. 3.21). The pendulum is actuated by a DC motor through a gear box. The angular position of the motor shaft is denoted by θ_m, and that of the pendulum is denoted by θ. The nonlinear model of the system is given by

$$\left[(m_e + \frac{M}{3})L^2 + J\right]\ddot{\theta} + (m_e + \frac{M}{2})gL\sin(\theta) = \tau \qquad (3.71)$$

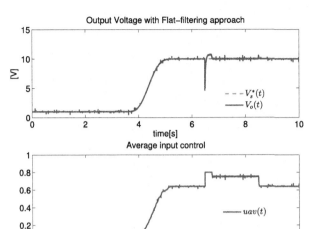

FIGURE 3.19 Trajectory tracking response for the output voltage of the Buck converter with Flat-filtering approach and variable input average voltage with dynamic load change.

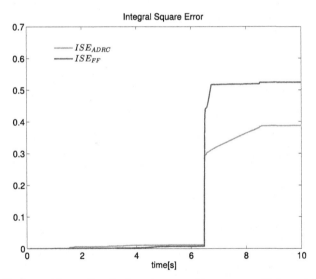

FIGURE 3.20 Integral Square Error for the trajectory tracking response with dynamic load change.

where J includes the moments of inertia of the gear train and the rotor shaft of the DC motor. The parameter N stands for the gear ratio. The pendulum angle θ is related via $\theta = \theta_m / N$ to the motor shaft displacement and the torque applied to the pendulum $\tau = \tau_m N$.

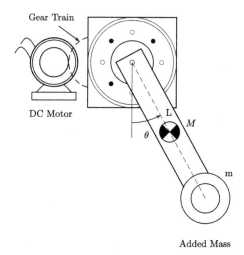

Gear Train

DC Motor

L

M

θ

m

Added Mass

FIGURE 3.21 Schematic of the nonlinear pendulum.

Some terms as viscous and Coulomb friction, dead zones, and backslash in the geared motor are considered as unmodeled dynamics. In this case, these perturbation inputs and the nonlinear gravitational effects are assumed to be unknown and, according to our proposal, considered to be a bounded perturbation input signal lumped into a single additive time-varying function $\xi(t)$. The simplified model of the pendulum system, in terms of the pendulum angle, is given by

$$\ddot{\theta} = \alpha\tau + \xi(t) \qquad (3.72)$$

with $\alpha = 1/(m_e + \frac{M}{3})L^2$ neglecting the parameter J. The control gain α is assumed to be known. The problem formulation is the following: *Consider the perturbed nonlinear system (3.71) and its simplified representation (3.72). It is desired to track a given angular position reference trajectory $\theta^*(t)$ by means of the proposed robust GPI controller scheme, in spite of the nonmodeled dynamics, the lack of knowledge of some nonlinearities, and possible arising of external disturbances, lumped as the time function $\xi(t)$.*

By (3.72) the unperturbed feedforward control input is $u^*(t) = \frac{1}{\alpha}\ddot{\theta}^*(t)$. Thus the input error e_u is given as follows:

$$e_u(t) = u(t) - u^*(t) = u(t) - \frac{1}{\alpha}\ddot{\theta}^*(t) \qquad (3.73)$$

By the Active Disturbance Rejection Philosophy the effects of the generalized disturbance input $\xi(t)$ are locally modeled by means of the following approximation:

$$\ddot{e}_\theta = \alpha e_u + z_1$$
$$\dot{z}_i = z_{i+1}, i = 1, 2, \ldots, m - 1 \qquad (3.74)$$
$$\dot{z}_m = 0$$
$$e_\theta = \theta - \theta^*$$

where $m \in \mathbb{Z}^+$ indicates the degree of approximation of the ultralocal model. Here, to illustrate the effect of the approximation degree of the ultralocal model, we consider the following cases: $m = 2, 3, 4, 5$ with $n = 2$.

The flat-filtering feedback controllers are given as follows:

$$\dot{\zeta}_i = \zeta_{i+1}, \qquad i = 1, \cdots, m$$
$$\dot{\zeta}_{m+1} = -\lambda_{m+2}\zeta_{m+1} + e_\theta$$
$$e_{\theta_f} = \zeta_1$$
$$e_{um} = \frac{1}{\alpha}((\lambda_{m+1}\lambda_{m+2} - \lambda_m)\zeta_{m+1} - \lambda_{m-1}\zeta_m - \lambda_{m-2}\zeta_{m-1}, \cdots,$$
$$\qquad -\lambda_1\zeta_2 - \lambda_0\zeta_1 - \lambda_{m+1}e_\theta) \qquad (3.75)$$

For $m = 2$,

$$\dot{\zeta}_i = \zeta_{i+1}, i = 1, 2$$
$$\dot{\zeta}_3 = -k_4\zeta_3 + e_\theta$$
$$e_{\theta_f} = \zeta_1$$
$$e_{u2} = \frac{1}{\alpha}((k_3k_4 - k_2)\zeta_3 - k_1\zeta_2 - k_0\zeta_1 - k_3e_\theta) \qquad (3.76)$$

where e_{θ_f} is the flat output. The characteristic polynomial of the linear dominant dynamics of the closed-loop tracking error is

$$p_{cm}(s) = s^{m+3} + \lambda_{m+2}s^{m+2} + \lambda_{m+1}s^{m+1} + \cdots + \lambda_1 s + \lambda_0 \qquad (3.77)$$

For $m = 3$,

$$\dot{\zeta}_i = \zeta_{i+1}, i = 1, 2, 3$$
$$\dot{\zeta}_4 = -k_5\zeta_4 + e_\theta$$
$$e_{\theta_f} = \zeta_1$$
$$e_{u3} = \frac{1}{\alpha}((k_4k_5 - k_3)\zeta_4 - k_2\zeta_3 - k_1\zeta_2 - k_0\zeta_1 - k_4e_\theta)$$
$$p(s) = s^6 + k_5s^5 + k_4s^4 + k_3s^3 + k_2s^2 + k_1s + k_0 \qquad (3.78)$$

FIGURE 3.22 Pendulum system prototype.

For $m = 4$,

$$\dot{\zeta}_i = \zeta_{i+1}, i = 1, 2, \cdots, 4$$
$$\dot{\zeta}_5 = -k_6\zeta_5 + e_\theta$$
$$e_{\theta_f} = \zeta_1$$
$$e_{u4} = \frac{1}{\alpha}((k_5k_6 - k_4)\zeta_5 - k_3\zeta_4 - k_2\zeta_3 - k_1\zeta_2 - k_0\zeta_1 - k_5e_\theta)$$
$$p(s) = s^7 + k_6s^6 + k_5s^5 + k_4s^4 + k_3s^3 + k_2s^2 + k_1s + k_0 \qquad (3.79)$$

For $m = 5$,

$$\dot{\zeta}_i = \zeta_{i+1}, i = 1, 2, \cdots, 5$$
$$\dot{\zeta}_6 = -k_7\zeta_6 + e_\theta$$
$$e_{\theta_f} = \zeta_1$$
$$e_{u5} = \frac{1}{\alpha}((k_6k_7 - k_5)\zeta_6 - k_4\zeta_5 - k_3\zeta_4 - k_2\zeta_3 - k_1\zeta_2 - k_0\zeta_1 - k_6e_\theta)$$
$$p(s) = s^8 + k_7s^7 + k_6s^6 + k_5s^5 + k_4s^4 + k_3s^3 + k_2s^2 + k_1s + k_0 \qquad (3.80)$$

3.5.1 Experimental Results

The experimental prototype consists of a DC motor from Electro-Craft servo products (model E586), which drives the pendulum system via a synchronous belt a pulley and a gear box with a ratio $N = 16:1$ as is shown in Fig. 3.22.

The angular position of the pendulum is measured using an incremental optical encoder of 2500 pulses per revolution.

A power amplifier Copley Controls model 412CE drives the motor in torque mode. The data acquisition is carried out through a data card, model PCI6221 National Instruments. The control strategy was implemented in the Matlab-Simulink platform. The sampling time is set to be 0.001 [s]. The parameters of the pendulum system are $M = 0.095$ [kg], $m_e = 0.190$ [kg], and $L = 0.3$ [m].

The initial condition of the pendulum position is $\theta = 0$ radians. The GPI control gain parameters were chosen according with controller design parameter m as:

For $m = 2$, $\varsigma_2 = 5$, $\omega_2 = 28$, and $p_2 = 28$,

$$p_{c2}(s) = (s + p_2)(s^2 + 2\varsigma_2\omega_2 s + \omega_2^2)^2 \tag{3.81}$$

For $m = 3$, $\varsigma_3 = 7$, and $\omega_3 = 14$,

$$p_{c3}(s) = (s^2 + 2\varsigma_3\omega_3 s + \omega_3^2)^3 \tag{3.82}$$

For $m = 4$, $\varsigma_4 = 5$, $\omega_4 = 16$, and $p_4 = 16$,

$$p_{c4}(s) = (s + p_4)(s^2 + 2\varsigma_4\omega_4 s + \omega_4^2)^3 \tag{3.83}$$

and, for $m = 5$, $\varsigma_5 = 5$, and $\omega_5 = 12$,

$$p_{c5}(s) = (s^2 + 2\varsigma_5\omega_5 s + \omega w_5^2)^3 \tag{3.84}$$

The first test is shown in Fig. 3.23 for the tracking results on a rest-to-rest trajectory, obtained with the help of a Bézier polynomial of degree 9. The pendulum starts on the initial position $\theta(0) = 0$, and then it is moved to a first rest position $\theta(1.6) = \frac{\pi}{2}$ [rad] during the interval $[1, 1.6]$ [s]. Once the pendulum is stabilized, it is taken, at time $t = 3.1$ [s], toward a final position $\theta(3.8) = -\frac{\pi}{2}$ [rad]. In this test, all the controllers had accurate tracking results. Fig. 3.23 (bottom) depicts a zoom for trajectory tracking. The tracking error is shown in Fig. 3.24, which is bounded in all cases in spite of the nonlinear terms present in the system as the Coulomb friction, gravitational forces, viscous friction, etc. Fig. 3.25 depicts the torque control input for each approximation. A performance index consisting on the integral square of the trajectory tracking error (ISI) was set for assessing the effect of the approximation parameter m. Consider the following performance index:

$$ISI = \epsilon \int_0^t e_\theta^2(\sigma)d\sigma \tag{3.85}$$

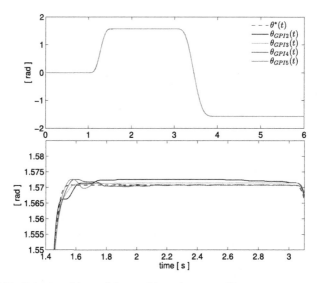

FIGURE 3.23 Behavior of the pendulum position trajectory tracking.

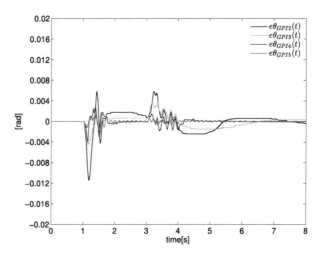

FIGURE 3.24 Trajectory tracking error.

with $\epsilon \in \mathbb{R}^+$; in this case, $\epsilon = 2 \times 10^5$. Fig. 3.26 shows how the increase of m allows a better trajectory tracking, illustrated in the performance index.

The second test consisting in tracking the same desired trajectory used in the first test was carried out. This time the system is perturbed by adding an extra mass of 0.095 [kg] (see Fig. 3.27) increasing the inertia of the pendulum and the gravitational force. Here, the best result is obtained using an approximation parameter $m = 5$ while compensating the effects of disturbances and unmod-

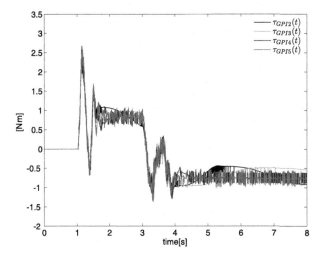

FIGURE 3.25 Torque control input.

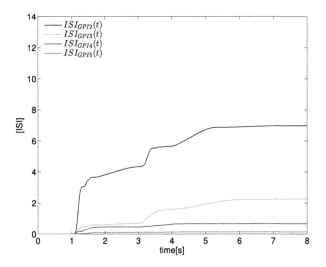

FIGURE 3.26 Performance index behavior.

eled dynamics. As shown in the results (Fig. 3.28), the trajectory tracking error for the pendulum position when the added mass is applied remain restricted to the interval of $[-0.002, 0.002]$ [rad] (see Fig. 3.29). These results show a remarkable robustness of the GPI controller and how the use of nested integration compensation terms improves the controller performance (see Fig. 3.30). Fig. 3.31 shows the evolution of the control input torques τ. The experimental results illustrate the fact that the applied torque on the DC motor is bounded. On

FIGURE 3.27 Pendulum system prototype with unknown added mass.

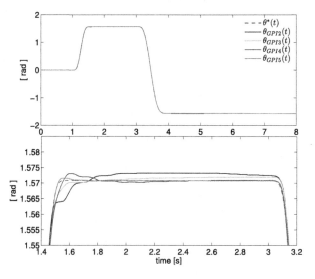

FIGURE 3.28 Behavior of the pendulum position trajectory tracking.

the other hand, the corrective reaction of the robust GPI control against parameter variations is performed within a reasonable interval.

3.6 TWO-LINK ROBOT MANIPULATOR

Consider the two-link planar robot manipulator shown in Fig. 3.32, which consists of two actuated links that move in the plane. The generalized coordinates of the system are given by the angular positions $\mathbf{q} = [\theta \ \phi]^T$, and the dynamic

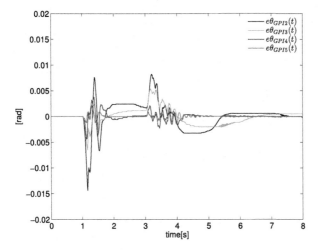

FIGURE 3.29   Trajectory tracking error.

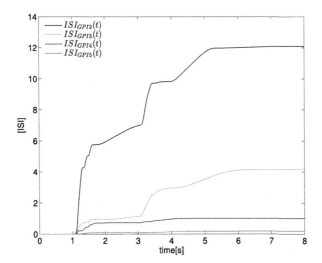

FIGURE 3.30   Performance index behavior for different approximations.

model of the robot manipulator [19], [20], [21] is given as

$$\mathbf{M}(\mathbf{q})\ddot{\mathbf{q}} + \mathbf{C}(\mathbf{q}, \dot{\mathbf{q}})\dot{\mathbf{q}} + \mathbf{G}(\mathbf{q}) = \boldsymbol{\tau} - \mathbf{D}\dot{\mathbf{q}} \qquad (3.86)$$

where the matrices

$$\mathbf{M}(\mathbf{q}) = \begin{bmatrix} m_1 l_{c1}^2 + m_2 l_1^2 + m_2 l_{c2}^2 + 2m_2 l_1 l_{c2} \cos(\phi) + I_1 + I_2 & m_2 l_{c2}^2 + m_2 l_1 l_{c2} \cos(\phi) + I_2 \\ m_2 l_{c2}^2 + m_2 l_1 l_{c2} \cos(\phi) + I_2 & m_2 l_{c2}^2 + I_2 \end{bmatrix}$$

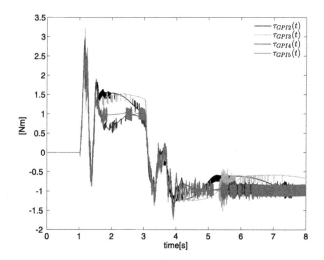

FIGURE 3.31   Torque control input.

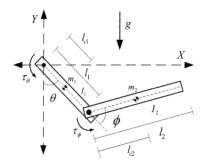

FIGURE 3.32   2-DOF robot manipulator diagram.

$$\mathbf{D} = \begin{bmatrix} d_1 & 0 \\ 0 & d_2 \end{bmatrix}$$

$$\mathbf{C(q, \dot{q})} = \begin{bmatrix} -2m_2 l_1 l_{c2} \sin(\phi)\dot{\phi} & -m_2 l_1 l_{c2} \sin(\phi)\dot{\phi} \\ m_2 l_1 l_{c2} \sin(\phi)\dot{\theta} & 0 \end{bmatrix}$$

$$\mathbf{G(q)} = \begin{bmatrix} (m_1 l_{c1} + m_2 l_1)g \sin(\theta) + m_2 l_{c2}g \sin(\theta + \phi) \\ m_2 l_{c2}g \sin(\theta + \phi) \end{bmatrix}$$

where $\boldsymbol{\tau} = \begin{bmatrix} \tau_\theta & \tau_\phi \end{bmatrix}^T$ represents the generalized input torque vector, θ and ϕ are the angular displacements related to each joint, l_1 and l_2 are the distances of each link, l_{c1} and l_{c2} are the distances to the center of mass, m_1 and m_2 are the link masses, respectively, the inertia is represented with I_1 and I_2, and g is the gravitational constant. The control variables τ_θ and $\tau_p hi$ are the control inputs, and $\mathbf{D}$ is the matrix of the friction viscous with coefficients d_1 and d_2.

3.6.1 The Two-Link Manipulator Robot Tracking Trajectory Problem

It is desired that the end effector of the robot manipulator located at the position

$$
\begin{aligned}
x_e &= l_1 \sin(\theta) + l_2 \sin(\theta + \phi) \\
y_e &= -l_1 \cos(\theta) - l_2 \cos(\theta + \phi)
\end{aligned}
\tag{3.87}
$$

tracks a trajectory specified by the position reference $x^*(t)$, $y^*(t)$. Using the inverse kinematic of planar robot manipulator, we can process the desired Cartesian trajectories to desired articular trajectories:

$$
\cos(\phi) = \frac{\left[x^*(t)\right]^2 + \left[y^*(t)\right]^2 - l_1^2 - l_2^2}{2 l_1 l_2}
$$

$$
\sin(\phi) = \sqrt{1 - (\cos(\phi))^2}
$$

$$
\cos(\theta) = \frac{(l_1 + l_2 \cos(\phi)) y^*(t) - l_2 \sin(\phi) x^*(t)}{[x^*(t)]^2 + \left[y^*(t)\right]^2}
$$

$$
\sin(\theta) = \frac{(l_1 + l_2 \cos(\phi)) x^*(t) - l_2 \sin(\phi) x^*(t)}{[x^*(t)]^2 + \left[y^*(t)\right]^2}
$$

$$
\theta_2^*(t) = \arctan\left(\frac{\sin(\phi)}{\cos(\phi)}\right)
$$

$$
\theta_1^*(t) = \arctan\left(\frac{\sin(\theta)}{\cos(\theta)}\right) + \frac{\pi}{2}
\tag{3.88}
$$

3.6.2 Robot Manipulator GPI Controller

Given the dynamic of the robot (3.86), the nonlinear Coriolis and gravity forces and friction viscous terms are considered unknown and lumped into time-varying vector $\tilde{\boldsymbol{\xi}}(t) = [\tilde{\xi}_\theta(t)\ \tilde{\xi}_\phi(t)]^T$:

$$
\begin{bmatrix} \ddot{\theta}_1 \\ \ddot{\theta}_2 \end{bmatrix} = \mathbf{M}(\mathbf{q})^{-1} \begin{bmatrix} \tau_1 \\ \tau_2 \end{bmatrix} + \begin{bmatrix} \tilde{\xi}_\theta(t) \\ \tilde{\xi}_\phi(t) \end{bmatrix}
\tag{3.89}
$$

with $\tilde{\xi}(t) = -\mathbf{M}(\mathbf{q})^{-1}\left(\mathbf{C}(\mathbf{q},\dot{\mathbf{q}})\dot{\mathbf{q}} + \mathbf{G}(\mathbf{q}) + \mathbf{D}\dot{\mathbf{q}}\right)$. We define the tracking trajectory errors $e_\theta = \theta - \theta^*(t)$ and $e_\phi = \phi - \phi^*(t)$ by

$$
\begin{bmatrix} \ddot{e}_\theta \\ \ddot{e}_\phi \end{bmatrix} = \mathbf{M}(\mathbf{q})^{-1}\begin{bmatrix} \tau_\theta \\ \tau_\phi \end{bmatrix} - \mathbf{M}(\mathbf{q}^*(t))^{-1}\begin{bmatrix} \tau_\theta^*(t) \\ \tau_\phi^*(t) \end{bmatrix} + \begin{bmatrix} \tilde{\xi}_\theta(t) \\ \tilde{\xi}_\phi(t) \end{bmatrix} \tag{3.90}
$$

We rewrite the error dynamic of system (3.90) as a perturbed double chain of integers

$$
\begin{bmatrix} \ddot{e}_\theta \\ \ddot{e}_\phi \end{bmatrix} = \mathbf{M}(\mathbf{q})^{-1}\begin{bmatrix} e_{\tau\theta} \\ e_{\tau\phi} \end{bmatrix} + \begin{bmatrix} \xi_\theta(t) \\ \xi_\phi(t) \end{bmatrix} \tag{3.91}
$$

with $\dot{e}_{\tau\theta} = \tau_\theta - \tau_\theta^*(t)$, $\dot{e}_{\tau\phi} = \tau_\phi - \tau_\phi^*(t)$, and the disturbance vector

$$
\begin{bmatrix} \xi_\theta(t) \\ \xi_\phi(t) \end{bmatrix} = \mathbf{M}(\mathbf{q})^{-1}\begin{bmatrix} \tau_\theta^*(t) \\ \tau_\phi^*(t) \end{bmatrix} - \mathbf{M}(\mathbf{q}^*(t))^{-1}\begin{bmatrix} \tau_\theta^*(t) \\ \tau_\phi^*(t) \end{bmatrix} + \begin{bmatrix} \tilde{\xi}_\theta(t) \\ \tilde{\xi}_\phi(t) \end{bmatrix} \tag{3.92}
$$

The perturbed system (3.91) can be expressed using the Active Disturbance Rejection Control philosophy. The effects of the generalized disturbance vector input $\begin{bmatrix} \xi_\theta(t) & \xi_\phi(t) \end{bmatrix}^T = \begin{bmatrix} z_{\theta 1} & z_{\phi 1} \end{bmatrix}^T$ are locally modeled by means following approximation:

$$
\begin{bmatrix} \ddot{e}_\theta \\ \ddot{e}_\phi \end{bmatrix} = \mathbf{M}(\mathbf{q})^{-1}\begin{bmatrix} e_{\tau\theta} \\ e_{\tau\phi} \end{bmatrix} + \begin{bmatrix} z_\theta \\ z_\phi \end{bmatrix}
$$

$$
\begin{bmatrix} \dot{z}_{\theta i} \\ \dot{z}_{\phi i} \end{bmatrix} = \begin{bmatrix} z_{\theta i+1} \\ z_{\phi i+1} \end{bmatrix}, \quad i = 1,2\ldots,m-1 \tag{3.93}
$$

$$
\begin{bmatrix} \dot{z}_{\theta m} \\ \dot{z}_{\phi m} \end{bmatrix} = \begin{bmatrix} 0 \\ 0 \end{bmatrix}
$$

where m indicates the degree of approximation of the ultralocal model. For this example, $n = 2$, and we choose $m = 3$. We propose the flat-filter feedback controller

$$
\begin{bmatrix} \dot{\zeta}_{\theta i} \\ \dot{\zeta}_{\phi i} \end{bmatrix} = \begin{bmatrix} \zeta_{\theta(i+1)} \\ \zeta_{\phi(i+1)} \end{bmatrix}, \quad i = 1,2,3
$$

$$
\begin{bmatrix} \dot{\zeta}_{\theta 3} \\ \dot{\zeta}_{\phi 3} \end{bmatrix} = \begin{bmatrix} -k_{\theta 5}\zeta_{\theta 4} + e_\theta \\ -k_{\phi 5}\zeta_{\phi 4} + e_\phi \end{bmatrix} \tag{3.94}
$$

$$\begin{bmatrix} e_{\theta f} \\ e_{\phi f} \end{bmatrix} = \begin{bmatrix} \zeta_\theta \\ \zeta_\phi \end{bmatrix}$$

$$\begin{bmatrix} \tau_\theta \\ \tau_\phi \end{bmatrix} = \mathbf{M}(\mathbf{q}) \begin{bmatrix} \ddot{x}^*(t) \\ \ddot{y}^*(t) \end{bmatrix}$$

$$+ \mathbf{M}(\mathbf{q}) \begin{bmatrix} (k_{\theta 4}k_{\theta 5} - k_{\theta 3})\zeta_{\theta 4} - k_{\theta 2}\zeta_{\theta 3} - k_{\theta 1}\zeta_{\theta 2} - k_{\theta 0}\zeta_{\theta 1} - k_{\theta 4}e_\theta \\ (k_{\phi 4}k_{\phi 5} - k_{\phi 3})\zeta_{\phi 4} - k_{\phi 2}\zeta_{\phi 3} - k_{\phi 1}\zeta_{\phi 2} - k_{\phi 0}\zeta_{\phi 1} - k_{\phi 4}e_\phi \end{bmatrix}$$

$$(3.95)$$

The values of the controller design parameters $\{k_{\theta 5}, k_{\theta 4}, k_{\theta 3}, k_{\theta 2}, k_{\theta 1}, k_{\theta 0}\}$ and $\{k_{\phi 5}, k_{\phi 4}, k_{\phi 3}, k_{\phi 2}, k_{\phi 1}, k_{\phi 0}\}$ are chosen so that the closed-loop characteristic polynomials

$$\begin{bmatrix} s^6 + k_{\theta 5}s^5 + k_{\theta 4}s^4 + k_{\theta 3}s^3 + k_{\theta 2}s^2 + k_{\theta 1}s + k_{\theta 0} \\ s^6 + k_{\phi 5}s^5 + k_{\phi 4}s^4 + k_{\phi 3}s^3 + k_{\phi 2}s^2 + k_{\phi 1}s + k_{\phi 0} \end{bmatrix}$$

$$= \begin{bmatrix} (s^2 + 2\zeta_\theta \omega_\theta s + \omega_\theta^2) \\ (s^2 + 2\zeta_\phi \omega_\phi s + \omega_\phi^2) \end{bmatrix}$$

$$(3.96)$$

have all roots in the left half of the complex plane using a desired stable closed-loop characteristic polynomials.

3.6.3 Simulation Results for Robot Manipulator

Simulations were performed for the following reference trajectories: The robot manipulator starts with initial condition $x_e(0) = 0$, $y_e(0) = -0.6$, which implies that $[\theta(0) = 0 \; \phi(0) = 0]$; at the time $t = 1.5$ [s], the end effector moves with a smooth trajectory to $x_e(4) = 0.45$, $y_e(4) = 0$ in 2.5 [s] and remains in this position for 2 [s], when at time 7 [s], the end effector starts a smooth circle trajectory (3.97) with radius $R = 0.15$ [m] and smooth angle $\alpha^*(t)$ that is accomplished in 10 [s]; when the end effector gets one turn, it remains in this position to the end of the simulation.

$$x^*(t) = R\cos(\alpha^*(t))$$
$$y^*(t) = R\sin(\alpha^*(t)) \qquad (3.97)$$

The inverse kinematic (3.88) is used to compute for the articular desired positions $\theta(t)^*$ and $\phi(t)^*$. The parameters of system (3.86) used in the simulations

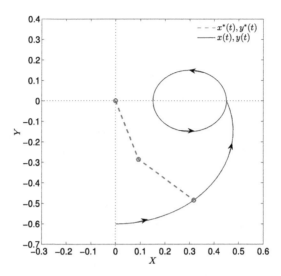

FIGURE 3.33 Robot manipulator trajectory tracking performance.

were set to be

$$m_1 = 0.263 \text{ [kg]} \qquad m_2 = 0.1306 \text{ [kg]} \qquad I_1 = 0.002 \text{ [Nm}^2]$$
$$I_2 = 0.00098 \text{ [Nm}^2] \qquad l_1 = 0.3 \text{ [m]} \qquad l_2 = 0.3 \text{ [Ns/m]}$$
$$l_{c1} = 0.15 \text{ [m]} \qquad l_{c2} = 0.15 \text{ [m]} \qquad g = 9.81 \text{ [m/s}^2]$$
$$d_1 = 0.03 \text{ [Nms/rad]} \quad d_2 = 0.005 \text{ [Nms/rad]}$$

The controller gains were chosen in accordance with the desired closed-loop characteristic polynomial for each position coordinate dynamics as follows: $\varsigma_\theta = 4$, $\omega_\theta = 10$, $\varsigma_\phi = 4$, $\omega_\phi = 10$. Fig. 3.33 depicts the time evolution of the end effector position coordinates $x_e(t)$ and $y_e(t)$, along with the desired reference trajectories. Fig. 3.34 shows the trajectory tracking of the first link position $\theta(t)$, and Fig. 3.35 depicts the trajectory tracking of the second link position $\phi(t)$. The time evolution of the GPI controller inputs $\tau_\theta(t)$ and $\tau_\phi(t)$ are shown in Figs. 3.36 and 3.37, respectively.

3.7 NONHOLONOMIC WHEELED CAR

Consider the following nonholonomic car [22], [23], [24] shown in Fig. 3.38. The kinematic model is described as

$$\dot{x} = V\cos(\theta) \tag{3.98}$$
$$\dot{y} = V\sin(\theta) \tag{3.99}$$
$$\dot{\theta} = \omega \tag{3.100}$$

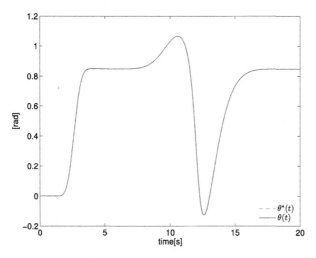

FIGURE 3.34 Controlled evolution of the first link.

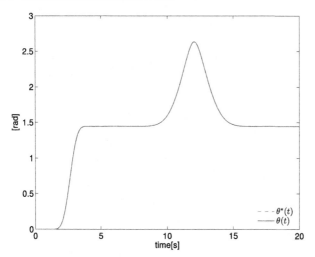

FIGURE 3.35 Controlled evolution of the second link.

The coordinates (x, y) are the plane coordinates, and θ is the angular displacement related to the mobile orientation with respect to the positive x coordinate axis. The control variables (V, ω) are the control inputs representing, respectively, the velocity and the velocity of change of orientation.

3.7.1 The Nonholonomic Car Tracking Trajectory Problem

It is desired that the robot tracks a trajectory specified by the position reference signals $x^*(t), y^*(t)$. The kinematic car model (3.98)–(3.100) is differentially

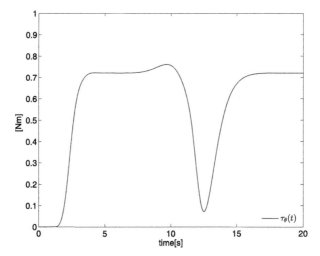

FIGURE 3.36 First link control torque input.

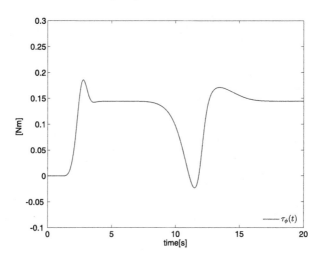

FIGURE 3.37 Second link control torque input.

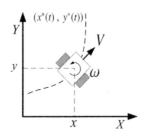

FIGURE 3.38 Nonholonomic car diagram.

flat [25], and all system variables may be differentially parameterized by the output variables x, y and a finite number of their time derivatives. Indeed,

$$\theta = \tan^{-1}\left(\frac{\dot{y}}{\dot{x}}\right) \tag{3.101}$$

$$\omega = \frac{\ddot{y}\dot{x} - \ddot{x}\dot{y}}{\dot{x}^2 + \dot{y}^2} \tag{3.102}$$

$$V = \sqrt{\dot{x} + \dot{y}} \tag{3.103}$$

From this differential parameterization we can immediately see that the control input variable V requires a first-order extension in order to have a well-defined relative degree and achieve invertibility of the matrix defining the input-to-output highest time derivative relation. The input–output model with $\dot{V}$ as new auxiliary control input variables yields

$$\begin{bmatrix} \ddot{x} \\ \ddot{y} \end{bmatrix} = \begin{bmatrix} \dfrac{\dot{x}}{\sqrt{\dot{x}^2 + \dot{y}^2}} & -\dot{y} \\ \dfrac{\dot{y}}{\sqrt{\dot{x}^2 + \dot{y}^2}} & \dot{x} \end{bmatrix} \begin{bmatrix} \dot{V} \\ \omega \end{bmatrix} \tag{3.104}$$

3.7.2 Nonholonomic Car GPI Controller

Consider the tracking trajectory errors $e_x = x - x^*(t)$ and $e_y = y - y^*(t)$. Using the differential flatness parameterization of the car (3.104), we obtain the trajectory tracking error dynamic

$$\begin{bmatrix} \ddot{x} - \ddot{x}^*(t) \\ \ddot{y} - \ddot{y}^*(t) \end{bmatrix} = \begin{bmatrix} \dfrac{\dot{x}}{\sqrt{\dot{x}^2 + \dot{y}^2}} & -\dot{y} \\ \dfrac{\dot{y}}{\sqrt{\dot{x}^2 + \dot{y}^2}} & \dot{x} \end{bmatrix} \begin{bmatrix} \dot{V} \\ \omega \end{bmatrix}$$
$$- \begin{bmatrix} \dfrac{\dot{x}^*(t)}{\sqrt{[\dot{x}^*(t)]^2 + [\dot{y}^*(t)]^2}} & -\dot{y}^*(t) \\ \dfrac{\dot{y}^*(t)}{\sqrt{[\dot{x}^*(t)]^2 + [\dot{y}^*(t)]^2}} & \dot{x}^*(t) \end{bmatrix} \begin{bmatrix} \dot{V}^*(t) \\ \omega^*(t) \end{bmatrix} \tag{3.105}$$

We simplify the model (3.105) as a perturbed double chain of integers

$$\begin{bmatrix} \ddot{e}_x \\ \ddot{e}_y \end{bmatrix} = \begin{bmatrix} \dfrac{\dot{x}}{\sqrt{\dot{x}^2 + \dot{y}^2}} & -\dot{y} \\ \dfrac{\dot{y}}{\sqrt{\dot{x}^2 + \dot{y}^2}} & \dot{x} \end{bmatrix} \begin{bmatrix} \dot{e}_V \\ e_\omega \end{bmatrix} + \begin{bmatrix} \varphi_x \\ \varphi_y \end{bmatrix} \tag{3.106}$$

with $\dot{e}_V = \dot{V} - \dot{V}^*(t)$, $e_\omega = \omega - \omega^*(t)$, and the disturbance vector

$$\begin{bmatrix} \varphi_x \\ \varphi_y \end{bmatrix} = \begin{bmatrix} \frac{\dot{x}}{\sqrt{\dot{x}^2+\dot{y}^2}} & -\dot{y} \\ \frac{\dot{y}}{\sqrt{\dot{x}^2+\dot{y}^2}} & \dot{x} \end{bmatrix} \begin{bmatrix} \dot{V}^*(t) \\ \omega^*(t) \end{bmatrix}$$
$$- \begin{bmatrix} \frac{\dot{x}^*(t)}{\sqrt{[\dot{x}^*(t)]^2+[\dot{y}^*(t)]^2}} & -\dot{y}^*(t) \\ \frac{\dot{y}^*(t)}{[\dot{x}^*(t)]^2+[\dot{y}^*(t)]^2} & \dot{x}^*(t) \end{bmatrix} \begin{bmatrix} \dot{V}^*(t) \\ \omega^*(t) \end{bmatrix} \qquad (3.107)$$

From the Active Disturbance Rejection Control philosophy we can reexpress the perturbed system (3.106), where the effects of the generalized disturbance vector input $\begin{bmatrix} \varphi_x(t) & \varphi_y(t) \end{bmatrix}^T = \begin{bmatrix} z_{x1} & z_{y1} \end{bmatrix}^T$ are locally modeled by means of the following approximation:

$$\begin{bmatrix} \ddot{e}_x \\ \ddot{e}_y \end{bmatrix} = \begin{bmatrix} \frac{\dot{x}}{\sqrt{\dot{x}^2+\dot{y}^2}} & -\dot{y} \\ \frac{\dot{y}}{\sqrt{\dot{x}^2+\dot{y}^2}} & \dot{x} \end{bmatrix} \begin{bmatrix} \dot{e}_V \\ e_\omega \end{bmatrix} + \begin{bmatrix} z_{x1} \\ z_{y1} \end{bmatrix}$$
$$\begin{bmatrix} \dot{z}_{xi} \\ \dot{z}_{yi} \end{bmatrix} = \begin{bmatrix} z_{xi+1} \\ z_{yi+1} \end{bmatrix}, \quad i = 1, 2 \ldots, m-1 \qquad (3.108)$$
$$\begin{bmatrix} \dot{z}_{xm} \\ \dot{z}_{ym} \end{bmatrix} = \begin{bmatrix} 0 \\ 0 \end{bmatrix}$$

where m indicates the degree of approximation of the ultralocal model. For this example, $n = 2$, and we choose $m = 2$. The flat-filtering feedback controllers are given as follows:

$$\begin{bmatrix} \dot{\zeta}_{xi} \\ \dot{\zeta}_{yi} \end{bmatrix} = \begin{bmatrix} \zeta_{xi+1} \\ \zeta_{yi+1} \end{bmatrix}, \quad i = 1, 2 \qquad (3.109)$$

$$\begin{bmatrix} \dot{\zeta}_{x3} \\ \dot{\zeta}_{y3} \end{bmatrix} = \begin{bmatrix} -k_{x4}\zeta_{x3} + e_x \\ -k_{y4}\zeta_{y3} + e_y \end{bmatrix} \qquad (3.110)$$

$$\begin{bmatrix} e_{xf} \\ e_{yf} \end{bmatrix} = \begin{bmatrix} \zeta_{x1} \\ \zeta_{y1} \end{bmatrix} \qquad (3.111)$$

$$\begin{bmatrix} \dot{V} \\ \omega \end{bmatrix} = \begin{bmatrix} \frac{\dot{x}}{\sqrt{\dot{x}^2+\dot{y}^2}} & \frac{\dot{y}}{\sqrt{\dot{x}^2+\dot{y}^2}} \\ -\frac{\dot{y}}{\dot{x}^2+\dot{y}^2} & \frac{\dot{x}}{\dot{x}^2+\dot{y}^2} \end{bmatrix} \begin{bmatrix} \ddot{x}^*(t) \\ \ddot{y}^*(t) \end{bmatrix}$$

$$
+ \begin{bmatrix} \dfrac{\dot{\hat{x}}}{\sqrt{\dot{\hat{x}}^2+\dot{\hat{y}}^2}} & \dfrac{\dot{\hat{y}}}{\sqrt{\dot{\hat{x}}^2+\dot{\hat{y}}^2}} \\[3mm] -\dfrac{\dot{\hat{y}}}{\dot{\hat{x}}^2+\dot{\hat{y}}^2} & \dfrac{\dot{\hat{x}}}{\dot{\hat{x}}^2+\dot{\hat{y}}^2} \end{bmatrix} \begin{bmatrix} (k_{x3}k_{x4}-k_{x2})\zeta_{x3}-k_{x1}\zeta_{x2}-k_{x0}\zeta_{x1}-k_{x3}e_x \\ (k_{y3}k_{y4}-k_{y2})\zeta_{y3}-k_{y1}\zeta_{y2}-k_{y0}\zeta_{y1}-k_{y3}e_y \end{bmatrix}
$$

(3.112)

The velocities $\dot{x}$ and $\dot{y}$ used in the gain matrix are estimated with (3.104):

$$
\begin{bmatrix} \dot{\hat{x}} \\ \dot{\hat{y}} \end{bmatrix} = \begin{bmatrix} \int^{(1)}\ddot{x} \\ \int^{(1)}\ddot{y} \end{bmatrix} = \int^{(1)} \begin{bmatrix} \dfrac{\dot{\hat{x}}}{\sqrt{\dot{\hat{x}}^2+\dot{\hat{y}}^2}} & -\dot{\hat{y}} \\[3mm] \dfrac{\dot{\hat{y}}}{\sqrt{\dot{\hat{x}}^2+\dot{\hat{y}}^2}} & \dot{\hat{x}} \end{bmatrix} \begin{bmatrix} \dot{V} \\ \omega \end{bmatrix}
$$

(3.113)

The values of the controller design parameters $\{k_{x4}, k_{x3}, k_{x2}, k_{x1}, k_{x0}\}$ and $\{k_{y4}, k_{y3}, k_{y2}, k_{y1}, k_{y0}\}$ are chosen so that the closed-loop characteristic polynomials

$$
\begin{bmatrix} s^5 + k_{x4}s^4 + k_{x3}s^3 + k_{x2}s^2 + k_{x1}s + k_{x0} \\ s^5 + k_{y4}s^4 + k_{y3}s^3 + k_{y2}s^2 + k_{y1}s + k_{y0} \end{bmatrix}
$$
$$
= \begin{bmatrix} (s+p_x)(s^2+2\zeta_x\omega_x s + \omega_x^2)^2 \\ (s+p_y)(s^2+2\zeta_y\omega_y s + \omega_y^2)^2 \end{bmatrix}
$$

(3.114)

have all roots in the left half of the complex plane using a desired stable closed-loop characteristic polynomial.

3.7.3 Simulation Results for Nonholonomic Car

Simulations were performed for the following reference trajectories describing a lemniscate on the plane:

$$
R^*(t) = \cos(\frac{2\pi}{T}t)
$$
$$
\theta^*(t) = \sin(\frac{2\pi}{T}t)
$$
$$
x^*(t) = 2R^*(t)\cos(\theta^*(t))
$$
$$
y^*(t) = 4R^*(t)\sin(\theta^*(t))
$$

(3.115)

where $R^*(t)$ and $\theta^*(t)$ are the radius and the angle of the lemniscate, respectively, the desired trajectory period was set $T = 20$ [s], and the controller gains were chosen in accordance with the desired closed-loop characteristic polynomial for each position coordinate dynamics as follows: $\zeta_x = 3$, $\omega_x = 4$, $p_x = 4$, $\zeta_y = 3$, $\omega_y = 4$, and $p_y = 4$. The initial condition for the nonholonomic car was

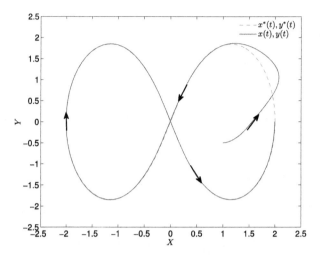

FIGURE 3.39 Nonholonomic trajectory tracking performance.

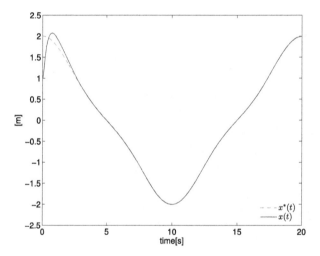

FIGURE 3.40 Controlled evolution of the horizontal position coordinate.

set $[x(0) = 1,\ y(0) = -0.5,\ \theta(0) = 0.5]$, and for the estimation of velocities, $[\hat{\dot{x}}(0) = 0.01,\ \hat{\dot{y}}(0) = 0.01]$. Fig. 3.39 depicts the time evolution of the position coordinates $x(t)$ and $y(t)$ for the nonholonomic car along with the desired reference trajectories. Fig. 3.40 shows the trajectory tracking of the horizontal car position $x(t)$, and Fig. 3.41 depicts the trajectory tracking of the vertical car position $y(t)$. The angular position $\theta(t)$ of the car is shown in Fig. 3.42, and the time evolutions of the GPI controller inputs $V(t)$ and $\omega(t)$ are shown in Figs. 3.43 and 3.44, respectively.

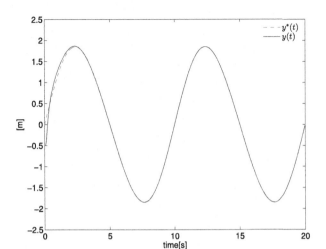

FIGURE 3.41 Controlled evolution of the vertical position coordinate.

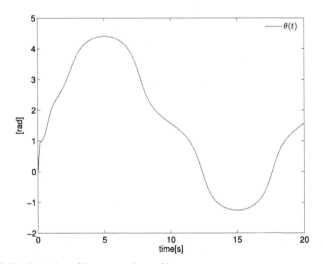

FIGURE 3.42 Evolution of the car angular position.

3.8 TWO-MASS RECTILINEAR MECHANISM

3.8.1 Mathematical Model

A two-mass rectilinear mechanism consists of two masses suspended on anti-friction linear bearings. The mathematical model of the mechanical part of the system is as follows:

$$m_1\ddot{x}_1 + c_1\dot{x}_1 + (K_1 + K_2)x_1 - K_2x_2 = u \tag{3.116}$$

$$m_2\ddot{x}_2 + c_2\dot{x}_2 - K_2x_1 + (K_2 + K_3)x_2 = 0 \tag{3.117}$$

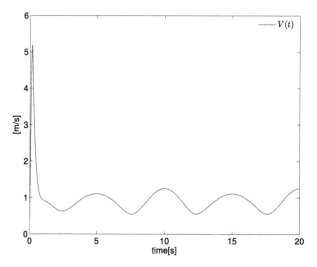

FIGURE 3.43 Control velocity input.

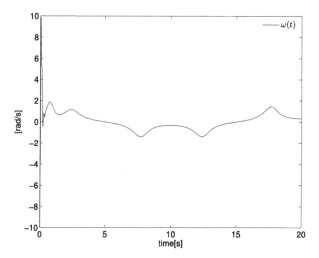

FIGURE 3.44 Angular velocity input.

The first mass is driven by a linear actuator, as is shown in Fig. 3.45, and the second mass is used as an underactuated simple mass; it is connected to the first mass by a spring with coefficient K_2, and both masses are constrained by springs that are attached to the base with coefficients K_1, K_3, respectively. The variables x_1 and x_2 are mass positions, u represents the input control signal of the system, and c_1 and c_2 are the friction viscous coefficients. We neglect the friction viscous forces, which are considered as an external input perturbation.

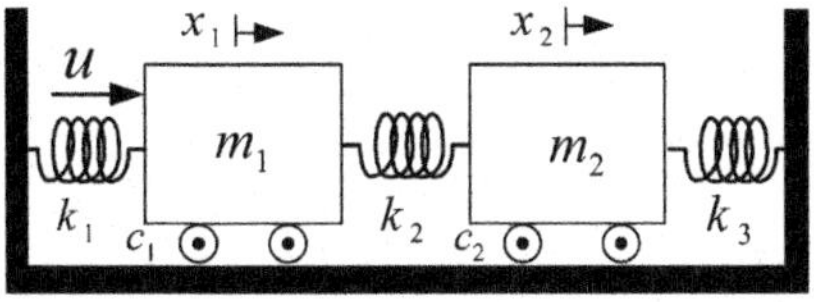

FIGURE 3.45 Two mass rectilinear mechanism diagram.

Then system (3.116)–(3.117) is simplified as follows:

$$m_1\ddot{x}_1 + (K_1 + K_2)x_1 - K_2 x_2 = u \tag{3.118}$$

$$m_2\ddot{x}_2 - K_2 x_1 + (K_2 + K_3)x_2 = 0 \tag{3.119}$$

In the state space model the system can be represented as

$$\dot{\mathbf{x}} = \mathbf{A}\mathbf{x} + \mathbf{B}u$$

$$y = \mathbf{C}\mathbf{x} \tag{3.120}$$

where

$$\mathbf{A} = \begin{bmatrix} 0 & 1 & 0 & 0 \\ -\dfrac{K_1+K_2}{m_1} & 0 & \dfrac{K_2}{m_1} & 0 \\ 0 & 0 & 0 & 1 \\ \dfrac{K_2}{m_2} & 0 & -\dfrac{K_1+K_3}{m_2} & 0 \end{bmatrix}, \quad \mathbf{x} = \begin{bmatrix} x_1 \\ \dot{x}_1 \\ x_2 \\ \dot{x}_2 \end{bmatrix}$$

$$\mathbf{B} = \begin{bmatrix} 0 \\ \dfrac{1}{m_1} \\ 0 \\ 0 \end{bmatrix} \quad \mathbf{C} = \begin{bmatrix} 0 & 0 & 1 & 0 \end{bmatrix}$$

3.8.2 Problem Formulation

Considering the two-mass Rectilinear Mechanism with dynamic model (3.118)–(3.119), we require a robust control law such that it forces the position of Mass-2, which is x_2, to track a point-to-point smooth trajectory. The task consists in taking the Mass-2 from the initial value $x_2(0) = 0$ toward the final position $x_2(t_f) = X$ within a prescribed time interval $[0, t_f]$.

3.8.3 Differential Flatness of Two-Mass Rectilinear Mechanism

To obtain the flat output, we obtain the controllability matrix of the linear system (3.118)–(3.119) given by $\mathbf{C}_k = \begin{bmatrix} \mathbf{B} & \mathbf{AB} & \mathbf{A}^2\mathbf{B} & \mathbf{A}^3\mathbf{B} \end{bmatrix}$, and the system under consideration is controllable with $\det(\mathbf{C}_k) = \frac{K_2^2}{m_1^4 m_2^2}$, and hence it is flat [16]. The flat output can be obtained as

$$F = \begin{bmatrix} 0 & 0 & 0 & 1 \end{bmatrix} \mathbf{C}_K^{-1} \mathbf{x} = \frac{m_1 m_2}{K_2} x_2 \tag{3.121}$$

We chose the flat output as $x_\gamma = \epsilon F$, where $\epsilon = \frac{K_2}{m_1 m_2}$, and this flat output coincides with the position x_2. We readily obtain all state variables as differential functions of the flat output x_γ:

$$x_1 = \frac{K_1 + K_3}{K_2} x_\gamma + (\frac{K_3}{K_1} + \frac{m_2}{K_2}) \ddot{x}_\gamma \tag{3.122}$$

$$\dot{x}_1 = \frac{K_1 + K_3}{K_2} \dot{x}_\gamma + (\frac{K_3}{K_1} + \frac{m_2}{K_2}) x_\gamma^{(3)} \tag{3.123}$$

$$x_2 = x_\gamma \tag{3.124}$$

$$\dot{x}_2 = \dot{x}_\gamma \tag{3.125}$$

$$u = \frac{m_1 m_2}{K_2} x_\gamma^{(4)} + \frac{1}{K_2}(K_1 m_1 + K_2 m_1 + K_2 m_2 + K_3 m_1) \ddot{x}_\gamma$$

$$+ \frac{1}{K_2}(K_1 K_2 + K_1 K_3 + K_2 K_3) x_\gamma \tag{3.126}$$

3.8.4 Two-Mass Rectilinear Mechanism GPI Controller Approach

To obtain GPI controller, a disturbed chain of integrator representation is necessary. The input–flat output system form (3.126) can be computed as follows:

$$x_\gamma^{(4)} = \frac{K_2}{m_1 m_2} u - \frac{1}{m_1 m_2}(K_1 K_2 + K_1 K_3 + K_2 K_3) x_\gamma$$

$$- \frac{1}{m_1 m_2}(K_1 m_2 + K_2 m_1 + K_2 m_2 + K_3 m_1) \ddot{x}_\gamma + d(t) \tag{3.127}$$

where $d(t)$ contains the uncertainty disturbances, such as friction viscous, nonlinear terms, and other disturbances that can be neglected in (3.118)–(3.119). We define the tracking error as

$$e_x = x_\gamma - x_\gamma^*(t)$$

The fourth time derivative of the tracking error can be represented as a linear perturbed dynamic model, which is the key step in flatness-based ADRC controllers:

$$e_x^{(4)} = \frac{K_2}{m_1 m_2} e_u + \varphi(t) \tag{3.128}$$

where the unperturbed feedforward control is $u(t)^* = \frac{m_1 m_2}{K_2} x_\gamma^{(4)*}(t)$, and thus the input error $e_u = u(t) - \frac{m_1 m_2}{K_2} x_\gamma^{(4)*}(t)$, and $\varphi(t)$ is the total disturbance [4], [5] which lumped the effects of neglected internal linear dynamic disturbances, non-linear dynamic disturbances, and unmodeled external disturbances:

$$\varphi(t) = -\frac{1}{m_1 m_2}(K_1 K_2 + K_1 K_3 + K_2 K_3)e_x$$
$$-\frac{1}{m_1 m_2}(K_1 m_2 + K_2 m_1 + K_2 m_2 + K_3 m_1)\ddot{e}_x + d(t) \qquad (3.129)$$

The generalized disturbance input $\varphi(t)$ is locally modeled as the following approximation:

$$e_x^{(4)} = \frac{K_2}{m_1 m_2}e_u + z_1$$
$$\dot{z}_i = z_{i+1} \qquad (3.130)$$
$$\dot{z}_m = 0$$

We propose the following flat-filter feedback controller with $n = 4$ and $m = 2$:

$$\dot{\zeta}_i = \zeta_{i+1}, i = 1, 4$$
$$\dot{\zeta}_5 = -k_8 \zeta_5 - k_7 \zeta_4 - k_6 \zeta_3 + e_x \qquad (3.131)$$
$$e_{xf} = \zeta_1$$
$$u(t) = \frac{m_1 m_2}{K_2} x_\gamma^{(4)*}(t) + \frac{m_1 m_2}{K_2}\Big((k_5 k_8 - k_4)\zeta_5 + (k_5 k_7 - k_3)\zeta_4$$
$$+ (k_5 k_6 - k_2)\zeta_3 - k_1 \zeta_2 - k_0 \zeta_1 - k_5 e_{xf}\Big) \qquad (3.132)$$

where e_{xf} is the flat filtering output. The controller gains $\{k_8, k_7, k_6, \ldots, k_1, k_0\}$ are chosen matching the closed-loop tracking error characteristic polynomial with a desired stable polynomial with all roots in the left of the complex plane:

$$s^9 + k_8 s^8 + k_7 s^7 + \cdots + k_1 s + k_0 = (s + p_x)(s^2 + 2\varsigma_x \omega_x s + \omega_x^2)^4 \quad (3.133)$$

3.8.5   Simulation Results for Two-Mass Rectilinear Mechanism

The parameters of system (3.116)–(3.130) used in the simulations were set to be

$$m_1 = 2.34 \text{ [kg]} \quad m_2 = 2 \text{ [kg]} \quad K_1 = 450 \text{ [N/m]} \quad K_2 = 450 \text{ [N/m]}$$
$$K_3 = 175 \text{ [N/m]} \quad c_1 = 10 \text{ [Ns/m]} \quad c_2 = 10 \text{ [Ns/m]}$$

The controller parameters were chosen as $\varsigma_x = 3$, $\omega_x = 65$, and $p_x = 65$. The test starts with initial conditions of the masses at the point $[x_1(0) = 0,$

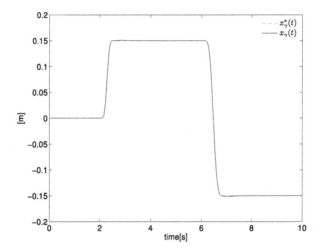

FIGURE 3.46 Flat output x_γ trajectory tracking.

$x_2(0) = 0$], which implies that flat output is $x_\gamma(0) = 0$. Fig. 3.46 shows system performance using the flat-filter scheme: the flat output $x_\gamma(t)$ follows the desired point-to-point smooth trajectory $x_\gamma^*(t)$ as is depicted in Fig. 3.47 with an efficient result, the GPI controller-based differential flatness allows the second mass position $x_\gamma(t) = x_2$ carry out tracking trajectories avoiding undesired oscillations that could affect the system due to springs, the mass position x_1 adjusts its path allowing flat output make the desired task, and the control force input $u(t)$ is depicted in Fig. 3.48; it has to overcome the forces provided by springs that do not allow the masses to move freely as well as unmodeled dynamics in the case of viscous friction.

3.9 REMARKS

In this chapter, we have introduced a remarkable new property of flatness, summarized as follows: Any linear controllable system whose flat output (or Brunovsky output) coincides with the measurable system output can be controlled in stabilization or trajectory tracking tasks by means of a linear well-tuned stable filter alone. The filter processes, respectively, the system output or the system output trajectory tracking error. A linear combination of the obviously measurable internal states of the filter conforms the required stabilizing feedback law. The coefficients of the linear filter, which turns out to be also flat in a filtering context, are uniquely determined in a rather trivial manner involving only an overall pole placement effort for closed-loop stability. The flat filter constitutes a reinterpretation of the classical compensation networks for linear systems. The classical compensation networks are shown to be totally equivalent

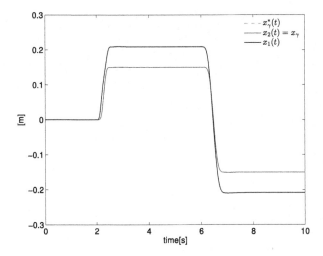

FIGURE 3.47 Flat-filter controller closed-loop system performance.

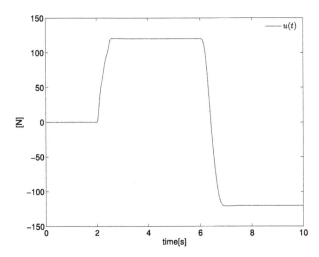

FIGURE 3.48 Force input control.

to the well-known Generalized Proportional Integral (GPI) controllers based on integral reconstructors. GPI controllers are widely known to efficiently evade the need for asymptotic state observers. The striking fact, coincident in a dual manner with Extended State Observer-based Active Disturbance Rejection Control schemes, is that the proposed linear dynamic controllers may be used for the control of nonlinear, uncertain flat systems. Whereas the tracking performance is approximate in nature, it is realized with a remarkable precision even in experimental environments. The approach easily overcomes a classical limitation of nonlinear system trajectory tracking and stabilizing control via their tangent linearization model, provided that such a model is controllable. The local validity of the linearized controllable model is overcome by the annihilating features of the dynamic controller with respect to the excited nonlinearities and neglected terms in the first-order approximation. This was illustrated here in the trajectory tracking task of a classic nonlinear pendulum system. In this particular example, it was shown that increasing the degree of approximation of the family of time polynomials that locally model the disturbance input can improve the tracking performance.

REFERENCES

[1] H. Sira-Ramírez, S.K. Agrawal, Differentially Flat Systems, Marcel Dekker, 2004.
[2] J. Levine, Analysis and Control of Nonlinear Systems: a Flatness-Based Approach, Springer Science & Business Media, 2009.
[3] M. Fliess, C. Join, H. Sira-Ramirez, Non-linear estimation is easy, International Journal of Modelling, Identification and Control 4 (1) (2008) 12–27.
[4] Z. Gao, Active disturbance rejection control: a paradigm shift in feedback control system design, in: 2006 American Control Conference, IEEE, 2006, 7 pp.
[5] J. Han, From PID to active disturbance rejection control, IEEE Transactions on Industrial Electronics 56 (3) (2009) 900–906.
[6] Z. Chen, D. Xu, Output regulation and active disturbance rejection control: unified formulation and comparison, Asian Journal of Control 18 (5) (2016) 1–11.
[7] L. Dong, Y. Zhang, Z. Gao, A robust decentralized load frequency controller for interconnected power systems, ISA Transactions 51 (3) (2012) 410–419.
[8] B.Z. Guo, J.J. Liu, A.S. Al-Fhaid, A.M. Younas, A. Asiri, The active disturbance rejection control approach to stabilisation of coupled heat and ODE system subject to boundary control matched disturbance, International Journal of Control 88 (8) (2015) 1554–1564.
[9] G.A. Ramos-Fuentes, J.A. Cortés-Romero, Z. Zou, R. Costa-Castelló, K. Zhou, Power active filter control based on a resonant disturbance observer, Power Electronics, IET 8 (4) (2015) 554–564.
[10] S. Zhao, Z. Gao, An active disturbance rejection based approach to vibration suppression in two-inertia systems, Asian Journal of Control 15 (2) (2013) 350–362.
[11] Q. Zheng, H. Richter, Z. Gao, Active disturbance rejection control for piezoelectric beam, Asian Journal of Control 16 (6) (2014) 1612–1622.
[12] R. Madoński, P. Herman, Model-free control or active disturbance rejection control? On different approaches for attenuating the perturbation, in: Control & Automation (MED), 2012 20th Mediterranean Conference on, IEEE, 2012, pp. 60–65.
[13] M. Fliess, R. Marquez, Continuous-time linear predictive control and flatness: a module-theoretic setting with examples, International Journal of Control 73 (2000) 606–623.
[14] M. Fliess, R. Marquez, E. Delaleau, H. Sira-Ramírez, Correcteurs proportionnels-intégraux généralisés, ESAIM: Control, Optimisation and Calculus of Variations 7 (2002) 23–41.
[15] E.W. Zurita-Bustamante, J. Linares-Flores, E. Guzmán-Ramírez, H. Sira-Ramírez, A comparison between the GPI and PID controllers for the stabilization of a DC–DC Buck converter: a field programmable gate array implementation, IEEE Transactions on Industrial Electronics 58 (11) (2011) 5251–5262.
[16] M. Fliess, J. Lévine, P. Martin, P. Rouchon, Flatness and defect of non-linear systems: introductory theory and applications, International Journal of Control 61 (1995) 1327–1361.

[17] H. Sira-Ramírez, A. Luviano-Juárez, M. Ramírez-Neria, R. Garrido-Moctezuma, Flat filtering: a classical approach to robust control of nonlinear systems, in: Proceedings of the 2016 American Control Conference, IEEE, 2016, pp. 3844–3849.

[18] R.A. McCann, M.S. Islam, I. Husain, Application of a sliding-mode observer for position and speed estimation in switched reluctance motor drives, IEEE Transactions on Industry Applications 37 (1) (2001) 51–58.

[19] H. Asada, J.J. Slotine, Robot Analysis and Control, John Wiley & Sons, 1986.

[20] J.J. Craig, Introduction to Robotics: Mechanics and Control, vol. 3, Pearson Prentice Hall, Upper Saddle River, 2005.

[21] M.W. Spong, M. Vidyasagar, Robot Dynamics and Control, John Wiley & Sons, 2008.

[22] B. D'Andréa Novel, G. Campion, G. Bastin, Control of nonholonomic wheeled mobile robots by state feedback linearization, International Journal of Robotics Research 14 (6) (1995) 543–559.

[23] A. De Luca, G. Oriolo, C. Samson, Feedback control of a nonholonomic car-like robot, in: Robot Motion Planning and Control, Springer, 1998, pp. 171–253.

[24] W. Dixon, D.M. Dawson, E. Zergeroglu, A. Behal, Nonlinear Control of Wheeled Mobile Robots, Springer-Verlag, ISBN 978-1-85233-414-7, 2001.

[25] H. Sira-Ramírez, C. López-Uribe, M. Velasco-Villa, Trajectory-tracking control of an input delayed omnidirectional mobile robot, in: Electrical Engineering Computing Science and Automatic Control (CCE), 2010 7th International Conference on, IEEE, 2010, pp. 470–475.

Chapter 4

Extensions of ADRC

Chapter Points

- In this chapter, we present some extensions of the Active Disturbance Rejection Control for multivariable systems, its combination with sliding-mode control, sampled time systems, and time delay systems.
- The multivariable treatment deals with two schemes, the integral reconstruction approach and the use of a quadratic Lyapunov function to prove its convergence.
- The sampled time systems are treated by means of the delta operator approach, which is an alternative tool to analyze fast-sampled time systems.
- A class of control-input delayed systems in trajectory tracking tasks is approached from the classic Smith Predictor viewpoint, but, in this case, the prediction is proposed from a Taylor series approximation.

4.1 INTRODUCTION

Over the years, ADRC has been applied to different classes of systems including the linear and nonlinear system cases, the single- and multivariable system structures, the finite- and infinite-dimensional system cases. Formulations of ADRC are found, which are cast in the continuous-time framework and in the discrete-time environment. Recent surveys [2,3] indicate that the field of applications of ADRC has been steadily expanding to each new class of dynamic system framework, or to each new class of problems, which, for a variety of reasons, becomes a "contagious" research trend in the field of automatic control.

In this chapter, we explore some extensions of ADRC control to classes of systems of general interest. We first present an integral reconstruction of states in multivariable systems, just to complete the integral reconstruction-based GPI control schemes proposed in previous chapters in the realm of observer-free ADRC for simple chains of integration systems. Enhanced robustness and effectiveness can be bestowed on Sliding-Mode control schemes by using disturbance observers. This evades the need for adaptive switching "amplitudes," representing binary control values characterizing typical control input limitations in sliding-mode control. We then move to present ADRC results to discrete-time systems in the context of delta transforms, a successful technique to deal with the efficient control of sampled-data dynamical systems [4,5]. The chapter also

Active Disturbance Rejection Control of Dynamic Systems. http://dx.doi.org/10.1016/B978-0-12-849868-2.00004-6
Copyright © 2017 Elsevier Inc. All rights reserved.

includes an excursion into the ADRC control of systems exhibiting input delays. The topic is also addressed in more detail via a specific laboratory example in Chapter 5. The chapter closes with an illustrative example pointing toward an extension of ADRC control to a class of fractional-order systems.

4.2 INTEGRAL RECONSTRUCTORS OF MIMO LINEAR SYSTEMS

In this section, we complete our treatment of integral reconstructor-based ADRC of Chapters 2 and 3. The results in this section can be found in the work of Fliess et al. [1]. We present those results with little modification.

Consider the observable time-invariant linear system of m inputs and p outputs

$$\dot{x} = Ax + Bu, \ x(0) = x_0, \quad y = Cx \tag{4.1}$$

where $x \in \mathbb{R}^n$ is the state vector, $x_0 \in \mathbb{R}^n$ represents the initial state, $y \in \mathbb{R}^l$ is the set of outputs, and $u \in \mathbb{R}^m$ is the set of inputs. Also, $A \in \mathbb{R}^{n \times n}$, $B \in \mathbb{R}^{n \times m}$, and $C \in \mathbb{R}^{l \times n}$.

Integrating (4.1), in the frequency domain, we have

$$x = A\frac{x}{s} + B\frac{u}{s} \tag{4.2}$$

Iterating this functional relation, we have

$$x(t) = A^2 \frac{x}{s^2} + AB\frac{u}{s^2} + B\frac{u}{s} \tag{4.3}$$

Continuing with the functional iteration $n - 1$ times, we have

$$x(s) = A^{n-1}\left(\frac{x(s)}{s^{n-1}}\right) + \sum_{i=1}^{n-1} A^{i-1} B \frac{u(s)}{s^i} \tag{4.4}$$

On the other hand, consider the output and its successive time derivatives:

$$\begin{pmatrix} I \\ sI \\ \vdots \\ s^{(n-1)}I \end{pmatrix} y(s) = \begin{pmatrix} C \\ CA \\ \vdots \\ CA^{n-1} \end{pmatrix} x(s)$$

$$+ \begin{pmatrix} 0 & 0 & \cdots & 0 \\ CB & 0 & \cdots & 0 \\ \vdots & \vdots & \ddots & 0 \\ CA^{n-2}B & \cdots & \cdots & CB \end{pmatrix} \begin{pmatrix} I \\ sI \\ \vdots \\ s^{(n-2)}I \end{pmatrix} u(s) \qquad (4.5)$$

Integrating in the frequency domain $n - 1$ times, we have

$$\begin{pmatrix} \dfrac{I}{s^{n-1}} \\ \dfrac{I}{s^{n-2}} \\ \vdots \\ I \end{pmatrix} y(s) = \begin{pmatrix} C \\ CA \\ \vdots \\ CA^{n-1} \end{pmatrix} \dfrac{x(s)}{s^{n-1}} + \mathcal{M} \begin{pmatrix} \dfrac{I}{s^{n-1}} \\ \dfrac{I}{s^{n-2}} \\ \vdots \\ \dfrac{I}{s} \end{pmatrix} u(s) \qquad (4.6)$$

Thanks to the assumed observability of the system,

$$\dfrac{x(s)}{s^{n-1}} = [\mathcal{O}^T \mathcal{O}]^{-1} \mathcal{O}^T \left[\begin{pmatrix} \dfrac{I}{s^{n-1}} \\ \dfrac{I}{s^{n-2}} \\ \vdots \\ I \end{pmatrix} y(s) - \mathcal{M} \begin{pmatrix} \dfrac{I}{s^{n-1}} \\ \dfrac{I}{s^{n-2}} \\ \vdots \\ \dfrac{I}{s} \end{pmatrix} u(s) \right] \qquad (4.7)$$

We can now combine this expression with the preceding one:

$$x(s) = A^{n-1} \left(\dfrac{x(s)}{s^{n-1}} \right) + \sum_{i=1}^{n-1} A^{i-1} B \dfrac{u(s)}{s^i} \qquad (4.8)$$

Finally, combining frequency domain and time domain notations, we have

$$x(t) = \mathcal{P}(s^{-1}) y(t) + \mathcal{Q}(s^{-1}) u(t) \qquad (4.9)$$

where the functional dependence on s^{-1} indicates a polynomial dependence on powers of such an operator, representing finite linear combinations of integrations of the involved quantities being operated on.

We address the expression, which does not take into account the influence of the initial states,

$$\widehat{x}(t) = \mathcal{P}(s^{-1}) y(t) + \mathcal{Q}(s^{-1}) u(t) \qquad (4.10)$$

as the *integral state reconstructor* based on finite linear combinations of iterated integrals of inputs and outputs.

The integral state reconstructor may be used, in principle, on any linear state feedback control law $u = -k^T x(t)$ as long as it is complemented with additional integral output or input error compensation, which counteracts the effects of the neglected initial conditions:

$$u = -k^T \left[\mathcal{P}(s^{-1}) y(t) + \mathcal{Q}(s^{-1}) u(t) \right] + v \qquad (4.11)$$

Such a compensation term v only requires iterated integrations of the outputs (i.e., output tracking or stabilization errors) or of the inputs (nominal input tracking errors).

4.2.1   Example: Planar Vertical Take-Off and Landing Aircraft (PVTOL)

This section illustrates the computation of integral reconstructors in a multivariable system and their use on a GPI controller. The use of the reconstructed states in an ADRC controller is not addressed and is left as an exercise for the interested reader.

Consider the normalized model of the PVTOL system

$$\ddot{x} = -u_1 \sin\theta + \epsilon u_2 \cos\theta$$
$$\ddot{z} = u_1 \cos\theta + \epsilon u_2 \sin\theta - g$$
$$\ddot{\theta} = u_2 \qquad (4.12)$$

A desired equilibrium for the system is given by

$$x = \overline{x}, \quad z = \overline{z}, \quad \theta = 0, \quad u_1 = g, \quad u_2 = 0 \qquad (4.13)$$

Linearization of the nonlinear system of equations around the equilibrium point yields

$$\ddot{x}_\delta = -g\theta_\delta + \epsilon u_{2\delta}$$
$$\ddot{z}_\delta = u_{1\delta}$$
$$\ddot{\theta}_\delta = u_{2\delta} \qquad (4.14)$$

It is easy to verify that x_δ is a nonminimum phase output. Indeed, the zero dynamics associated to $x_\delta = 0$ is

$$\ddot{\theta}_\delta = (g/\epsilon)\theta_\delta \qquad (4.15)$$

which exhibits an unstable saddle-point type of equilibrium.

The flat outputs may be considered to be

$$F = x_\delta - \epsilon\theta_\delta, \quad L = z_\delta \tag{4.16}$$

The system is equivalent to two chains of integrations

$$u_{1\delta} = \ddot{L}, \quad u_{2\delta} = -\frac{1}{g}F^{(4)} \tag{4.17}$$

A pole assignment-based compensator is readily obtained as

$$u_{1\delta} = -\gamma_2\dot{L} - \gamma_1 L$$
$$u_{2\delta} = -\frac{1}{g}\left[-k_5 F^{(3)} - k_4\ddot{F} - k_3\dot{F} - k_2 F\right] \tag{4.18}$$

Take the system outputs simply as

$$y_{1\delta} = x_\delta, \quad y_{2\delta} = z_\delta \tag{4.19}$$

The integral state reconstructors are obtained in the following manner:

$$\theta_\delta = \left(\int\int u_{2\delta}\right), \quad x_\delta = y_{1\delta}, \quad z_\delta = y_{2\delta}$$
$$\dot{\theta}_\delta = \left(\int u_{2\delta}\right), \quad \dot{x}_\delta = -g\left(\int\int\int u_{2\delta}\right) + \epsilon\left(\int u_2\right),$$
$$\dot{z}_\delta = \left(\int u_{1\delta}\right) \tag{4.20}$$

A feedback control law based on the integral reconstructors is readily obtained as

$$u_{1\delta} = -\gamma_2\left(\int u_{1\delta}\right) - \gamma_1 y_{2\delta} - \gamma_0\left(\int y_{2\delta}\right)$$
$$u_{2\delta} = -\frac{1}{g}\left[-k_5\widehat{F}^{(3)} - k_4\widehat{\ddot{F}} - k_3\widehat{\dot{F}} - k_2\widehat{F} - k_1\left(\int y_{1\delta}\right) - k_0\left(\int\int y_{1\delta}\right)\right] \tag{4.21}$$

This yields the following GPI controller:

$$u_{1\delta} = -\int[\gamma_2 u_{1\delta} - \gamma_0 y_{2\delta}] - \gamma_1 y_{2\delta}$$
$$u_{2\delta} = -\frac{1}{g}\left[(k_5 g u_{2\delta} - k_1 y_{1\delta}) - k_2 y_{1\delta} + \int\int[(k_4 g + k_2\epsilon)u_{2\delta} - k_0 y_{1\delta}]\right.$$
$$\left.+ \int\int\int (k_3 g u_2 - k_2 y_{1\delta})\right] \tag{4.22}$$

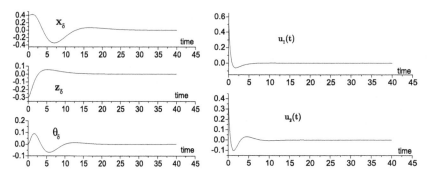

FIGURE 4.1 Performance of PVTOL model controlled by a GPI multivariable controller.

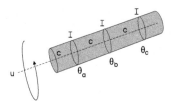

FIGURE 4.2 A flexible shaft system.

The performance of the incremental controller on the nonlinear model of the PVTOL is depicted in Fig. 4.1.

4.2.2 A Monovariable Example: A Flexible Shaft

Consider the following simplified model of a flexible shaft (see Fig. 4.2) acted upon by an external torque T and applied on the boundary.

The shaft is discretized to three representative points a, b, c characterized by lumped moments of inertia I, connected by springs, and angular deflections θ_a, θ_b, and θ_c satisfying the following set of differential equations:

$$
\begin{aligned}
I\frac{d^2\theta_a}{dt^2} &= k(-\theta_a + \theta_b) + T \\
I\frac{d^2\theta_b}{dt^2} &= k(\theta_a - 2\theta_b + \theta_c) \\
I\frac{d^2\theta_c}{dt^2} &= k(\theta_b - \theta_c)
\end{aligned}
\tag{4.23}
$$

Normalizing the system via the following time scale and input coordinate transformation we obtain

$$
d\tau = \sqrt{\frac{k}{I}}dt, \quad u = \frac{T}{I}
\tag{4.24}
$$

The normalized system is flat, with flat output given by $F = \theta_c$. The differential parameterization of all system variables follows immediately as

$$\theta_c = F$$
$$\theta_b = \ddot{F} + F \tag{4.25}$$
$$\theta_a = F^{(4)} + 3\ddot{F} + F$$
$$u = F^{(6)} + 4F^{(4)} + 3\ddot{F} \tag{4.26}$$

If the linear endogenous terms of the system are overlooked or purposely neglected, then the system is simplified to the following perturbed chain of integrations:

$$F^{(6)} = u + \xi(t) \tag{4.27}$$

A pole placement controller design, for having the flat output track a desired output reference trajectory $F^*(t)$, is obtained for u in the usual manner:

$$u = [F^*(t)]^{(6)} - k_5\left(F^{(5)} - [F^*(t)]^{(5)}\right) - k_4\left(F^{(4)} - [F^*(t)]^{(4)}\right)$$
$$- k_3\left(F^{(3)} - [F^*(t)]^{(3)}\right) - k_2\left(\ddot{F} - \ddot{F}^*(t)\right)$$
$$- k_1\left(\dot{F} - \dot{F}^*(t)\right) - k_0\left(F - F^*(t)\right) \tag{4.28}$$

with the following state-dependent relations replaced on the higher-order flat output time derivatives:

$$F = \theta_c$$
$$\dot{F} = \dot{\theta}_c$$
$$\ddot{F} = (\theta_b - \theta_c)$$
$$F^{(3)} = \dot{\theta}_b - \dot{\theta}_c$$
$$F^{(4)} = (\theta_a - 3\theta_b + 2\theta_c)$$
$$F^{(5)} = (\dot{\theta}_a - 3\dot{\theta}_b + 2\dot{\theta}_c)$$

We picked the following rest-to-rest trajectory for the flat output nominal trajectory $F^*(t)$:

$$F^*(t) = F_{ini} + (F_{fin} - F_{ini})\psi(t, t_0, t_f) \tag{4.29}$$

with $F_{ini} = 0$, $F_{fin} = 1$, and

$$\psi(t, t_0, t) = \tau^8\left[r_1 - r_2\tau + \cdots - r_8\tau^7 + r_9\tau^8\right] \tag{4.30}$$

where

$$r_1 = 12{,}870, \ r_2 = 91{,}520, \ r_3 = 288{,}288,$$
$$r_4 = 524{,}160, \ r_5 = 600{,}600, \ r_6 = 443{,}520,$$
$$r_7 = 205{,}920, \ r_8 = 54{,}912, \ r_9 = 6435$$

and $\tau = \dfrac{t - t_i}{t_f - t_i}$.

The linear feedback controller coefficients were set as coming from the desired characteristic polynomial,

$$(s + a)^6, \ a < 0 \tag{4.31}$$

The system is observable from θ_c since this is the flat output. We iterate twice on the second-order normalized equations:

$$
\begin{aligned}
\ddot{\theta}_a &= -\theta_a + \theta_b + u \\
\ddot{\theta}_b &= \theta_a - 2\theta_b + \theta_c \\
\ddot{\theta}_c &= \theta_b - \theta_c \\
y &= \theta_c
\end{aligned}
\tag{4.32}
$$

Integrating twice, we obtain

$$
\begin{aligned}
\theta_a &= -\left(\int\int \theta_a\right) + \left(\int\int \theta_b\right) + \left(\int\int u\right) \\
\theta_b &= \left(\int\int \theta_a\right) - 2\left(\int\int \theta_b\right) + \left(\int\int y\right) \\
\theta_c &= \left(\int\int \theta_b\right) - \left(\int\int y\right)
\end{aligned}
\tag{4.33}
$$

Iterating on the expressions for θ_a and θ_b once, we have

$$
\begin{aligned}
\theta_a &= 2\left(\int^{[4]} \theta_a\right) - 3\left(\int^{[4]} \theta_b\right) + \left(\int^{[4]} y\right) + \left(\int^{[4]} u\right) \\
\theta_b &= -3\left(\int^{[4]} \theta_a\right) + 5\left(\int^{[4]} \theta_b\right) - 2\left(\int^{[4]} y\right) + \left(\int^{[4]} u\right)
\end{aligned}
\tag{4.34}
$$

From the flatness of θ_c we have the following differential parameterizations for θ_a and θ_b:

$$
\begin{aligned}
\theta_a &= y^{(4)} + 3\ddot{y} + 3y \\
\theta_b &= \ddot{y} + y
\end{aligned}
\tag{4.35}
$$

Hence,

$$\left(\int^{[4]} \theta_a\right) = y + 3\left(\int\int y\right) + 3\left(\int^{[4]} y\right) \tag{4.36}$$

$$\left(\int^{[4]} \theta_b\right) = \left(\int\int y\right) + \left(\int^{[4]} y\right)$$

Substituting these expressions on the intermediate parameterizations of θ_a and θ_b, in terms of fourth iterated integrals, we obtain

$$\theta_a = 2y + 3\left(\int\int y\right) + 4\left(\int^{[4]} y\right) + \left(\int^{[4]} u\right) \tag{4.37}$$

$$\theta_b = -3y - 4\left(\int\int y\right) - 4\left(\int^{[4]} y\right)$$

From the system equations we obtain a parameterization for the flat output velocity,

$$\dot{\theta}_c = \left(\int \theta_b\right) - \left(\int y\right) \tag{4.38}$$

Then this yields

$$\dot{\theta}_c = -3y - \left(\int y\right) - 4\left(\int\int y\right) - 4\left(\int^{[4]} y\right) \tag{4.39}$$

Similarly, from the system equations,

$$\dot{\theta}_a = -\left(\int \theta_a\right) + \left(\int \theta_b\right) + \left(\int u\right) \tag{4.40}$$

$$\dot{\theta}_b = \left(\int \theta_a\right) - 2\left(\int \theta_b\right) + \left(\int \theta_c\right)$$

we readily obtain

$$\dot{\theta}_a = -5\left(\int y\right) - 7\left(\int^{[3]} y\right) - 8\left(\int^{[5]} y\right) - \left(\int^{[5]} u\right) + \left(\int u\right) \tag{4.41}$$

$$\dot{\theta}_b = 9\left(\int y\right) + 14\left(\int^{[3]} y\right) + 12\left(\int^{[5]} y\right) + \int^{[5]} u$$

The following integral input–output parameterization of all system variables allows us to design a GPI controller for this system:

$$\theta_a = 2y + 3\left(\int\int y\right) + 4\left(\int^{[4]} y\right) + \left(\int^{[4]} u\right)$$

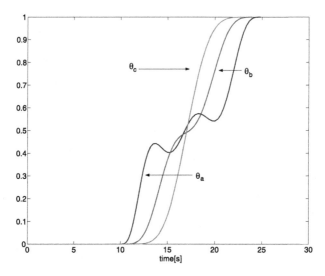

FIGURE 4.3 Performance of flexible bar subject to a linear feedback controller.

$$\theta_b = -3y - 4(\int\int y) - 4(\int^{[4]} y)$$

$$\theta_c = y$$

$$\dot{\theta}_a = -5(\int y) - 7(\int^{[3]} y) - 8(\int^{[5]} y) - (\int^{[5]} u) + (\int u)$$

$$\dot{\theta}_b = 9(\int y) + 14(\int^{[3]} y) + 12(\int^{[5]} y) + \int^{[5]} u \qquad (4.42)$$

$$\dot{\theta}_c = -3y - (\int y) - 4(\int\int y) - 4(\int^{[4]} y)$$

These estimates are off by at most an $o(t^4)$ function, as seen from the iterated integrations of the system equations. We use a controller with fifth-order iterated integral compensations. In total, we adjust an 11th-order characteristic polynomial.

Fig. 4.3 shows the performance of the controller on a rest to rest trajectory tracking task. The control input $u(t)$ is shown in Fig. 4.4 and a 3D mesh reconstruction using the variables θ_a, θ_b, and θ_c is provided in Fig. 4.5.

4.3 A LYAPUNOV APPROACH FOR A CLASS OF NONLINEAR MULTIVARIABLE SYSTEMS

In this section, the ADRC problem is tackled for a class of interconnected uncertain second-order systems, which can model a large class of electromechanical

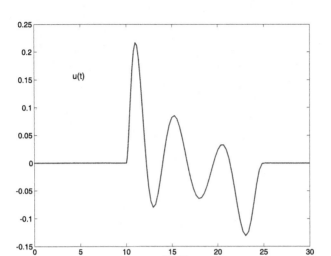

FIGURE 4.4 Control input trajectory generated by the linear controller.

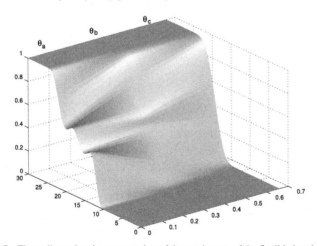

FIGURE 4.5 Three-dimensional representation of time trajectory of the flexible bar deformations.

systems formed by a set of inertial terms. Also, an alternative way of modeling the lumped disturbance inputs is discussed, and, finally, a quadratic Lyapunov function is provided for the stability of the closed-loop system. The theoretical part of this section is provided in [6].

Consider the following class of coupled second-order nonlinear systems with uncertain structure:

$$\dot{x}_a = x_b(t)$$
$$\dot{x}_b = f(x_a, x_b) + g(x_a(t))u(t) + \xi(x, t) \qquad (4.43)$$

We denote by $x = [x_a^\top, x_b^\top]^\top$, $x \in \mathbb{R}^{2n}$, the state vector of the robotic system; x_a is the set of position coordinates, and x_b represents the corresponding velocities. The initial condition of system (4.43) is given by $x(0) = x_0$, $\|x_0\| < \infty$.

Let X be a set such that $x \in X \subset \mathbb{R}^{2n}$. The bounded function $u \in \mathbb{R}^n$ is the control input function.

The nonlinear Lipschitz function $f : \mathbb{R}^{2n} \to \mathbb{R}^n$ is composed of n-uncertain nonlinear functions that describe the drift term of (4.43).

The matrix function $g : \mathbb{R}^n \to \mathbb{R}^{n \times n}$ is invertible and satisfies the following constraint:

$$0 < g^- \le \|g(\cdot)\| \le g^+, \quad g^-, g^+ \in \mathbb{R}^+ \tag{4.44}$$

The term $\xi(x, t)$ contains the combined effects of perturbations and uncertainties that can also include, for example, parameter variations and modeling errors. We consider the following assumption regarding $\xi(x, t)$.

Assumption 1. The class of uncertainties $\xi(x, t)$ satisfies the following inequality:

$$\|\xi(x, t)\|_{\Lambda_\xi}^2 \le \gamma_0 + \gamma_1 \|x\|^2, \quad \forall t \ge 0$$
$$\gamma_0, \gamma_1 \in \mathbb{R}^+; \ 0 < \Lambda_\xi = \Lambda_\xi^\top \in \mathbb{R}^{n \times n} \tag{4.45}$$

System (4.43) can be rewritten as follows:

$$\dot{x} = Ax(t) + G(x))U(t) + F(x, t) \tag{4.46}$$

$$A = \begin{bmatrix} 0_{n \times n} & I_{n \times n} \\ 0_{n \times n} & 0_{n \times n} \end{bmatrix}, \ G(x) = \begin{bmatrix} 0_{n \times n} & 0_{n \times n} \\ 0_{n \times n} & g(x_a) \end{bmatrix}, \ U(t) = \begin{bmatrix} 0_n \\ u(t) \end{bmatrix} \tag{4.47}$$

$$F(x, t) := \begin{bmatrix} 0_n^\top & f^\top(x) + \xi^\top(x, t) \end{bmatrix}^\top$$

The function $F(x)$ satisfies the following inequality for all $t \ge 0$: $\|F(x, t)\|^2 \le f_0 + f_1 \|x\|^2$, $f_0, f_1 \in \mathbb{R}^+$. The following assumption concerns the nature of the lumped disturbance according to the chain of integrators structure needed to propose the controller design.

Assumption 2. There exists a constant vector a such that the function $f(x) + \xi(x(t), t)$ evaluated along the system trajectories, that is, $x = x(t)$, can be represented as $f(x) + \xi(x, t) = a^\top \kappa(x) + \tilde{f}(x, t)$, where the vector $a \in \mathbb{R}^{(p+1) \times n}$ is formed by a set of constant parameters a_k that must be adjusted to improve the approximation of $f(x) + \xi(x, t)$. According to the structure of the GPI extended state observers, the vector $\kappa \in \mathbb{R}^{p+1}$ is composed as $\kappa = [1, t, \cdots, t^p]$. This is a regular decomposition of $f(x) + \xi(x, t)$ in terms of a finite number of elements that form a basis. In particular, the last set of polynomials is considered here. The term $\tilde{f}(x, t)$ is called the modeling error produced by the approximation of $f(x) + \xi(x, t)$ by a finite number p of elements in the basis.

The so-called nominal model $a^\top \kappa(t)$ can be represented as $a^\top \kappa(t) = a_0 + a_1 t + a_2 t^2 + \cdots + a_p t^p$. Each term a_j is a vector, i.e., $a_j \in \mathbb{R}^n$, $j = 0, \ldots, p$. This representation is similar to the one used throughout this book. However, the underlying approximation method requires the differentiability of the term $f(x) + \xi(x, t)$, which is considered to be a strong restriction. In this section, an alternative is proposed, in which the function $a^\top \kappa(x(t))$ can be represented as a chain of integrators. Thus, the approximation presented above states that $f(x) + \xi(x, t)$ must be the solution of integration of an uncertain function plus the approximation error. This condition relaxes the constraint, which is usually a drawback in this class of extended state observers. Thus, the previous equation is written as

$$a^\top \kappa(t) = a_0 + \int_{\tau_1=0}^{t} a_1 d\tau_1 + \int_{\tau_1=0}^{t} \int_{\tau_2=0}^{\tau_1} 2a_2 d\tau_2 d\tau_1 + \cdots$$
$$+ \int_{\tau_1=0}^{t} \cdots \int_{\tau_p=0}^{\tau_{p-1}} p! a_p d\tau_p \cdots d\tau_1 \qquad (4.48)$$

The last equation can be expressed in a differential form:

$$a^\top \kappa(t) = \rho_0(t) \qquad (4.49)$$
$$\frac{d\rho_i(t)}{dt} = \rho_{i+1}(t), \ i = 0, \ldots, p-1$$
$$\frac{d\rho_p(t)}{dt} = 0 \qquad \rho_j(0) = a_j, j = 0, 1, \ldots, p$$

The problem formulation given above can be rephrased as follows: Given a vector output reference trajectory x^* for system (4.43), design an output feedback controller that, regardless of the unknown unmodeled dynamics or external disturbances (both lumped in an additive signal $F(x)$), forces the outputs x to track, asymptotically, the desired reference trajectories with the tracking error restricted to evolve on a small neighborhood near the origin and proportional to a power of the uncertainties and perturbations. The first stage in solving this problem consists in designing an extended state observer to reconstruct the unmeasurable part of the state.

4.3.1   The High-Order Extended State Observer

The extended state observer for system (4.43) with the corresponding feedback controller using the observer states $u(t) := u(\hat{x}(t))$ is proposed as

$$\dot{\hat{x}}(t) = A\hat{x}(t) + G(x_a(t))U(t) + B\hat{\rho}_0(t) + He(t) \qquad (4.50)$$

$$\dot{\hat{\rho}}(t) = \Phi\hat{\rho}(t) + Le(t)$$

$$\Phi = \begin{bmatrix} 0_{n\times n} & I_{n\times n} & 0_{n\times n} & \cdots & 0_{n\times n} \\ 0_{n\times n} & 0_{n\times n} & I_{n\times n} & \cdots & 0_{n\times n} \\ \vdots & \vdots & \vdots & \cdots & \vdots \\ 0_{n\times n} & 0_{n\times n} & 0_{n\times n} & \cdots & I_{n\times n} \\ 0_{n\times n} & 0_{n\times n} & 0_{n\times n} & \cdots & 0_{n\times n} \end{bmatrix} \Bigg\} p \text{ times}$$

$$\hat{\rho}_0(t) = D^{\top}\hat{\rho}(t); \quad e(t) = x_a(t) - C^{\top}\hat{x}(t)$$

Here the vectors $\hat{x}$ and $\hat{\rho}$ are the estimated states for x and ρ, respectively. The matrix $H \in \mathbb{R}^{2n\times n}$ is the state observer gain. The matrix $L \in \mathbb{R}^{n(p+1)\times n}$ is the extended observer gain. The matrix $D \in \mathbb{R}^{n(p+1)\times p}$ is represented as $D^{\top} = [I_{n\times n} \ 0_{n\times n} \ \cdots \ 0_{n\times n}]$. The matrix $\Phi \in \mathbb{R}^{n(p+1)\times n(p+1)}$ is called the extended state self-matrix. The output matrix $C \in \mathbb{R}^{2n\times n}$ is given by $C^{\top} = [I_{n\times n} \ 0_{n\times n}]$. Finally, the matrix $B \in \mathbb{R}^{2n\times n}$ is represented as $B^{\top} = [0_{n\times n} \ I_{n\times n}]$.

On the basis of the extended state observer proposed in (4.50), the output-based controller can then be designed using the following structure:

$$u(t) = -g(x_a(t))^{-1}\left[K^{\top}(\hat{x}(t) - x^*(t)) + \hat{\rho}_0(t) - s(x^*(t))\right] \qquad (4.51)$$

where $s(x^*(t))$ is the feedforward input. In space state terms, the reference trajectory vector $x^* \in \mathbb{R}^{2n}$ satisfies

$$\dot{x}^*(t) = Ax^*(t) + Bs(x^*(t)), \ x^*(0) = x_0^* \in \mathbb{R}^{2n} \qquad (4.52)$$

where $s : \mathbb{R}^n \to \mathbb{R}^n$ is a Lipschitz function. The reference trajectory is designed to be bounded for all $t \geq 0$, that is, $\|x^*(t)\|^2 \leq [x_{max}^*]^2$, $x_{max}^* \in \mathbb{R}^+$.

The controller gain matrix K is designed such that the matrix $A - BK$ is Hurwitz. To solve this part of the problem, all the methodologies to tune a desired characteristic polynomial of a dominant linear dynamics can be applied.

The main result is given as follows.

Theorem 1. *Consider the state observer given in (4.50) and the output feedback controller proposed in (4.51) (with the gain K adjusted so that the linear closed-loop dominant dynamics is Hurwitz) for the class of nonlinear uncertain systems (4.43) fulfilling condition (4.44) with incomplete information and* *affected by perturbations that obey the constraints given in (4.45). Define the matrices $\Pi(H, L, K)$, R, and Q described by*

$$\Pi(H, L, K) = \begin{bmatrix} A - HC^\top & BD^\top & 0_{2n \times 2n} \\ -LC^\top & \Phi & 0_{n(p+1) \times n(p+1)} \\ HC^\top & 0_{2n \times 2n} & A - BK \end{bmatrix} \quad (4.53)$$

$$R = \Lambda + \begin{bmatrix} 2f_1 I & 0 & 0 \\ 0 & 0 & 0 \\ 0 & 0 & 4f_1 I \end{bmatrix} \quad (4.54)$$

where $\Lambda = \Lambda^\top \in \mathbb{R}^{N_p \times N_p}$, $\Lambda > 0$, $Q = Q_0$, $Q_0 = Q_0^\top \in \mathbb{R}^{N_p \times N_p}$, $Q_0 > 0$, and $N_p = 4n + n(p + 1)$. If there is a positive definite matrix Q_0 such that the algebraic Riccati equation $Ric(P) = 0$ with

$$Ric(P) = P\Pi(H, L, K) + \Pi(H, L, K)^\top P + PRP + Q \quad (4.55)$$

has a positive definite symmetric solution $P \in \mathbb{R}^{N_p \times N_p}$, then the tracking trajectory error $\varepsilon = x - x^*$ is ultimately bounded with an upper bound given by $\beta = 2\dfrac{f_0 + 4f_1 \left[x_{max}^* \right]^+}{\alpha_Q}$ with $\alpha_Q := \lambda_{min} \left\{ P^{-1/2} Q_0 P^{-1/2} \right\}$.

Proof. The estimation error Δ satisfies

$$\dot{\Delta}(t) = \left[A - HC^\top \right] \Delta(t) + F(x(t)) - B\hat{\rho}_0(t) \quad (4.56)$$

The approximation proposed in (4.49) makes it possible to transform the previous differential equation into

$$\dot{\Delta}(t) = [A - HC^\top]\Delta(t) + B\tilde{f}(x(t), t) + BD^\top \left[\rho(t) - \hat{\rho}(t) \right] \quad (4.57)$$

On the other hand, the extended state error, defined as $\delta = \rho - \hat{\rho}$, and the tracking error σ satisfy

$$\dot{\delta}(t) = \Phi\delta(t) - LC^\top \Delta(t) \quad (4.58)$$

$$\dot{\sigma}(t) = A\hat{x}(t) + G(x_a(t))u(t) + B\hat{\rho}_0(t) + He(t) - s(x^*(t)) \quad (4.59)$$

Substituting the control action described by (4.51) into the previous differential equation, we get

$$\dot{\sigma}(t) = (A - BK)\sigma(t) + HC^\top \Delta(t) \quad (4.60)$$

To prove that the equilibrium point of (4.57) and (4.58) is stable, we propose the following Lyapunov function candidate:

$$V(z) = z^\top P z \tag{4.61}$$

where the vector $z \in \mathbb{R}^{N_p}$ is constructed as $z^\top = [\Delta^\top \ \delta^\top \ \sigma^\top]$. Taking the Lie derivative of $V(z)$ with respect to time, we have

$$\dot{V}(t) = 2z^\top(t) P \dot{z}(t) \tag{4.62}$$

By (4.57) and (4.58) the time derivative of z is

$$\dot{z}(t) = \begin{bmatrix} A - HC^\top & BD^\top & 0_{2n \times 2n} \\ -LC^\top & \Phi & 0_{n(p+1) \times n(p+1)} \\ HC^\top & 0_{2n \times 2n} & A - BK \end{bmatrix} z(t) + \begin{bmatrix} B\tilde{f}(x(t),t) \\ 0_{n(p+1)} \\ 0_n \end{bmatrix} \tag{4.63}$$

Using this result in the previous differential equation, we get

$$\dot{V}(t) = 2z^\top(t) P \Pi(H,L,K) z(t) + 2z^\top(t) P \eta(t) \tag{4.64}$$

where η represents all the internal uncertainties and external perturbations.

The direct application of the matrix inequality $X^\intercal Y + Y^\intercal X \leq X^\intercal N X + Y^\intercal N^{-1} Y$, which is valid for any $X, Y \in \mathbb{R}^{r \times s}$ and any $0 < N = N^\intercal \in \mathbb{R}^{s \times s}$ [7], yields

$$\dot{V}(t) \leq z^\top(t) \left(P\Pi(H,L,K) + \Pi(H,L,K)^\top P + P\Lambda P \right) z(t) + \eta^\top(t)\Lambda^{-1}\eta(t) \tag{4.65}$$

An algebraic manipulation involving the term $z^\top(t) Q_0 z(t)$ yields

$$\dot{V}(t) \leq z^\top(t)(P\Pi(H,L,K) + \Pi(H,L,K)^\top P + P\Lambda P + Q_0)z(t) + \eta^\top(t)\Lambda^{-1}\eta(t) - z^\top(t)Q_0 z(t) \tag{4.66}$$

By taking the upper bound presented in (4.45) the term $\eta^\top \Lambda^{-1} \eta$ for all $t \geq 0$ is bounded from above by

$$\eta^\top \Lambda^{-1} \eta \leq f_0 + 2f_1\|\Delta\|^2 + 4f_1\|\sigma\|^2 + 4f_1\|x^*\|^2 \tag{4.67}$$

On the basis of this inequality, the differential inclusion presented in (4.66) is transformed into

$$\dot{V}(t) \leq z^\top(t) Ric(P) z(t) - z^\top(t) Q_0 z(t) + \beta_0 \tag{4.68}$$

where $\beta_0 = f_0 + 4f_1\left[x^*\right]^+$. Using the assumption regarding the existence of a positive definite solution to the Riccati equation $Ric(P) = 0$, the previous differential inclusion is modified to

$$\dot{V}(t) \leq -z^\top(t)Q_0 z(t) + \beta_0 \qquad (4.69)$$

The right-hand side of the previous differential inclusion can be bounded from above as

$$\dot{V}(t) = -\lambda_{min}\left\{P^{-1/2}Q_0 P^{-1/2}\right\}z^\top(t)Pz(t) + \beta_0 \qquad (4.70)$$

The last inequality was obtained using the so-called Rayleigh inequality and the Cholesky decomposition [7]. According to the definition of β, we have the following differential inequality:

$$\dot{V}(t) \leq -\alpha_Q V(t) + \beta_0 \qquad (4.71)$$

Taking the equality in the previous differential inclusion, we have

$$\dot{V}^{eq} = -\alpha_Q V^{eq} + \beta_0 \qquad (4.72)$$

The solution of the previous differential equation is

$$V^{eq}(t) = e^{-\alpha_Q t}V^{eq}(0) + \frac{\beta_0}{\alpha_Q}(1 - e^{-\alpha_Q t}) \qquad (4.73)$$

Applying the comparison principle [8] to the previous equation and using the fact that $V(t) \geq 0$, we have

$$V(t) \leq e^{-\alpha_Q t}V(0) + \frac{\beta_0}{\alpha_Q}(1 - e^{-\alpha_Q t}) \qquad (4.74)$$

Taking the upper limit of the previous inequality, we finally obtain the main result presented in the theorem. $\qquad\qquad\qquad\qquad\qquad\qquad\qquad\qquad\square$

4.3.2 Example: A Cylindrical Robot

Consider a cylindrical robotic system (Fig. 4.6). The mathematical model is given as follows:

$$\begin{bmatrix} J + m_3 q_3^3 & 0 & 0 \\ 0 & m_2 + m_3 & 0 \\ 0 & 0 & m_3 \end{bmatrix}\begin{bmatrix} \ddot{q}_1 \\ \ddot{q}_2 \\ \ddot{q}_3 \end{bmatrix} + \begin{bmatrix} m_3 q_3 \dot{q}_3 & 0 & m_3 q_3 \dot{q}_1 \\ 0 & 0 & 0 \\ -m_3 q_3 \dot{q}_1 & 0 & 0 \end{bmatrix}\begin{bmatrix} \dot{q}_1 \\ \dot{q}_2 \\ \dot{q}_3 \end{bmatrix}$$

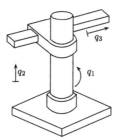

FIGURE 4.6 Cylindrical Robot.

$$+ \begin{bmatrix} 0 \\ (m_2 + m_3)g \\ 0 \end{bmatrix} = \begin{bmatrix} u_1 \\ u_2 \\ u_3 \end{bmatrix} \tag{4.75}$$

where J is the moment of inertia of the first link (revolute), m_2 and m_3 are the masses of the prismatic links, g is the gravity acceleration, q_1, q_2, and q_3 are the joint coordinates that represent the rotational movement (q_1), the vertical motion (q_2), and the radial motion of the end effector (q_3). The control problem is to design a multivariable control $u(t)$ such that the joint position coordinates q are forced to track a reference trajectory q^*, in spite of the lack of knowledge of the Coriolis and gravity terms and possible external disturbances within an ultimate bounded tracking behavior.

The last model can be seen as

$$D(q)\ddot{q} + C(q,\dot{q})\dot{q} + \bar{G}(q) = u \tag{4.76}$$

where $D(q) \in \mathbb{R}^{3\times 3}$ is the inertia matrix, $C(q,\dot{q}) \in \mathbb{R}^{3\times 3}$ is the matrix of Coriolis effects, $\bar{G}(q) \in \mathbb{R}^3$ is the gravity vector, $\ddot{q} = \begin{bmatrix} \ddot{q}_1 & \ddot{q}_2 & \ddot{q}_3 \end{bmatrix}^\mathsf{T} \in \mathbb{R}^3$ is the joint acceleration vector, $\dot{q} = \begin{bmatrix} \dot{q}_1 & \dot{q}_2 & \dot{q}_3 \end{bmatrix}^\mathsf{T} \in \mathbb{R}^3$ is the joint velocity vector, $q = \begin{bmatrix} q_1 & q_2 & q_3 \end{bmatrix}^\mathsf{T} \in \mathbb{R}^3$ is the joint position vector (measurable), and $u = \begin{bmatrix} u_1 & u_2 & u_3 \end{bmatrix}^\mathsf{T} \in \mathbb{R}^3$ stands for the input vector.

Assuming the presence of external forces, denoted by $\eta := \begin{bmatrix} \eta_1 & \eta_2 & \eta_3 \end{bmatrix}^\mathsf{T}$, we have, according to the structure (4.46), the following system:

$$\dot{x} = Ax + G(x_a)u + F(x,t) \tag{4.77}$$

where

$$x_a = \begin{bmatrix} q_1 & q_2 & q_3 \end{bmatrix}^\mathsf{T}, \quad x_b = \begin{bmatrix} \dot{q}_1 & \dot{q}_2 & \dot{q}_3 \end{bmatrix}^\mathsf{T}, \quad G(x_a) = \begin{bmatrix} 0_{3\times 3} & 0_{3\times 3} \\ 0_{3\times 3} & D^{-1}(x_a) \end{bmatrix}$$

$$\tag{4.78}$$

$$g(x_a) = D^{-1}(x_a), \quad F(x,t) = C(x)x_b + \bar{G}(x_a) + \eta \tag{4.79}$$

From the bounds of the inertia matrix property [9] we achieve condition (4.44) and assume that each component of η lies in the class defined in Assumption 1.

For the approximation of F, we propose a family of 3rd-degree time polynomials ($p = 1$), which leads to the following extended state observer form:

$$\dot{\hat{x}}(t) = A\hat{x}(t) + G(x_a)U(t) + B\hat{\rho}_0(t) + He(t) \tag{4.80}$$
$$\dot{\hat{\rho}}(t) = \Phi\hat{\rho}(t) + Le(t)$$

with

$$\Phi = \begin{bmatrix} 0_{3\times3} & I_{3\times3} & 0_{3\times3} & 0_{3\times3} \\ 0_{3\times3} & 0_{3\times3} & I_{3\times3} & 0_{3\times3} \\ 0_{3\times3} & 0_{3\times3} & 0_{3\times3} & I_{3\times3} \\ 0_{3\times3} & 0_{3\times3} & 0_{3\times3} & 0_{3\times3} \end{bmatrix}, \quad D^\mathsf{T} = \begin{bmatrix} I_{3\times3} & 0_{3\times3} & 0_{3\times3} & 0_{3\times3} \end{bmatrix}$$

The matrices H and L are proposed as

$$H = \begin{bmatrix} \lambda_{51} & 0 & 0 \\ 0 & \lambda_{52} & 0 \\ 0 & 0 & \lambda_{53} \\ \lambda_{41} & 0 & 0 \\ 0 & \lambda_{42} & 0 \\ 0 & 0 & \lambda_{43} \end{bmatrix}, \quad L = \begin{bmatrix} \lambda_{31} & 0 & 0 \\ 0 & \lambda_{32} & 0 \\ 0 & 0 & \lambda_{33} \\ \lambda_{21} & 0 & 0 \\ 0 & \lambda_{22} & 0 \\ 0 & 0 & \lambda_{23} \\ \lambda_{11} & 0 & 0 \\ 0 & \lambda_{12} & 0 \\ 0 & 0 & \lambda_{13} \\ \lambda_{01} & 0 & 0 \\ 0 & \lambda_{02} & 0 \\ 0 & 0 & \lambda_{03} \end{bmatrix}$$

such that the linear dominant injection dynamics of each joint coordinate in terms of the Laplace operator is given as follows:

$$s^6 + \lambda_{5i}s^5 + \lambda_{4i}s^4 + \lambda_{3i}s^3 + \lambda_{2i}s^2 + \lambda_{1i}s + \lambda_{0i} \tag{4.81}$$

with $i = 1, 2, 3$. The trajectory tracking control is given as

$$u(t) = -g(x_a(t))^{-1}\left[K^\mathsf{T}(\hat{x}(t) - x^*(t)) + \hat{\rho}_0(t) - s(x^*(t)) \right] \tag{4.82}$$

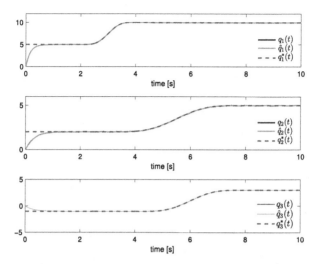

FIGURE 4.7 Closed-loop trajectory tracking results and position estimation.[1]

4.3.3 Numerical Results

A trajectory tracking consisting in a rest-to-rest reference function for the three joints was carried out for the robotic system. The robot parameters were $J = 0.02$ [kgm^2], $m_2 = 0.2$ [kg], $m_3 = 0.15$ [kg], $g = 9.81$ [m/s^2]. The desired characteristic polynomials of the dominant dynamics (4.81) of the observer were set to have the form $(s^2 + 2\zeta_{1i}\omega_{1i} + \omega_{1i}^2)(s^2 + 2\zeta_{2i}\omega_{2i} + \omega_{2i}^2)(s^2 + 2\zeta_{3i}\omega_{3i} + \omega_{3i}^2)$ with $\zeta_{i1} = 2$, $\zeta_{2i} = 4$, $\zeta_{3i} = 5$, $\omega_{1i} = 22$, $\omega_{2i} = 25$, $\omega_{3i} = 30$, $i = 1, 2, 3$. The linear control input was set to match the closed-loop error characteristic polynomial of $s^2 + 2\zeta_{ic}\omega_{ic}^2$ with $\zeta_{ic} = 3$ and $\omega_{ic} = 20$, $i = 1, 2, 3$. Additionally, the states of a Chen chaotic system were used to disturb each joint, forcing the extended state observer to estimate both nonmodeled dynamics and external disturbances. Fig. 4.7 shows the tracking results, where the three reference trajectories are tracked with a bounded error. The same figure shows the position estimation of the observer. Finally, Fig. 4.8 shows the lumped disturbance inputs and the control inputs, where the chaotic behavior of the disturbances is present in the extended state estimator and the control inputs.

4.4 COMBINED SLIDING-MODE–ADRC CONTROLLERS

4.4.1 Generalities About Sliding Regimes

Consider the following n-dimensional nonlinear single-input–single-output system:

---

1. For interpretation of the references to color in this and the following figures, the reader is referred to the web version of this chapter.

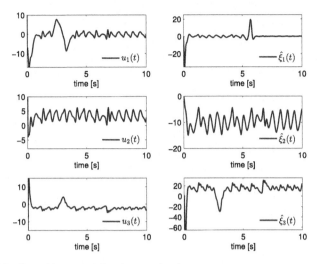

FIGURE 4.8 Control inputs and disturbance estimation.

$$\dot{x} = f(x) + g(x)u, \quad \sigma = h(x) \tag{4.83}$$

where f and g are known smooth vector fields on $T\mathbb{R}^n$. The scalar control input function u is assumed to take values on the closed interval $[-U, U]$. The function $h(x)$ is a smooth function $h : \mathbb{R}^n \to \mathbb{R}$. The zero level set for the scalar output σ,

$$S = \{x \in \mathbb{R}^n \mid \sigma = h(x) = 0\} \tag{4.84}$$

represents a smooth $(n - 1)$-dimensional manifold acting as a sliding surface. We say that σ is the *sliding surface coordinate function*. Let $x^0 \notin S$ be an arbitrary operating point in $\mathbb{R}^n$. We restrict our considerations to an open neighborhood $\mathcal{N}(x^0) \subset \mathbb{R}^n$ exhibiting a nonempty intersection with the sliding surface S.

Suppose that it is desired to locally create a sliding motion on the sliding surface S from an operating point x^0. Furthermore, assume that the directional derivative $L_g h(x)$ is perfectly known strictly positive and known to be bounded away from zero in $\mathcal{N}(x^0)$. This is known as the *transversality condition*. A necessary and sufficient condition for the existence of a local sliding regime on the smooth surface S from a point x^0 is given by the well-known existence condition (see Utkin [10]):

$$\sigma\dot{\sigma} < 0 \tag{4.85}$$

For points $x \in \mathcal{N}(x^0)$ where $\sigma(x) > 0$, the control input choice $u = u^+(x)$ must be such that

$$\dot{\sigma} = L_f h(x) + L_g h(x)u^+(x) < 0 \tag{4.86}$$

whereas for points where $\sigma(x) < 0$, the control input choice $u = u^-(x)$ must satisfy the condition

$$\dot{\sigma} = L_f h(x) + L_g h(x) u^-(x) > 0 \qquad (4.87)$$

Suppose that we adopt the following *switching policy*: $u^+(x) = -W$, $u^-(x) = W$, i.e., $u = -W \operatorname{sign}(\sigma)$ with $W < U$. Then

$$-W L_g h(x) < L_f h(x) < L_g h(x) W \qquad (4.88)$$

Dividing out by the negative quantity $-L_g h(x)$, we obtain, for points in the vicinity of the sliding surface S, a necessary and sufficient condition for the existence of a sliding regime translated into

$$-W < -\frac{L_f h(x)}{L_g h(x)} < W \qquad (4.89)$$

Ideal sliding motions are defined as the motions locally and smoothly taking place on the sliding surface S under the influence of the *virtual*, smooth, equivalent control input function, denoted as $u_{eq}(x)$, for initial state conditions belonging to S. The ideal sliding motions are characterized by the *Invariance conditions* $\sigma = 0$, $\dot{\sigma} = 0$. These conditions define the equivalent control as

$$u_{eq}(x) = -\left.\frac{L_f h(x)}{L_g h(x)}\right|_{\sigma=0} \qquad (4.90)$$

The ideal sliding motions are described by the following n-dimensional redundant dynamics, valid on the constraint manifold S:

$$\dot{x} = \left[I - \frac{g(x)}{L_g h(x)}\frac{\partial h}{\partial x^T}\right] f(x) := M(x) f(x) \qquad (4.91)$$

It can be verified that since $dh(x)M(x) = 0$ for $x \in S \cap \mathcal{N}(x^0)$, it follows that $M(x)f(x) \in T_x S$, whereas $M^2(x) = M(x)$ and $M(x)g(x) = 0$. The matrix $M(x)$ is, hence, a *projection operator onto the tangent space* $T_x S$ *along the* span$\{g\}$.

We have the following result:

Given the smooth nonlinear system $\dot{x} = f(x) + g(x)u$ with $u \in [-U, U]$ and $L_g h(x)$ a strictly positive scalar function on $\mathcal{N}(x^0)$, the switched feedback control policy $u = -W \operatorname{sign}(\sigma)$ with $W < U$ locally creates a sliding regime on $S = \{x \in \mathbb{R}^n \mid \sigma = h(x) = 0\}$ if and only if for points x on $S \cap \mathcal{N}(x^0)$,

$$-W < u_{eq}(x) = -\frac{L_f h(x)}{L_g h(x)} < W \qquad (4.92)$$

The proof is immediate upon realizing that, due to the smoothness of the equivalent control, the condition $-W < u_{eq}(x) < W$ remains valid on an open neighborhood of $x \in S$. Hence, a sliding regime locally exists on $S \cap \mathcal{N}(x^0)$.

4.4.2 Main Results

Consider the following n-dimensional nonlinear single-input–single-output system

$$\dot{x} = f(x) + g(x)u, \quad \sigma = h(x) \tag{4.93}$$

where the drift vector field $f(x)$ is a smooth, but uncertain, vector field on $T\mathbb{R}^n$. The case of an additive exogenously perturbed input field of the form $\gamma(x)\theta(t)$ follows quite a similar treatment.

4.4.3 Assumptions

- The scalar control input u is assumed to take values on the finite closed interval $[-U, U]$ and the amplitude W of the switching part of the control input u satisfies $W < U$.
- The directional derivative $L_f h(x)$ and a finite number of its time derivatives, say $(L_f h(x(t)))^{(k)}$, $k = 0, 1, 2, \ldots, m$, for a sufficiently large m, are assumed to be uniformly absolutely bounded for any feedback control input stabilizing the sliding surface coordinate dynamics.[2]
- $L_g h(x)$ is perfectly known and locally strictly positive.

4.4.4 A GPI Observer Approach to Sliding-Mode Creation

The state-dependent sliding surface dynamics is characterized by

$$\dot{\sigma} = L_f h(x) + L_g h(x)u \tag{4.94}$$

Under the uncertain but bounded nature of $L_f h(x)$, the existence condition $\sigma \dot{\sigma} < 0$ for $u = -W \text{sign}(\sigma)$ adopts the more conservative form

$$-W \inf_t L_g h(x) < \sup_t \left| L_f h(x) \right| < W \inf_t L_g h(x) \tag{4.95}$$

which transforms into

$$-W < -\frac{\sup_t \left| L_f h(x) \right|}{\inf_t L_g h(x)} < W \tag{4.96}$$

---

2. This assumption is to be understood in the "almost everywhere" sense.

It is clear that in comparison with the existence condition (4.92), for the nonuncertain case, the control input switching amplitude W may be superseded by the expression in (4.96).

Note, however, that an accurate, although possibly approximate, online estimation of the scalar uncertain quantity $L_f h(x)$ in the form $\widehat{L_f h}(x)$ may result in a substantially enhanced possibility for the creation of a sliding motion via the active disturbance cancellation strategy,

$$u = -\widehat{L_f h}(x) - W\text{sign}(\sigma). \tag{4.97}$$

The observer-based controller (4.97) produces the following closed-loop sliding surface dynamics:

$$\dot{\sigma} = \left[L_f h(x) - \widehat{L_f h}(x)\right] - WL_g h(x)\text{sign}(\sigma) \tag{4.98}$$

which would require a smaller control input switching amplitude W than in the case where the observer is not used:

$$\dot{\sigma} = L_f h(x) - WL_g h(x)\text{sign}(\sigma) \tag{4.99}$$

We have the following result.

Theorem 2. *Consider the following sliding surface coordinate function GPI observer:*

$$
\begin{aligned}
\dot{\hat{\sigma}} &= z_1 + L_g h(x)u + \gamma_m(\sigma - \hat{\sigma}) \\
\dot{z}_1 &= z_2 + \gamma_{m-1}(\sigma - \hat{\sigma}) \\
\dot{z}_2 &= z_3 + \gamma_{m-2}(\sigma - \hat{\sigma}) \\
&\vdots \\
\dot{z}_{m-1} &= z_m + \gamma_1(\sigma - \hat{\sigma}) \\
\dot{z}_m &= \gamma_0(\sigma - \hat{\sigma})
\end{aligned}
\tag{4.100}
$$

with m being a sufficiently large integer ([11], where m is set to 1). Given the uncertain sliding surface dynamics

$$\dot{\sigma} = L_f h(x) + L_g h(x)u \tag{4.101}$$

the sliding surface estimation error $e_\sigma = \sigma - \hat{\sigma}$ and its time derivatives, $e_\sigma^{(j)}$, $j = 1, 2, \ldots, m$, asymptotically converge, in an exponentially dominated manner, toward a small ball $\mathcal{B}(0, \rho)$ of radius ρ, centered around the origin of the estimation error phase space $\chi = (e_\sigma, \dot{e}_\sigma, \ldots, e_\sigma^{(m)})$. The radius ρ can be made *as small as desired, provided that the roots of the polynomial in the complex variable s,*

$$p(s) = s^{m+1} + \gamma_m s^m + \cdots + \gamma_1 s + \gamma_0 \tag{4.102}$$

are located further into the left half of the complex plane via the specification of the observer design coefficients $\{\gamma_0, \cdots, \gamma_m\}$. Moreover, the observer variable z_1 estimates, in an arbitrarily close fashion, the time signal representing the uncertain directional derivative $L_f h(x(t))$.

Proof. Notice that the estimation error variable $e_\sigma = \sigma - \hat{\sigma}$ evolves governed by

$$
\begin{aligned}
\dot{e}_\sigma &= (L_f h(x(t))) - z_1 - \gamma_m e_\sigma \\
\dot{z}_1 &= z_2 + \gamma_{m-1} e_\sigma \\
\dot{z}_2 &= z_3 + \gamma_{m-2} e_\sigma \\
&\vdots \\
\dot{z}_{m-1} &= z_m + \gamma_1 e_\sigma \\
\dot{z}_m &= \gamma_0 e_\sigma
\end{aligned}
\tag{4.103}
$$

Therefore, the sliding surface estimation error e_σ evolves according to the linearly dominated perturbed dynamics

$$e_\sigma^{(m+1)} + \gamma_m e_\sigma^{(m)} + \cdots + \gamma_1 \dot{e}_\sigma + \gamma_0 e_\sigma = \eta(t) \tag{4.104}$$

where $\eta(t) = (L_f h(x(t)))^{(m)}$. Since $(L_f h(x(t)))^{(m)}$ is assumed to be uniformly bounded, there exists a positive constant κ_m such that $\sup_t |\eta(t)| < \kappa_m$. The phase variables χ of the injected dynamics evolve governed by $\dot{\chi} = A\chi + b\eta(t)$ with $A \in \mathbb{R}^{(m+1)\times(m+1)}$, in companion form, with complex eigenvalues exhibiting strictly negative real parts. The $(m+1)$-dimensional vector b is a vector of zero entries except for the last one, which is equal to 1. Let P be a positive definite symmetric matrix. Then consider the positive definite Lyapunov function candidate $V(\chi) = \frac{1}{2}\chi^T P \chi$ defined in the estimation error phase space. Since $\frac{1}{2}(A^T P + PA)$ is a symmetric matrix with $m+1$ real negative eigenvalues, written in decreasing order fashion as $-\lambda_{max}, \ldots, -\lambda_{min}$, it is immediate to verify, that, along the solutions of the estimation error dynamics, we have

$$
\begin{aligned}
\dot{V}(\chi) &= \frac{1}{2}\chi^T(A^T P + PA)\chi + b^T P \chi \eta(t) \\
&\leq -\lambda_{max}\|P\|\|\chi\|^2 + \kappa_m\|\chi\| \\
&= -\lambda_{max}\|P\|\|\chi\| \left(\|\chi\| - \frac{\kappa_m}{\lambda_{max}} \right)
\end{aligned}
\tag{4.105}
$$

The trajectories of the tracking error phase vector starting outside the ball

$$\mathcal{B}(0, \rho) = \{\chi \mid \|\chi\| \leq \rho, \ \rho = \frac{\kappa_m}{\lambda_{max}} \|P\|\} \tag{4.106}$$

asymptotically converge toward the boundary of the ball. Those trajectories starting inside the ball $\mathcal{B}(0, \rho)$ will never be able to abandon it. Then, as the complex eigenvalues of the matrix A are chosen further to the left of the complex plane, the radius ρ of the ball $\mathcal{B}(0, \rho)$ becomes as small as desired. The sliding surface estimation error phase variables $\chi(t)$ will be ultimately uniformly bounded by the ball $\mathcal{B}(0, \rho)$.

From the first equation it is clear that $L_f h(x(t)) - z_1 = -\dot{e}_\sigma - \gamma_m e_\sigma$. Therefore, from the fact that $\dot{e}_\sigma$ and e_σ are ultimately uniformly bounded inside $\mathcal{B}(0, \rho)$, the difference $z_1 - L_f h(x(t))$ is asymptotically convergent toward a small interval around the origin, which is as small as desired. The observer signal z_1 estimates, in an arbitrarily close fashion, the uncertain directional derivative $L_f h(x(t))$. The observer variable z_1 may be properly denoted as the estimate $\widehat{L_f h(x)}$. $\qquad\Box$

Let q be an even integer, and $\epsilon > 0$ be a small real number ($\epsilon \approx \kappa_m^{-1}\rho = \lambda_{max}^{-1}$). We define $\phi_s(t)$ as the "smoothed" version of a time signal $\phi(t)$ whenever $\phi_s(t) = s_f(t)\phi(t)$ with $s_f(t)$ given by

$$s_f(t) = \begin{cases} \sin^q(\frac{\pi t}{2\epsilon}) & \text{for } t \leq \epsilon \\ 1 & \text{for } t > \epsilon \end{cases} \tag{4.107}$$

Corollary 1. *Let z_{1s} be a smoothing of the observer variable $z_1 = \widehat{L_f h(x(t))}$ during a small time interval $[0, \epsilon]$. Under all the previous assumptions, the discontinuous active disturbance cancelation feedback controller*

$$u = \frac{1}{L_g h(x)}\left[-z_{1s} - W \text{sign}(\sigma)\right] \tag{4.108}$$

locally creates a sliding regime on $S \cap \mathcal{N}(x^0)$ for any amplitude W satisfying $W > \rho = \kappa_m/\lambda_{max}$.

Proof. The closed-loop sliding surface dynamics is given by

$$\dot{\sigma} = L_f h(x) - z_{1s} - W \text{sign}(\sigma) \tag{4.109}$$

Since, for $t > \epsilon$, the disturbance estimation error $L_f h(x) - z_{1s} = L_f h(x) - \widehat{L_f h}(x)$ asymptotically converges toward an arbitrary small interval around the origin, and the local existence of a sliding regime on $\sigma = 0$ is guaranteed even if W is rather small. An estimate of the required W is just given by

$$W > \rho = \frac{\kappa_m}{\lambda_{max}} \tag{4.110}$$

where $\kappa_m = \sup_t |L_f h(x)|$, and λ_{max} is the smallest, in absolute value, eigenvalue of the symmetric negative definite matrix $\frac{1}{2}(A + A^T)$. □

4.4.5   A Second-Order Example

Consider the uncertain nonlinear second-order system

$$\begin{aligned}
\dot{x}_1 &= x_2 \\
\dot{x}_2 &= \xi(x_1, x_2) + u
\end{aligned} \tag{4.111}$$

where $\xi(x_1, x_2)$ is a smooth unknown function of its arguments. We take, for instance,

$$\xi = -0.5 \sin(x_1) \cos(x_1) \tag{4.112}$$

Define the following stabilizing sliding surface:

$$S = \{x \in \mathbb{R}^2 \mid \sigma(x) = x_2 + ax_1 = 0, \ a > 0\} \tag{4.113}$$

with associated ideal sliding dynamics $\dot{x}_1 = -ax_1$ on S. Consider the following discontinuous feedback control law:

$$u = -z_s \, v - W \text{sign}(\sigma) \tag{4.114}$$

with $W = 0.2$ and with $v \in \{0, 1\}$ being an "activation function" for the disconnection, or active involvement, of the smoothed version of the variable z_1 arising from the following GPI observer:

$$\begin{aligned}
\dot{\hat{\sigma}} &= z_1 + u + \gamma_3(\sigma - \hat{\sigma}) \\
\dot{z}_1 &= z_2 + \gamma_2(\sigma - \hat{\sigma}) \\
\dot{z}_2 &= z_3 + \gamma_1(\sigma - \hat{\sigma}) \\
\dot{z}_3 &= \gamma_0(\sigma - \hat{\sigma})
\end{aligned} \tag{4.115}$$

The gains $\{\gamma_0, \cdots, \gamma_3\}$ are chosen so that the polynomial in the complex variable s,

$$p(s) = s^4 + \gamma_3 s^3 + \gamma_2 s^2 + \gamma_1 s + \gamma_0 \tag{4.116}$$

exhibits all its roots deeply in the left half portion of the complex plane.[3]

---

3. As a justification for such a low-order observer, recall J. Von Neumann's statement: "With four parameters, I can fit an elephant, and with five, I can make him wiggle his trunk!"

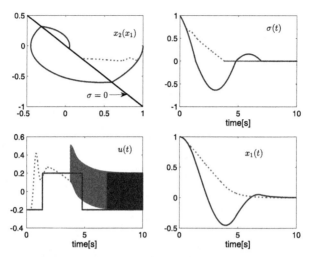

FIGURE 4.9 Comparison of the closed-loop performances of the GPI observer-assisted sliding-mode control scheme (dotted line) and the observer-less scheme (continuous line).

We set $p(s)$ identical to the polynomial $(s^2 + 2\zeta\omega_n s + \omega_n^2)^2$ with the damping factor ζ close to unity and ω_n chosen sufficiently large. We let

$$\gamma_3 = 4\zeta\omega_n, \ \gamma_2 = (2\omega_n^2 + 4\zeta^2\omega_n^2), \ \gamma_1 = 4\zeta\omega_n^3, \ \gamma_0 = \omega_n^4$$

The unknown function to be estimated via the GPI is simply $z = \xi(x_1, x_2) + ax_2$. Let q be an even integer. The smoothed version (or "clutched" version) of this observer variable is defined by $z_s = z_1 s_f(t)$ where $s_f(t)$ is a smoothing factor of the form (4.107).

4.4.5.1 Simulations

For simulation purposes, we have set $\zeta = 1$ and $\omega_n = 6$, $\epsilon = 1$, $q = 8$. The control input takes values on the set $[-U, U] = [-0.6, 0.6]$ with $W = 0.2$. The sliding surface is of the form

$$\{x \in \mathbb{R}^2 \mid x_2 + x_1 = 0\} \tag{4.117}$$

i.e., $a = 1$. Fig. 4.9 shows, in dotted lines, how the GPI observer-assisted sliding motion occurs much earlier than the traditional full force bang-bang sliding policy (in continuous line). The performance of the closed-loop system is clearly substantially improved.

Fig. 4.10 depicts the asymptotic estimation features of the GPI observer for the sliding surface coordinate function σ and for the estimation of the endogenous perturbation input signal $L_f h(x(t)) + ax_2(t)$, depicted as $\hat{z}(t)$.

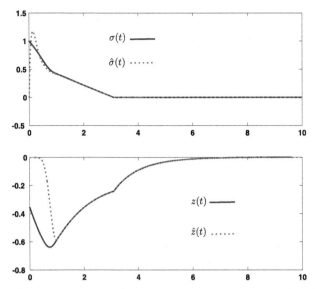

FIGURE 4.10 Sliding surface evolution $\sigma(t)$ and nonlinear state-dependent perturbation input $z(t) = L_f h(x(t)) + ax_2(t)$, along with their linear GPI observer-based estimates.

In order to test the robustness of the proposed approach, we now set the unknown function $\xi(x_1, x_2)$ to be given by $\xi(x_1, x_2) = -0.5\sin^2(x_1)$, making the uncontrolled system unstable. Fig. 4.11 depicts the fact that the GPI observer-assisted sliding-mode controller (in dotted line) locally creates a sliding regime with performance comparable to the last simulation example. However, the observer-less scheme (in continuous line) results in a totally unstable system.

Fig. 4.12 depicts the observer-assisted sliding-mode creation features ($v = 1$) for the unstable version of the illustrative system with controlled trajectories starting from two different initial conditions. It should be clear that sliding modes are being invariably created, in spite of control input switching authorities as low as $W = 0.05$. Fig. 4.13 depicts the same numerical experiment in the absence of the active disturbance estimation–rejection scheme ($v = 0$). The results clearly depict the beneficial effects of the active disturbance rejection approach based on linear high-gain GPI disturbance observers.

4.4.5.2 Control of Pendulum System

Consider the problem of controlling the angular position of a pendulum that was made with an aluminium bar and a concentrated mass at the tip at a distance L of pivot. As for the actuator, we use the DC motor with gear reduction ratio 16:1 shown in Fig. 4.14.

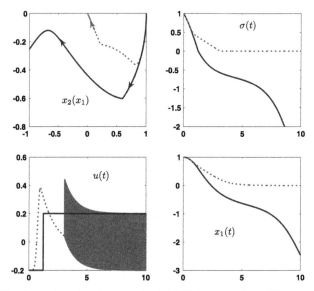

FIGURE 4.11 Comparison of performances for the GPI observer-assisted sliding-mode controller (dotted line) and the observer-less scheme (continuous line) for an unstable system.

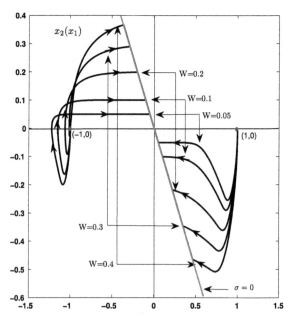

FIGURE 4.12 GPI observer-assisted sliding-mode creation for perturbed system using different switching control input authorities W.

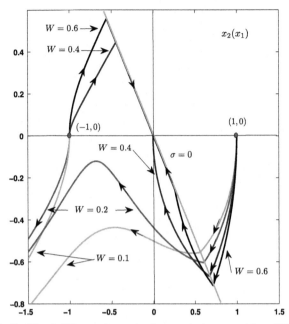

FIGURE 4.13 Traditional sliding-mode creation for perturbed system without GPI observer assistance, using different switching control input authorities W.

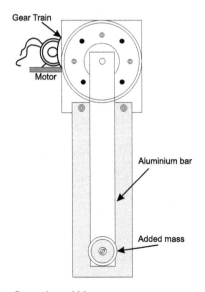

FIGURE 4.14 Pendulum, Gear train, and Motor.

The pendulum system is coupled to the DC motor through a gear train, there are viscous friction and Coulomb friction. The dynamics of this system is described by (4.118)

$$\left(\frac{1}{3}M + m\right)L^2\ddot{\theta} + \left(\frac{1}{2}M + m\right)gL\sin(\dot{\theta}) + B\dot{\theta} + F_c\text{sign}(\dot{\theta}) = \tau(t) \quad (4.118)$$

where τ is the input torque to the motor and gear train, θ, m, M, and L are the rotation angle, the mass of the bar, the added mass, and the length of the bar, respectively, B and F_c are viscous friction coefficient and Coulomb friction, and g denotes the gravity acceleration.

We simplify the nonlinear system model to the following linear perturbed system:

$$\ddot{\theta} = \frac{\tau(t)}{\left(\frac{1}{3}M + m\right)L^2} + \tilde{\varphi}(t) \quad (4.119)$$

For the system, we use a gear train between the motor and the pendulum system. The torque in the output of the gears train is given by

$$\tau(t) = \frac{K_1 N}{R_a}V(t) - \frac{K_1 K_2 N^2}{R_a}\dot{\theta} \quad (4.120)$$

We use equation (4.120) for the torque in the output of the gear train

$$\ddot{\theta} = \frac{\frac{K_1 N}{R_a}}{\left(\frac{1}{3}M + m\right)L^2}V(t) - \frac{\frac{K_1 K_2 N^2}{R_a}}{\left(\frac{1}{3}M + m\right)L^2}\dot{\theta} + \tilde{\varphi}(t) \quad (4.121)$$

We simplify the linear perturbed system:

$$\ddot{\theta} = \frac{K_1 N}{R_a\left(\frac{1}{3}M + m\right)L^2}V(t) + \varphi(t) \quad (4.122)$$

with $\varphi(t)$ including now the state-dependent nonlinearity

$$\varphi(t) = -\frac{K_1 K_2 N^2}{R_a\left(\frac{1}{3}M + m\right)L^2}\dot{\theta} - \frac{\left(\frac{1}{2}M + m\right)g\sin(y)}{\left(\frac{1}{3}M + m\right)L} - \frac{B\dot{y}}{\left(\frac{1}{3}M + m\right)L^2}$$
$$- \frac{F_c\text{sign}(\dot{y})}{\left(\frac{1}{3}M + m\right)L^2} \quad (4.123)$$

Define the following trajectory sliding surface:

$$S = \{x \in \mathbb{R}^2 \mid \sigma(x) = (\dot{y} - \dot{y}^*) + \lambda(y - y^*) = 0, \ \lambda > 0\} \quad (4.124)$$

where y^* is the reference trajectory $y = \theta$.

For the pendulum system, consider the following discontinuous feedback control law:

$$V(t) = -\frac{(\frac{1}{3}M + m)L^2}{\frac{K_1 N}{R_a}}(z_1 - \ddot{y}^*) - W\text{sign}(\sigma) \tag{4.125}$$

with $W = 1.5$ and $\lambda = 3$. We propose the GPI observer

$$\dot{\hat{\sigma}}_1 = \hat{\sigma}_2 + \gamma_7(\sigma - \hat{\sigma}_1) \tag{4.126}$$

$$\dot{\hat{\sigma}}_2 = z_1 + \frac{\frac{K_1 N}{R_a}}{(\frac{1}{3}M + m)L^2}V(t) + \gamma_6(\sigma - \hat{\sigma}_1)$$

$$\dot{z}_1 = z_2 + \gamma_5(\sigma - \hat{\sigma}_1)$$

$$\vdots$$

$$\dot{z}_6 = \gamma_0(\sigma - \hat{\sigma}_1) \tag{4.127}$$

The gains $\{\gamma_0, \cdots, \gamma_7\}$ are chosen so that the polynomial in the complex variable s,

$$p(s) = s^8 + \gamma_7 s^7 + \cdots + \gamma_0 \tag{4.128}$$

exhibits all its roots deeply in the left half portion of the complex plane.

We set $p(s)$ with the Hurwitz polynomial $(s^2 + 2\zeta \omega_n s + \omega_n^2)^4$ with the damping factor ζ close to unity and ω_n chosen sufficiently large.

Experimental Results

To measure the pendulum angular position, we used an incremental encoder of series CP-350-100-LD-1/4" of 1000 pulses per revolution. A Sensoray acquisition card, Model 620, with 6 counters and 4 D/A outputs, 16 A/D inputs, and 48 digital bidirectional I/O pins. The card is an interface for the Matlab Simulink Real-time Windows Target environment. The sampling time was set to 0.001 [s]. A power operational amplifier provides a current of 6 [A] and a voltage ±24 [V]. It receives the control signal and amplifies the motor current. The DC-motor used was an NC5475B NISCA MOTOR with parameters shown in Table 1; on the other hand, Fig. 4.15 shows the block diagram of the pendulum system.

The pendulum system exhibits the following parameter values:

$$M = 0.268 \ [\text{kg}], \quad m = 0.1 \ [\text{kg}] \quad L = 0.6 \ [\text{m}]$$

TABLE 1 NC5475B NISCA motor parameters

Parameter	Value
Torque constant	0.0724 [N/A]
FEM constant	0.0687 [Nm/rad/s]
Electrical resistance	2.983 [Ω]
Gear reduction ratio	16

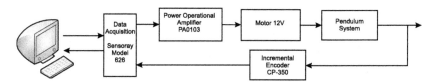

FIGURE 4.15   Block diagram of the implementation for the pendulum-CD motor system.

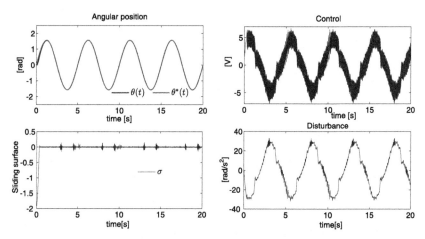

FIGURE 4.16   GPI observer-based trajectory tracking performance for pendulum-CD motor system.

Observer Gains:

$$\zeta_c = 2, \quad \omega_n = 10$$

Fig. 4.16 depicts the quality of the controller performance in a trajectory tracking task involving a sinusoidal signal reference trajectory for the angular position coordinate $\theta(t)$. The output reference trajectory $\theta(t)^*$ was specified to be

$$\theta(t)^* = \frac{\pi}{2} \sin(\omega t) \text{ [rad]}, \quad \omega = \frac{2\pi}{5} \left[\frac{\text{rad}}{\text{s}} \right] \qquad (4.129)$$

4.5   ADRC AND SAMPLED SYSTEMS: THE DELTA OPERATOR APPROACH

Most practical problems in control implementations deal with digital control, which implies a sampling process (discretization). The classic analysis of discretized systems is based on the time shift operator q, and the usual analysis for linear systems is the Z transform technique.

Current data acquisition technologies have allowed one to increase the sampling rate of the continuous-time systems, involving some advantages and drawbacks.

On the other hand, fast sampling induces several problems such as [12]:

- Vast amounts of data collected;
- Numerical ill-conditioning within algorithms, primarily caused by the discrete time-shift operator;
- Encroachment of the sample frequency onto the noise bandwidth causing aliasing/frequency folding.

This fact encouraged Goodwin and Middleton to develop a strategy capable of unifying both continuous- and discrete-time formulations [13]. Moreover, this approach overcomes the unstable sampling zero problem as analyzed in [14], and the procedure for the control gains is enhanced since the stability region increases as sampling time decreases. In this type of approach, the authors proposed to use an operator, called δ-operator, defined as follows:

$$\delta = \frac{q - 1}{\Delta} \tag{4.130}$$

where q is the shift operator in the time domain, and Δ is the sampling time. This approach has been used with good results regarding robustness and other features even in a class of nonlinear systems when fast sampled data are involved.

4.5.1   The Delta Operator Approach: A Theoretical Framework

In this section, we introduce some preliminary concepts regarding the δ operator and its properties. The reader is referred to [4,13] for detailed information.

Definition 1. The domain of possible nonnegative "times" $\Omega^+(\Delta)$ is defined as follows:

$$\Omega^+(\Delta) = \left\{ \begin{array}{ll} \mathbb{R}^+ \cup \{0\} & : \Delta = 0 \\ \{0, \Delta, 2\Delta, 3\Delta, \ldots\} & : \Delta \neq 0 \end{array} \right\} \tag{4.131}$$

where Δ denotes the sampling period in discrete time, or $\Delta = 0$ for a continuous-time framework.

Definition 2. A time function $x(t)$, $t \in \Omega^+(\Delta)$, is in general, simply a mapping from times to either the real or complex set, that is, $x(t) : \Omega^+ \to \mathbb{C}$.

Definition 3. The unified operator or δ operator is defined as follows:

$$\delta x(t) \triangleq \left\{ \begin{array}{ll} \frac{dx(t)}{dt} & : \Delta = 0 \\ \frac{(q-1)x(t)}{\Delta} = \frac{x(t+\Delta)-x(t)}{\Delta} & : \Delta \neq 0 \end{array} \right\} \tag{4.132}$$

where

$$\lim_{\Delta \to 0^+} \delta(x(t)) = \frac{dx(t)}{dt}$$

Definition 4. The integration operation $\mathbf{S}$ is given as follows:

$$\mathbf{S}_{t_1}^{t_2} x(\tau) d\tau = \left\{ \begin{array}{ll} \int_{t_1}^{t_2} x(\tau) d\tau & : \Delta = 0 \\ \Delta \sum_{l=t_1/\Delta}^{l=t_2/\Delta - 1} x(l\Delta) & : \Delta \neq 0 \end{array} \right\}, \quad t_1, t_2 \in \Omega^+ \tag{4.133}$$

The integration operator corresponds to the antiderivative operator.

Definition 5 (Generalized Matrix Exponential). In the case of the unified transform theory, the generalized exponential E is defined as follows:

$$E(A, t, \Delta) = \left\{ \begin{array}{ll} e^{At} & : \Delta = 0 \\ (I + A\Delta)^{t/\Delta} & : \Delta \neq 0 \end{array} \right\} \tag{4.134}$$

where $A \in \mathbb{C}^{n \times n}$, and I is the identity matrix. The Generalized Matrix Exponential satisfies to be the fundamental matrix of $\delta x = Ax$ and thus, the unique solution to

$$\delta x = Ax; \; x(0) = x_0 \tag{4.135}$$

is $x(t) = E(A, t, \Delta)x_0$. The general solution to

$$\delta x = Ax + Bu, \; x(0) = x_0 \tag{4.136}$$

is

$$x(t) = E(A, t, \Delta)x_0 + \mathbf{S}_0^t E(A, t - \tau - \Delta, \Delta) Bu(\tau) d\tau \tag{4.137}$$

Definition 6 (Stability Boundary). The solution of (4.135) is said to be asymptotically stable if and only if, for all x_0, $x(t) \to 0$ as time elapses. The stability arises if and only if $E(A, t, \Delta) \to 0$ as $t \to \infty$ if and only if every eigenvalue of A, denoted as λ_i, $i = 1, \ldots, n$, satisfies the following condition:

$$Re\{\lambda_i\} + \frac{\Delta}{2} |\lambda_i|^2 < 0 \tag{4.138}$$

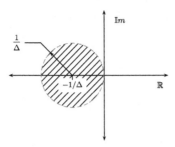

FIGURE 4.17 Stability region of the delta operator.

Therefore, the stability boundary is the circle with center $(-1/\Delta, 0)$ and radius $1/\Delta$ (see Fig. 4.17), which also can be described as

$$Re\{\lambda_i\} + \frac{\Delta}{2}|\lambda_i|^2 = 0 \qquad (4.139)$$

In particular, given the equation

$$\delta^n x + a_{n-1}\delta^{n-1}x + \ldots + a_1\delta x + a_0 x = 0 \qquad (4.140)$$

if all the roots $\lambda_i, i = 1\ldots, n$, of the polynomial

$$\lambda^n + a_{n-1}\lambda^{n-1} + \ldots + a_1\lambda x + a_0 = 0 \qquad (4.141)$$

satisfy condition (4.138), then the solution of (4.140) is asymptotically stable.

Definition 7. Suppose $V(t, x)$ is a positive definite function and x is a state of the system

$$\delta x(t) = f(x(t), t) \qquad (4.142)$$

Then $V(t, x)$ is called a Lyapunov Function Candidate (LFC) if $\delta V(t, x(t))$ exists.

Definition 8 (Lyapunov Stability). Suppose that, for some LFC $V(t, x)$ and some system (4.142), $\delta V(t, x) \leq 0$ for all t and x. Then the solution of (4.142) is stable in the Lyapunov sense. If $\delta V(t, x) < 0$, then the solution of (4.142) is asymptotically stable in the Lyapunov sense.

4.5.2 Problem Formulation

Consider the following continuous-time system, expressed as a disturbed integrator chain

$$y^{(n)} = u + \xi(t) \qquad (4.143)$$

The sampling process of the last system (with sampling time Δ) leads to the following delta operator approximation:

$$\delta^n y(t) = u + \bar{\xi}(t) \tag{4.144}$$

where $\bar{\xi}(t)$ includes the nonmodeled dynamics derived from the sampling process.

The problem statement is given as follows: Given a reference trajectory $y^*(t)$, devise a discrete-time controller u such that, in spite of the nonmodelled dynamics and possible arising of external disturbances, the output of system (4.143) is forced to track $y^*(t)$ within a bounded tracking error.

4.5.3 Delta Operator GPI Observer

Using the same methodology of the continuous-time ADRC control, the generalized disturbance input $\bar{\xi}(t)$ is proposed to be locally approximated by a family of polynomials of degree p. That implies the following space state representation of (4.144):

$$\begin{aligned}
\delta y_1 &= y_2 \\
\delta y_2 &= y_3 \\
&\vdots \\
\delta y_{n-1} &= y_n \\
\delta y_n &= u + \rho_0 \\
\delta \rho_0 &= \rho_1 \\
&\vdots \\
\delta \rho_{p-1} &= \rho_p \\
\delta \rho_p &= 0
\end{aligned} \tag{4.145}$$

where the state ρ_0 locally approximates $\bar{\xi}(t)$. The proposed ADRC control is given in terms of the delta operator as follows:

$$\begin{aligned}
u &= \delta^n y^* - \sum_{i=0}^{n-1} \kappa_i \left(\hat{y}_i - \delta^i y_i^* \right) - \hat{\rho}_0 \\
\delta \hat{y}_1 &= \hat{y}_2 + l_{n+p}(y_1 - \hat{y}_1) \\
\delta \hat{y}_2 &= \hat{y}_3 + l_{n+p-1}(y_1 - \hat{y}_1) \\
&\vdots
\end{aligned}$$

$$\delta \hat{y}_{n-1} = \hat{y}_n + l_{p+2}(y_1 - \hat{y}_1)$$
$$\delta \hat{y}_n = u + \hat{\rho}_0 + l_{p+1}(y_1 - \hat{y}_1) \qquad (4.146)$$
$$\delta \hat{\rho}_0 = \hat{\rho}_1 + l_p(y_1 - \hat{y}_1)$$
$$\vdots$$
$$\delta \hat{\rho}_{p-1} = \hat{\rho}_p + l_1(y_1 - \hat{y}_1)$$
$$\delta \hat{\rho}_p = l_0(y_1 - \hat{y}_1)$$
$$\hat{\rho}_0 = \hat{\bar{\xi}}(t)$$

Let us define the convergence error as $\varepsilon := y_1 - \hat{y}_1$. This error satisfies the following dynamics:

$$\delta^{n+p+1}\varepsilon + l_{n+p}\delta^{n+p}\varepsilon + \cdots + l_1\delta\varepsilon + l_0\varepsilon = \delta^p\bar{\xi} \qquad (4.147)$$

The characteristic equation of the linear dominant dynamics in the complex variable γ is

$$\gamma^{n+p+1} + l_{n+p}\gamma^{n+p} + \cdots + l_1\gamma + l_0 = 0 \qquad (4.148)$$

Thus, if the observer gains l_i, $i = 0, 1, \ldots, n+p$, are chosen such that the roots of (4.148), say γ_i, are located in the region $\frac{\Delta}{2}|\gamma_i|^2 + \mathbb{Re}\{\gamma_i\} < 0$, then the estimation error is restricted to a vicinity of the origin of the phase space.

The tracking error, defined as $e_y := y - y^*$, evolves according to the following disturbed linear dynamics:

$$\delta^n e_y + \kappa_{n-1}\delta^{n-1}e_y + \ldots + \kappa_0 e_y = \bar{\xi}(t) - \hat{\bar{\xi}}(t) - \varphi(\varepsilon, \delta\varepsilon, \ldots, \delta^{n-1}\varepsilon) \quad (4.149)$$

where $\varphi(\varepsilon, \delta\varepsilon, \ldots, \delta^{n-1}\varepsilon)$ is a linear function of the estimation error and its uniform time derivatives. By choosing κ_i, $i = 0, \ldots, n-1$, such that the polynomial

$$\gamma^n + \kappa_{n-1}\gamma^{n-1} + \ldots + \kappa_0 \qquad (4.150)$$

has its roots deep in the region $\frac{\Delta}{2}|\gamma_i|^2 + \mathbb{Re}\{\gamma_i\} < 0$, it is ensured that the tracking error and its time derivatives converge asymptotically to a vicinity of the origin of the phase state of the tracking error.

4.5.4 Gain Tuning via Continuous-Time Matching

One important aspect for a control practitioner in the use of discrete-time control systems is the gain tuning procedure. Since the stability region for a shift operator is the unitary circle, the tuning procedure based on the root location may be

TABLE 2 Zero-order-hold discrete-time equivalents (delta operator) for transfer functions

$G(s)$	$G'(\gamma)$
$\dfrac{1}{s}$	$\dfrac{1}{\gamma}$
$\dfrac{1}{s+a}$	$\dfrac{\frac{1}{a}(1-e^{-a\Delta})}{\gamma+\dfrac{1-e^{-a\Delta}}{\Delta}}$
$\dfrac{\omega}{(s+a)^2+\omega^2}$	$\dfrac{\left[\dfrac{\omega(1-e^{-a\Delta}\cos\omega\Delta)-ae^{-a\Delta}\sin\omega\Delta}{\Delta}\right]\gamma+\omega\left[\dfrac{1-2e^{-a\Delta}\cos\omega\Delta+e^{-2a\Delta}}{\Delta^2}\right]}{(a^2+\omega^2)\left(\gamma^2+\left[\dfrac{2-2e^{-a\Delta}\cos\omega\Delta}{\Delta}\right]\gamma+\left[\dfrac{1-2e^{-a\Delta}\cos\omega\Delta+e^{-2a\Delta}}{\Delta^2}\right]\right)}$
$\dfrac{s}{(s+a)^2+\omega^2}$	$\dfrac{\left[e^{-a\Delta}(\sin\omega\Delta/\omega\Delta)\right]\gamma}{\gamma^2+\left[\dfrac{2-2e^{-a\Delta}\cos\omega\Delta}{\Delta}\right]\gamma+\left[\dfrac{1-2e^{-a\Delta}\cos\omega\Delta+e^{-2a\Delta}}{\Delta^2}\right]}$

sensitive to the sampling time [14]. A direct connection between discrete- and continuous-time transfer function $(G(s))$, assuming a sampling process of the output of zero-order-hold input, is expressed as follows:

- Shift Operator (z)

$$G'(z) = \frac{z-1}{z}\mathbf{Z}\left(\mathcal{L}^{-1}\left\{\frac{1}{s}G(s)\right\}\right) \qquad (4.151)$$

- Delta Operator (γ)

$$G'(\gamma) = \frac{\gamma}{1+\gamma\Delta}\mathbf{T}\left(\mathcal{L}^{-1}\left\{\frac{1}{s}G(s)\right\}\right) \qquad (4.152)$$

Unlike (4.151), notice that in (4.152) the relation between the transfer function of $G'(\gamma)$ includes the sampling time. This fact can be an advantage in order to apply a tuning process based on a pole assignment rule in continuous time. In particular, Table 2 shows some zero-order-hold discrete-time equivalents, which are useful in the tuning procedure (for a comprehensive table, see [4]):

4.5.5 Example: Inertia Damper System

Consider the following Inertia Damper system with unknown disturbance input:

$$J\ddot{q} + B(\dot{q}) = u + \eta(t) \qquad (4.153)$$

where J represents the inertia of the system, B is a function representing the combination of viscous and Coulomb friction terms, u is the system input, and $\eta(t)$ stands for an external disturbance due to an external load.

Assuming a sampling process with sampling time Δ, the model in the delta operator is given by

$$\delta^2 q = -\frac{B(\delta q)}{J} + \frac{1}{J}u + \frac{1}{J}\eta(t) \tag{4.154}$$

Let us define the generalized lumped disturbance $\xi = -\frac{B(\delta q)}{J} + \frac{1}{J}\eta(t)$. We have

$$\delta^2 q = \frac{1}{J}u + \xi \tag{4.155}$$

We propose an approximation of $\xi(t)$ by a family of 2nd-degree time polynomials, that is, we propose a 3rd-order local approximation. The system including the ultramodel approximation is

$$\begin{aligned}
\delta q_1 &= q_2 \\
\delta q_2 &= \frac{1}{J}u + \rho_0 \\
\delta \rho_0 &= \rho_1 \\
\delta \rho_1 &= \rho_2 \\
\delta \rho_2 &= 0
\end{aligned} \tag{4.156}$$

The observer for system (4.156) is given as

$$\begin{aligned}
\delta \hat{q}_1 &= \hat{q}_2 + l_1(q_1 - \hat{q}_1) + l_4(q_1 - \hat{q}_1) \\
\delta \hat{q}_2 &= \frac{1}{J}u + \rho_0 + l_3(q_1 - \hat{q}_1) \\
\delta \hat{\rho}_0 &= \hat{\rho}_1 + l_2(q_1 - \hat{q}_1) \\
\delta \hat{\rho}_1 &= \hat{\rho}_2 + l_1(q_1 - \hat{q}_1) \\
\delta \hat{\rho}_2 &= l_0(q_1 - \hat{q}_1)
\end{aligned} \tag{4.157}$$

Some numerical simulations were carried out to test the estimation and control scheme. The sampling time in the control was 1 ms, and the tuning of the observer and controller was based on the use of the zero-order-hold equivalents. The closed-loop trajectory tracking characteristic polynomial was $s^2 + 2z_1\omega_{n1}s + w_{n1}^2$ with $z_1 = 0.5$ and $\omega_{n1} = 7$. The observer gains were selected such that the injection error dynamics matches the desired characteristic polynomial $(s^2 + 2z_{o1}\omega_{no1}s + w_{no1}^2)(s^2 + 2z_{o2}\omega_{no2}s + w_{no2}^2)(s + p_o)$ with

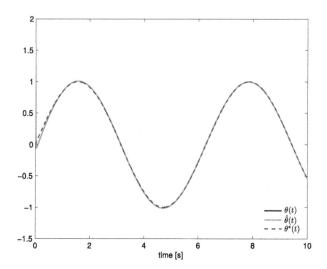

FIGURE 4.18 Trajectory tracking results of the sampled control.

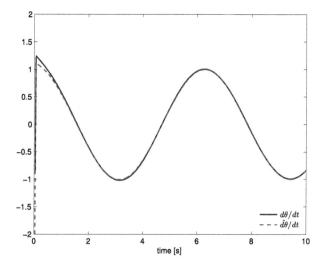

FIGURE 4.19 Time derivative estimation.

$z_{o1} = 0.5$, $z_{o1} = 0.8$, $\omega_{no1} = 5$, and $\omega_{no1} = 10$. Fig. 4.18 shows the tracking results for a sinusoidal reference trajectory. The time derivative estimation is shown in Fig. 4.19. The control input and the disturbance estimation are depicted in Fig. 4.20. Since the sampling time is reduced, the sampled time controller allows a similar behavior with respect to the continuous-time instance, and its digital implementation is direct.

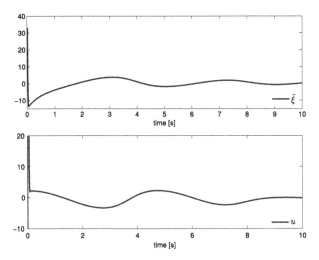

FIGURE 4.20 Disturbance estimation (top) and control input (bottom).

4.6 CONTROL OF TIME DELAY SYSTEMS

4.6.1 An Illustrative Example

Consider the following second-order linear delay system:

$$\ddot{y} = u(t - T) + \varphi(t) \qquad (4.158)$$

where T is a perfectly known strictly positive real number, and $\varphi(t)$ is a bounded perturbation of low-frequency character so that its time derivatives are negligible after certain order, say the fourth. This perturbation is, otherwise, unknown.

It is desired to stabilize the output of the system to a constant value y_{ref}.

Let $y_1 = y$ and consider the state representation of the given system

$$
\begin{aligned}
\dot{y}_1 &= y_2 \\
\dot{y}_2 &= u(t - T) + \varphi(t) \\
y &= y_1
\end{aligned}
\qquad (4.159)
$$

Consider the following GPI observer:

$$
\begin{aligned}
\dot{\hat{y}}_1 &= \hat{y}_2 + \gamma_{11}(y - \hat{y}) \\
\dot{\hat{y}}_2 &= u(t - T) + z_1 + \gamma_{10}(y - \hat{y}) \\
\dot{z}_1 &= z_2 + \gamma_9(y - \hat{y}) \\
\dot{z}_2 &= z_3 + \gamma_8(y - \hat{y}) \\
\dot{z}_3 &= z_4 + \gamma_7(y - \hat{y})
\end{aligned}
$$

$$\dot{z}_4 = z_5 + \gamma_6(y - \hat{y})$$
$$\dot{z}_5 = z_6 + \gamma_5(y - \hat{y}) \qquad (4.160)$$
$$\dot{z}_6 = z_7 + \gamma_4(y - \hat{y})$$
$$\dot{z}_7 = z_8 + \gamma_3(y - \hat{y})$$
$$\dot{z}_8 = z_9 + \gamma_2(y - \hat{y})$$
$$\dot{z}_9 = z_{10} + \gamma_1(y - \hat{y})$$
$$\dot{z}_{10} = \gamma_0(y - \hat{y})$$

It is easy to see that the output reconstruction error $e = y - \hat{y}$ evolves according to

$$e^{(12)} + \gamma_{11}e^{(11)} + \cdots + \gamma_1\dot{e} + \gamma_0 e = \varphi^{(10)}(t) \approx 0 \qquad (4.161)$$

The reconstruction error and its time derivatives converge toward a small neighborhood of zero in the reconstruction error state space. Similarly, the variable z_1 converges toward the unknown perturbation $\varphi(t)$, and $z_j \to \varphi^{(j-1)(t)}$ for $j = 2, 3, \ldots$.

We propose the following perturbation predictor computed at time t:

$$\hat{\varphi}(t + T) \approx z_1(t) + z_2(t)T + z_3(t)\frac{T^2}{2!} + z_4(t)\frac{T^3}{3!} + \cdots \qquad (4.162)$$

With this prediction of the perturbation input, we can certainly use the Smith predictor philosophy by considering the advance (or forward) system:

$$\ddot{y}_f = u(t) + \hat{\varphi}(t + T) \qquad (4.163)$$

By controlling this apparently noncausal system (recall that $\hat{\varphi}(t + T)$ is in fact available at time t) toward the desired reference value y_{ref} and sharing the designed state-feedback controller with the actual delay system, we control the delay system with a properly predicted state-feedback control law with predicted cancelation of the acting perturbation input.

We set the following proportional-derivative controller of the forward system:

$$u(t) = -\hat{\varphi}(t + T) - k_1\dot{y}_f - k_0(y_f - y_{ref}) \qquad (4.164)$$

The control scheme is shown in Figs. 4.21–4.22.

4.6.2 Another Smith Predictor-GPI-Observer-Based Controller

We now consider the following popular Smith predictor-based control scheme in the context of the following example. Consider a perturbed water heating

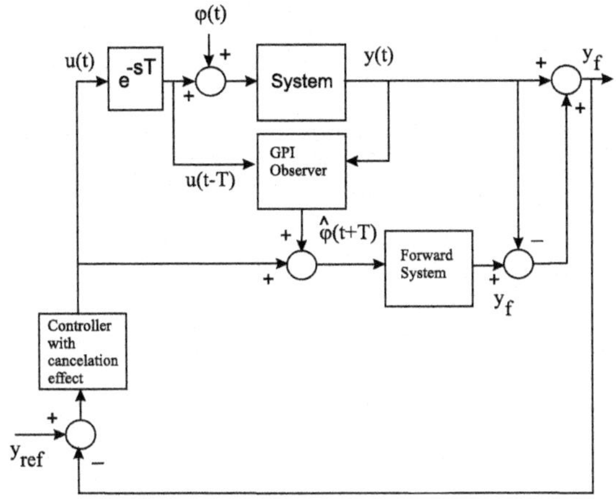

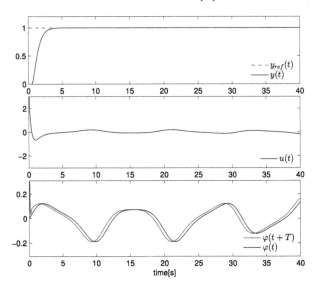

FIGURE 4.21   GPI observer-based controller for time delay systems.

FIGURE 4.22   Performance of Smith predictor plus GPI observer control scheme of the illustrative example.

system described by the following delayed differential equation:

$$10\dot{y} = -y + u(t - 20) + \tilde{\varphi}(t) \tag{4.165}$$

Since the delay $T = 20$ is very large compared to the time constant of the system $(1/10)$, we choose the following time normalization:

$$dt' = dt/20, \ u(t - 20) = u(20(t' - 1)) = v(t' - 1). \tag{4.166}$$

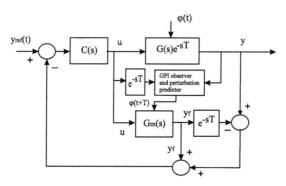

FIGURE 4.23 Smith predictor control scheme and GPI observer for perturbation prediction.

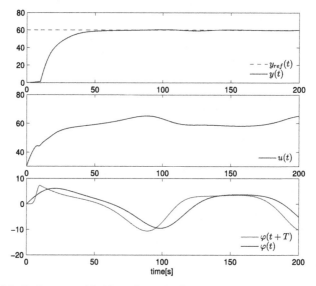

FIGURE 4.24 Performance of Smith predictor plus GPI observer control scheme in a perturbed water heating system.

We obtain the following normalized system:

$$\dot{y} = -2y + 2v(t' - 1) + \varphi(t') \tag{4.167}$$

Using the described Smith predictor scheme in combination with the perturbation prediction GPI observer (see Fig. 4.23 for the schematic of the proposal) we obtained the following simulation results, as shown in Fig. 4.24. We used the following normalized GPI observer:

$$\frac{d}{dt'}\hat{y} = -2\hat{y} + 2v(t' - 1) + z_1 + \gamma_3(y - \hat{y})$$

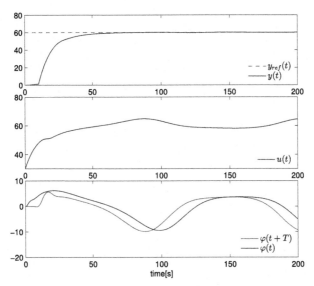

FIGURE 4.25 Performance of Smith predictor plus GPI observer control scheme in a perturbed water heating system with a higher-dimensional observer.

$$\frac{d}{dt'}z_1 = z_2 + \gamma_2(y - \hat{y}) \tag{4.168}$$
$$\frac{d}{dt'}z_2 = z_3 + \gamma_1(y - \hat{y})$$
$$\frac{d}{dt'}z_3 = \gamma_0(y - \hat{y})$$

The denormalized prediction of the perturbation is computed as

$$\hat{\varphi}(t + T) = z_1(t) + z_2(t)\frac{T}{1!} + z_3(t)\frac{T^2}{2!} \tag{4.169}$$

To obtain a better prediction of the perturbation input $\varphi(t)$, we proposed a higher-dimensional observer, thus obtaining a better estimate of the needed time derivatives (see Fig. 4.25). We continued to use the same predictor order:

$$\frac{d}{dt'}\hat{y} = -2\hat{y} + 2v(t' - 1) + z_1 + \gamma_7(y - \hat{y})$$
$$\frac{d}{dt'}z_1 = z_2 + \gamma_6(y - \hat{y})$$
$$\frac{d}{dt'}z_2 = z_3 + \gamma_5(y - \hat{y})$$
$$\frac{d}{dt'}z_3 = z_4 + \gamma_4(y - \hat{y})$$

$$\frac{d}{dt'}z_4 = z_5 + \gamma_3(y - \hat{y})$$

$$\frac{d}{dt'}z_5 = z_6 + \gamma_2(y - \hat{y})$$

$$\frac{d}{dt'}z_6 = z_7 + \gamma_1(y - \hat{y})$$

$$\frac{d}{dt'}z_7 = \gamma_0(y - \hat{y})$$

$$\hat{\varphi}(t + T) = z_1(t) + z_2(t)\frac{T}{1!} + z_3(t)\frac{T^2}{2!} \tag{4.170}$$

Now, let us consider a perturbed first-order system with unknown but bounded perturbation [15],

$$\dot{y} = u(t' - 5) + \tilde{\varphi}(t) \tag{4.171}$$

where t' stands for the actual time. It is desired to have $y \to y_{ref}$. First, consider, mainly to make the computer simulations feasible, the following time scaling of the system $t = t'/5$. We obtain

$$\dot{y} = 5u(t - 1) + \varphi(t) \tag{4.172}$$

where, with an abuse of notation, the dot notation stands for derivation with respect to the scaled time t.

To estimate the perturbation input $\varphi(t)$ and thus to be able to predict its future values, even if in an approximate manner, we propose the following GPI observer:

$$\dot{\hat{y}} = u(t - 1) + z_1 + \lambda_8(y - \hat{y})$$

$$\dot{z}_1 = z_2 + \lambda_7(y - \hat{y})$$

$$\dot{z}_2 = z_3 + \lambda_6(y - \hat{y})$$

$$\dot{z}_3 = z_4 + \lambda_5 * (y - \hat{y})$$

$$\dot{z}_4 = z_5 + \lambda_4(y - \hat{y}) \tag{4.173}$$

$$\dot{z}_5 = z_6 + \lambda_3(y - \hat{y})$$

$$\dot{z}_6 = z_7 + \lambda_2(y - \hat{y})$$

$$\dot{z}_7 = z_8 + \lambda_1(y - \hat{y})$$

$$\dot{z}_8 = \lambda_0(y - \hat{y})$$

The estimation error $e = y - \hat{y}$ evolves according to

$$e^{(9)} + \lambda_8 e^{(8)} + \cdots + \lambda_1 \dot{e} + \lambda_0 e = \varphi^{(8)}(t) \tag{4.174}$$

Our basic assumption is that the higher-order time derivatives of $\varphi(t)$ are negligible, i.e., $\varphi^{(8)}(t) \approx 0$.

Placing the poles of the corresponding characteristic polynomial

$$p(s) = s^{(9)} + \lambda_8 s^{(8)} + \cdots + \lambda_1 s + \lambda_0 \tag{4.175}$$

well into the left half of the complex plane, we further obtain that

$$e^{(j)}(t) \to \mathcal{N}(0), \ j = 1, 2, \ldots, 8 \tag{4.176}$$

$$z_k(t) \to \varphi^{(k-1)}(t) + \epsilon_k(t), \ k = 1, 2, \ldots$$

where $\mathcal{N}(0)$ is a small neighborhood of the origin of the reconstruction error state space $[e, \dot{e}, \ldots, e^{(8)}]$, and $\epsilon_k(t)$ is a scalar signal taking values in such a neighborhood $\mathcal{N}(0)$ for a reasonable set of consecutive values of k.

The accurate estimation of the perturbation input $\varphi(t)$ by means of the observer variable z_1 and its various time derivatives by the functions z_k, $k = 2, 3, \ldots$, prompts us to propose the following Taylor polynomial-based predictor for $\varphi(t + T)$:

$$\hat{\varphi}(t + T) = z_1 + \frac{1}{1!} z_2 T + \frac{1}{2!} z_3 T^2 + \frac{1}{3!} z_4 T^3 + \frac{1}{4!} z_5 T^4 \tag{4.177}$$

with $T = 1$. This predicted value of $\varphi(t + 1)$ is used in the following forward system:

$$\dot{y}_f = 5u(t) + \hat{\varphi}(t + 1) \tag{4.178}$$

The Smith predictor basic idea is to control, usually by means of a classical PI controller, this forward system so that it converges to the desired reference value y_{ref}. We propose

$$u = \frac{1}{5} \left[-\hat{\varphi}(t + 1) - k_1(y_f(t) - y_{ref}) - k_0 \int_0^t (y_f(\sigma) - y_{ref}) d\sigma \right] \tag{4.179}$$

The Smith predictor alternative proposes to share this synthesized controller for the forward system with the actual plant. However, since the plant has been completely bypassed, this controller is quite sensitive to the lack of a perfect prediction of the perturbation error and other modeling errors possibly committed in the forward system synthesis. Then the following modification of the above controller is usually enforced.

Let $e_{fr} = y - y_f(t - 1)$, i.e., the error between the actual plant output and the delayed output of the forward system. This quantity is used as an additive perturbation in the feedback signal trying to control the forward system. We thus propose

$$u = \frac{1}{5} \left[-\hat{\varphi}(t + 1) - k_1(y_f(t) - y_{ref} + e_{fr}) - k_0 \int_0^t (y_f(\sigma) - y_{ref} + e_{fr}) d\sigma \right] \tag{4.180}$$

Simulations

We set the observer gains by equating the 9th-degree characteristic polynomial $p(s)$, dominating the behavior of the reconstruction error, to the expression

$$(s^2 + 2\zeta\omega_n s + \omega_n^2)^4 (s + p) \tag{4.181}$$

$$
\begin{aligned}
\lambda_8 &= (p + 8\zeta\omega_n) \\
\lambda_7 &= (24\zeta^2\omega_n^2 + 8\zeta\omega_n p + 4\omega_n^2) \\
\lambda_6 &= (24\zeta^2\omega_n^2 p + 24\zeta\omega_n^3 + 32\zeta^3\omega_n^3 + 4\omega_n^2 p) \\
\lambda_5 &= 32\zeta^3\omega_n^3 p + 16\zeta^4\omega_n^4 + 48\zeta^2\omega_n^4 + 6\omega_n^4 + 24\zeta\omega_n^3 p \\
\lambda_4 &= 24\zeta\omega_n^5 + 48\zeta^2\omega_n^4 p + 32\zeta^3\omega_n^5 + 16\zeta^4\omega_n^4 p + 6\omega_n^4 p \\
\lambda_3 &= (24\zeta^2\omega_n^6 + 4\omega_n^6 + 32\zeta^3\omega_n^5 p + 24\zeta\omega_n^5 p) \\
\lambda_2 &= (8\zeta\omega_n^7 + 24\zeta^2\omega_n^6 p + 4\omega_n^6 p) \\
\lambda_1 &= (8\zeta\omega_n^7 p + \omega_n^8) \\
\lambda_0 &= \omega_n^8 p
\end{aligned}
\tag{4.182}
$$

with $\zeta = 1$, $\omega_n = 20$, $p = 20$.

The desired reference level for the output variable y was set to be $y_{ref} = 4.0$. The closed-loop poles for the PI controller were established as $k_1 = 2\zeta_c\omega_{nc}$, $k_0 = \omega_{nc}^2$ with $\zeta_c = 1$, $\omega_{nc} = 0.5$.

The perturbation input was synthesized as $\varphi(t) = e^{-\sin^2(0.6t)}\cos(0.3t)$. To avoid the large "peaking" that may arise from the high-gain observer, we propose to suitably attenuate certain signals used in the observer-based control scheme. We propose the use of the following "smoothing factor function" $s_f(t)$ during a certain time interval $[0, \delta]$ with $\delta > 0$:

$$
s_f(t) = \begin{cases} \sin^8(\frac{\pi t}{2\delta}) & \text{for } t \leq \delta \\ 1 & \text{for } t > \delta \end{cases}
\tag{4.183}
$$

We use this function to suppress the peaking behavior in the observer variables $z_1, z_2, \ldots, z_5$ used in the perturbation estimation and perturbation prediction. In the simulations presented next, δ was set to be 3. The perturbation estimation signal $z_1(t)$ and the perturbation prediction signal $\hat{\varphi}(t + 1)$ are all multiplied by the smoothing factor function $s_f(t)$. Fig. 4.26 shows the trajectory tracking performance as well as the disturbance prediction of the system. The control input as well as the cancelling by the prediction of the disturbance is illustrated in Fig. 4.27.

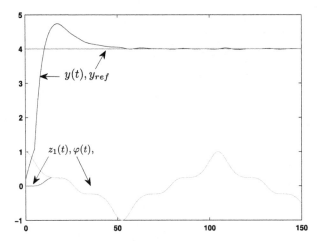

FIGURE 4.26   GPI observer-based controller for a first-order time delay system.

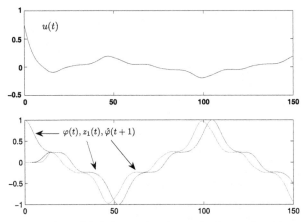

FIGURE 4.27   Performance of GPI observer-based controller in canceling input generation and perturbation estimation–prediction.

4.6.3   A Particularly Difficult Example

Consider the following delay system[4]:

$$\dot{y} = y(t - T_1) + u(t - T_2) \tag{4.184}$$

with $T_2 > T_1$. It is desired to control $y(t)$ toward a constant reference value y_{ref}. Our technique, combined with the Smith predictor, is capable of online estimat-

---

4. This example was suggested in the Nokia Lecture: Long Actuator Delays – Extending the Smith Predictor to Nonlinear systems by Miroslav Kristic, http://www.youtube.com/watch?v=0Nt1Pf5H3rA.

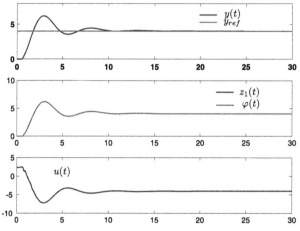

FIGURE 4.28 Performance of Smith predictor plus GPI observer control scheme on an unstable delay system with delayed state.

ing the delayed state $y(t - T_1)$ and to further predict the associated forward system perturbation $y(t - T_1 + T_2)$.

We consider the system as being just a perturbed system of the form

$$\dot{y} = u(t - T_2) + \varphi(t) \tag{4.185}$$

We will estimate $\varphi(t)$ by means of a GPI observer. From the observer variables we predict the value of $\varphi(t + T)$, i.e., $y(t - T_1 + T_2)$, and then control the forward perturbed system

$$\dot{y}_f = u(t) + \hat{\varphi}(t + T_2) \tag{4.186}$$

The problem reduces to the same class of problems we study along this manuscript.

The controller for the forward system is of the classical PI form with perturbation elimination and plant-retarded forward system stabilization error perturbation:

$$u = -\hat{\varphi}(t + T_2) + K_1(y_f - y_{ref} + y - y_f(t - T_2)) \tag{4.187}$$
$$+ K_0 \int_0^t (y_f - y_{ref} + y - y_f(\sigma - T_2))d\sigma$$

Fig. 4.28 shows the performance of the Smith predictor by means of the GPI observer in a regulation task. Fig. 4.29 shows the prediction of the state y by the estimation concept.

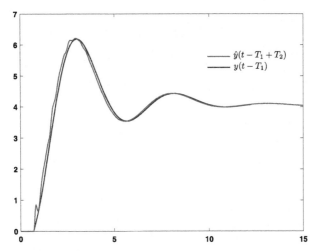

FIGURE 4.29 Performance of Smith predictor plus GPI observer control scheme on an unstable delay system with delayed state.

4.6.4 A Simple Nonlinear Delayed System

Consider the following normalized controlled pendulum with delayed inputs:

$$\ddot{y} = -\sin(y) + u(t - T) + p(t) \qquad (4.188)$$

where $p(t)$ is a time-varying perturbation of unknown but bounded nature. We simplify the nonlinear system model to the following linear perturbed system:

$$\ddot{y} = u(t - T) + \varphi(t) \qquad (4.189)$$

with $\varphi(t)$ including now the state-dependent nonlinearity $\varphi(t) = p(t) - \sin(y(t))$. Suppose, for simplicity, that we desire to control the angular position y toward a constant reference y_{ref}.

Letting $y_1 = y$, $y_2 = \dot{y}$, we postulate the following GPI observer for the linear perturbed system:

$$
\begin{aligned}
\dot{\hat{y}}_1 &= \hat{y}_2 + \lambda_7(y_1 - \hat{y}_1) \\
\dot{\hat{y}}_2 &= u(t - T) + z_1 + \lambda_6(y_1 - \hat{y}_1) \\
\dot{z}_1 &= z_2 + \lambda_5(y_1 - \hat{y}_1) \\
\dot{z}_2 &= z_3 + \lambda_4(y_1 - \hat{y}_1) \\
\dot{z}_3 &= z_4 + \lambda_3(y_1 - \hat{y}_1) \\
\dot{z}_4 &= z_5 + \lambda_2(y_1 - \hat{y}_1) \\
\dot{z}_5 &= z_6 + \lambda_1(y_1 - \hat{y}_1) \\
\dot{z}_6 &= \lambda_0(y_1 - \hat{y}_1)
\end{aligned}
\qquad (4.190)
$$

The reconstruction error $e = y_1 - \hat{y}_1 = y - \hat{y}_1$ satisfies the dynamics

$$e^{(8)} + \lambda_7 e^{(7)} + \lambda_6 e^{(6)} + \ldots + \lambda_1 \dot{e} + \lambda_0 e = \varphi^{(6)}(t) \quad (4.191)$$

Under the assumption that $\varphi^{(6)}(t)$ is negligible, we may assume that e and its time derivatives converge to a small neighborhood of zero. As a consequence, $z_1 \to \varphi(t)$, $z_2 \to \dot{\varphi}(t)$, $z_3 \to \ddot{\varphi}(t)$, etc. We use the following Taylor polynomial as an approximate prediction of the perturbation input:

$$\hat{\varphi}(t+T) = z_1(t) + \frac{1}{1!} z_2(t) T + \frac{1}{2!} z_3(t) T^2 \quad (4.192)$$

With this perturbation we consider, following the philosophy of the Smith Predictor controller methodology, the following forward or advanced system:

$$\ddot{y}_f = u(t) + \hat{\varphi}(t+T) \quad (4.193)$$

Let $e_f = y - y_f$ be the forward system-based output prediction error. Naturally, $\dot{e}_f = \dot{y} - \dot{y}_f$. The angular velocity $\dot{y}$ will be obtained from the observer as $\hat{y}_2$, and we assume that $\dot{y}_f$ is available by construction.

We propose the following proportional derivative controller for the forward system, with control input shared by the nonlinear delayed system:

$$u = -\hat{\varphi}_s(t+T) - k_1(\dot{y}_f + \dot{e}_f) - k_0(y + e_f - y_{\text{ref}}) \quad (4.194)$$

where $\hat{\varphi}_s(t+T)$ is a "smoothed" predicted state-dependent perturbation input. Let ϵ be a small positive real number. Let $s_f(t)$ denote the smoothing factor defined as

$$s_f(t) = \begin{cases} \sin^8(\frac{\pi t}{2\epsilon}) & \text{for } t \le \epsilon \\ 1 & \text{for } t > \epsilon \end{cases} \quad (4.195)$$

Thus, $\hat{\varphi}_s(t+T) = \hat{\varphi}(t+T) s_f(t)$.

Numerical Results

Some numerical simulations were implemented to show the effectiveness of the proposal. Fig. 4.30 shows the preformance of the GPI observer based Smith predictor control in a regulation task. Since the reference is not close to the initial condition of the output, there is a transient behavior during the first 3 [s], but this transient does not affect the regulation task as shown in the figure. On the other hand, Figs. 4.31, 4.32 show the same scheme for a trajectory tracking task. Fig. 4.31 shows the tracking results as well as the control input and the prediction scheme is depicted in Fig. 4.32.

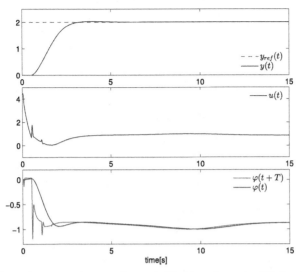

FIGURE 4.30   Performance trajectory and control of Smith predictor plus GPI observer controller.

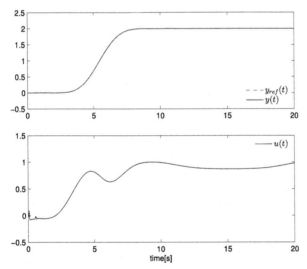

FIGURE 4.31   Performance for trajectory tracking of Smith predictor plus GPI observer control scheme in a perturbed one-degree-of-freedom perturbed pendulum with delayed control.

4.7   RELATIONS WITH MODEL-FREE CONTROL

Model-free control [16] is an alternative technique to control complex systems by using a simplified representation of the system (ultralocal model) and subsequent algebraic estimation techniques to design a simple, but effective, trajectory tracking controller. There are some similarities and differences between ADRC and the model-free approaches, which are listed as follows:

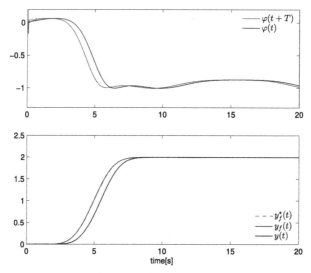

FIGURE 4.32 Performance for trajectory tracking of Smith predictor plus GPI observer control scheme in a perturbed one-degree-of-freedom perturbed pendulum with delayed control.

- The idea of using an ultralocal model is derived from the model-free control scheme. However, in the case of model-free control, the system order can be reduced, whereas in the case of ADRC, the order of the system is invariant. The use of the ultralocal model is then given for an alternative model of the lumped disturbance input.
- The nominal representation of a system in model-free control has the form

$$y^{(\nu)} = F + \alpha u \qquad (4.196)$$

where $\nu \in \mathbb{Z}^+$ is a designer choice, $\alpha \in \mathbb{R}$ is a constant gain, and F is a function to be estimated by means of algebraic estimation techniques, locally considered as piecewise constant.

- In ADRC, the control gain is assumed to be known, and a possible nonlinear function depends on space state. In general, there is an interval in which the controller works appropriately. In the case of the model-free approach, the control gain is obtained by trial-error, but it is not known from the beginning.

In the next subsection, we analyze an example by a model-free and ADRC scheme.

4.7.1 Example: A Van Der Pol Oscillator

Consider the following controlled van der Pol oscillator:

$$\dot{x}_1 = x_2$$

$$\dot{x}_2 = \mu(1 - x_1^2)x_2 - x_1 + u \qquad (4.197)$$

$$y = x_1$$

4.7.1.1 Intelligent PID Control

The ultramodel representation with $v = 2$ is given by

$$\dot{x}_1 = x_2$$
$$\dot{x}_2 = F + \alpha u \qquad (4.198)$$
$$y = x_1$$

This system with $F = 0$ is flat, with flat output y. The reference trajectory y^* leads to the nominal feed forward control $u^* = \ddot{y}^*$.

There are some ways to implement the model-free scheme by assuming an ultralocal model of order one or two (see [16] for a comprehensive analysis). In this section, we use the intelligent PID approach since the ultralocal model is similar to the one proposed in ADRC.

Assuming that the control gain is known ($\alpha = 1$), the intelligent proportional integral controller is defined as

$$u(t) = -\hat{F}(t) - \ddot{y}(t) + K_D(\dot{\hat{y}}(t) - \dot{y}^*(t)) + K_P e(t) + K_I \int_0^t e(\tau)d\tau \quad (4.199)$$

$$e(t) = y(t) - y^*(t)$$

F is estimated as follows:

$$\hat{F} = \dot{\hat{y}} - u \qquad (4.200)$$

4.7.1.2 Algebraic Time Derivative Scheme

The time derivative estimation is carried out through the algebraic derivative technique [17,18]. This method is based on the time derivative approximation by truncated expansion of the Taylor series of the signal, say y:

$$\tilde{y}(t) = y(t_i) + [\dot{y}(t_i)](t - t_i) \qquad (4.201)$$

The time derivative on the approximation expression (4.201) leads to the following linear model for the approximation:

$$\frac{d^2\tilde{y}}{dt^2} = 0 \qquad (4.202)$$

The procedure is generalized by adopting a higher-order model for the approximation of the output signal $y(t)$,

$$\tilde{y}(t) = \sum_{k=0}^{N-1} \frac{1}{k!} y^{(k)}(t_i)(t - t_i)^k, \quad N \in \mathbb{Z}^+ \qquad (4.203)$$

This approximation satisfies the following differential equation:

$$\frac{d^N \tilde{y}}{dt^N} = 0 \tag{4.204}$$

The problem of computing time derivatives of $y(t)$ is given as computing the states of the linear time-invariant linear homogeneous system of order N.

The linear approximation $y^{(N)}(t) = 0$ satisfies (in the s domain) the following relation:

$$\frac{d^N}{ds^N}\left[s^N Y(s)\right] = 0 \tag{4.205}$$

The expressions

$$s^{-k}\frac{d^N}{ds^N}\left[s^N Y(s)\right] = 0, \quad k = N-1, N-2, \cdots, N-k \tag{4.206}$$

contain implicit information on the kth derivatives of $y(t)$ in an approximate manner. In this case, we propose a fifth-order approximation that leads to the following expression in the s domain:

$$s^{-k}\frac{d^6}{ds^6}[s^6 \hat{y}(s)] = 0, \quad k = 5, 4, 3 \tag{4.207}$$

For $k = 5$, we have

$$720s^{-5}y(s) + 4320s^{-4}\frac{dy(s)}{ds} + 5400s^{-3}\frac{d^2 y(s)}{ds^2} + 2400s^{-2}\frac{d^3 y(s)}{ds^3}$$
$$+ 450s^{-1}\frac{d^4 y(s)}{ds} + 36\frac{d^5 y(s)}{ds^5} + s\frac{d^6 y(s)}{ds^6} = 0 \tag{4.208}$$

This expression relates the term $s\dfrac{d^6 y(s)}{ds^6}$, which in the time domain is

$$\frac{d}{dt}[t^6 y(t)] = 6t^5 y(t) + t^6 \dot{y}(t) \tag{4.209}$$

Expression (4.208) in the time domain is

$$720(\int^{(5)} y(t)) - 4320(\int^{(4)} t y(t)) + 5400(\int^{(3)} t^2 y(t)) - 2400(\int^{(2)} t^3 y(t))$$
$$+ 450\int t^4 y(t) - 36t^5 y(t) + \frac{d}{dt}t^6 y(t) = 0 \tag{4.210}$$

From (4.208) and (4.210) we have

$$
\dot{y}(t) = \frac{1}{t^6}\left[-720(\int^{(5)} y(t)) + 4320(\int^{(4)} ty(t)) - 5400(\int^{(3)} t^2 y(t)) \right.
$$

$$
\left. + 2400(\int^{(2)} t^3 y(t)) - 450 \int t^4 y(t) + 30t^5 y(t) \right] \tag{4.211}
$$

Since equation (4.211) is singular at $t = 0$ (the singularity disappears after time $t = \epsilon > 0$), we propose the following estimate of the first-order time derivative of $y(t)$ with respect to time:

$$
\dot{y}(t) = \begin{cases} \text{Arbitrary value, here, it is proposed } \dot{y}^*(t) & 0 \le t < \epsilon \\[2mm] \dfrac{1}{t^6}\left[-720(\int^{(5)} y(t)) + 4320(\int^{(4)} ty(t)) - 5400(\int^{(3)} t^2 y(t)) \right. \\[2mm] \left. + 2400(\int^{(2)} t^3 y(t)) - 450 \int t^4 y(t) + 30t^5 y(t) \right] & t \ge \epsilon \end{cases}
$$
$$\tag{4.212}$$

This computation can be expressed in the form of a time-varying linear filter of the form

$$
\dot{y}(t) = \begin{cases} \dot{y}^*(t) & 0 \le t < \epsilon \\[2mm] \dfrac{1}{t^6}\left[30t^5 y(t) + z_1 \right] & t \ge \epsilon \end{cases} \tag{4.213}
$$

where

$$
\begin{aligned}
\dot{z}_1 &= z_2 - 450t^4 y(t) \\
\dot{z}_2 &= z_3 + 2400t^3 y(t) \\
\dot{z}_3 &= z_4 - 5400t^2 y(t) \\
\dot{z}_4 &= z_5 + 4320t y(t) \\
\dot{z}_5 &= -720 y(t)
\end{aligned} \tag{4.214}
$$

To compute the second-order time derivative of $y(t)$, using $k = 4$ in (4.207), we have

$$
720s^{-4} y(s) + 4320s^{-3}\frac{dy(s)}{ds} + 5400s^{-2}\frac{d^2 y(s)}{ds^2} + 2400s^{-1}\frac{d^3 y(s)}{ds^3}
$$

$$
+ 450\frac{d^4 y(s)}{ds} + 36s\frac{d^5 y(s)}{ds^5} + s^2\frac{d^6 y(s)}{ds^6} = 0 \tag{4.215}
$$

The last two terms in the sum are, in time domain,

$$
-36\frac{d}{dt}[t^5 y(t)] + \frac{d^2}{dt^2}[t^6 y(t)] \tag{4.216}
$$

and allow us to compute the second-order time derivative of $y(t)$ in terms of the first-order time derivative of $y(t)$ and a finite sum of convolutions of $y(t)$ with powers of t. The resulting expression using the time-varying linear filter is

$$\ddot{y}(t) = \begin{cases} \ddot{y}^*(t) & 0 \le t < \epsilon \\ \dfrac{1}{t^6} \left(150t^4 y(t) + 24t^5 \dot{y}(t) + z_2\right) & t \ge \epsilon \end{cases} \tag{4.217}$$

with

$$\begin{aligned} \dot{z}_1 &= z_2 - 450t^4 y(t) \\ \dot{z}_2 &= z_3 + 2400t^3 y(t) \\ \dot{z}_3 &= z_4 - 5400t^2 y(t) \\ \dot{z}_4 &= z_5 + 4320t y(t) \\ \dot{z}_5 &= -720 y(t) \end{aligned} \tag{4.218}$$

4.7.1.3 Computing Resettings

Since this scheme is based on a local Taylor series approximation, the validity of the time derivative estimation is limited in the time horizon. For this reason, it is necessary to reinitialize the computations at some time $t_r > 0$. This time selection depends on the precision of the approximation, which is a function of the signal to differentiate, the number of approximation terms, etc.

In this case, under the assumption that $\epsilon \ll t_r$, $t_r < \delta$, the resetting strategy consists in using two identifiers in a resetting mode configuration such that the reinitialization of each identifier is separated in an interval of time $t_r/2$. This can be performed by defining two time lines t_{r1}, t_{r2} defined as follows:

$$\begin{aligned} t_1 &= t \mod t_r \\ t_2 &= t - t_r/2 \mod t_r \end{aligned} \tag{4.219}$$

where $\epsilon < t_r/2$.

4.7.1.4 Numerical Results

An intelligent PID was implemented for system (4.197) with $\mu = 0.5$. The time derivative estimation was set with a resetting time $t = 1.2$ [s]. The PID control gains were set to match the following dominant linear characteristic polynomial associated with the tracking error: $Pd(s) = (s^2 + 2\zeta \omega_n s + \omega_n^2)(s + p)$ with $\zeta = 1$, $\omega_n = 4$, $p = 5$. The reference trajectory was a sinusoidal wave with angular frequency of 0.5π [rad/s] and a unitary amplitude and a phase value equal to zero. Fig. 4.33 shows that the intelligent PID control is able to force the system output to track the reference trajectory within a second. Fig. 4.34 shows

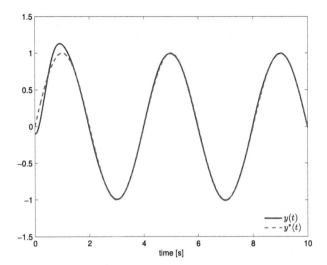

FIGURE 4.33   Closed-loop trajectory tracking results.

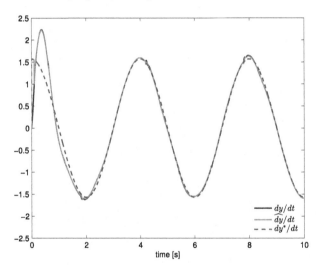

FIGURE 4.34   Time derivative algebraic estimation and reference signal.

that the time derivative estimation is accurate allowing an appropriate derivative feedback control component. In this case, the case of study is free of noise, but, as shown in the literature, this methodology is suitable in presence of zero-mean additive noises. Finally, Fig. 4.35 shows the generalized lumped disturbance estimation, which is clearly related to the control input compensation. As shown in the example, the ultralocal model is the same as that used in ADRC. The main difference is the fact that model-free control can consider a lower system order

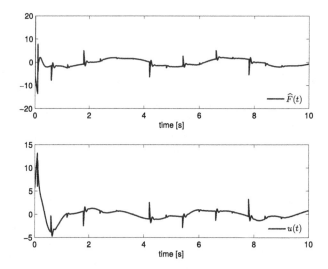

FIGURE 4.35 Algebraic disturbance estimation and control input.

(in this case, an intelligent PI control would be used instead). But the ideas of a simpler modelling and a compensation technique look for the same purposes.

4.8 FRACTIONAL-ORDER SYSTEMS

Consider the following Bagley–Torvik equation:

$$A_0 D_t^2 y + B_0 D_t^{\frac{3}{2}} y + C y = u \tag{4.220}$$

(where $A \neq 0$ and $B, C \in \mathbb{R}$), which arises in the modeling of the motion of a rigid plate immersed in a Newtonian fluid. It was originally proposed by Bagley and Torvik (see [19]) and is thoroughly discussed by Podlubny [20] and Trinks and Ruge [21].

Following the work of Podlubny, we assume homogeneous initial conditions corresponding to an equilibrium state at the beginning of the process: $y(0) = 0$ and $\dot{y} = 0$.

We desire to track a given smooth bounded reference trajectory, denoted by $y^*(t)$. For the control of the infinite-dimensional system, we use the following simplified perturbed model of the fractional-order system (we let $_0 D_t^2 y = \ddot{y}$ with homogeneous initial conditions):

$$\ddot{y} = \frac{1}{A} u + \xi(t) \tag{4.221}$$

$$y(0) = \dot{y} = 0 \tag{4.222}$$

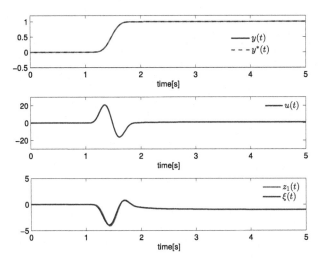

FIGURE 4.36   GPI observer-based trajectory tracking performance for the uncertain Bagley–Torvik system.

where

$$\xi(t) = \frac{B}{A_0} D_t^{\frac{3}{2}} y - \frac{C}{A} y \qquad (4.223)$$

is regarded as a bounded perturbation input of, otherwise, completely unknown nature. As usual, we locally model this perturbation input by means of a representative of a family of second-degree polynomials. We propose then the following high-gain GPI observer-based PID controller scheme:

$$\dot{\hat{y}}_1 = \hat{y}_2 + \lambda_2(y - \hat{y}_1)$$

$$\dot{\hat{y}}_2 = \frac{1}{A}u + z_1 + \lambda_1(y - \hat{y}_1)$$

$$\dot{z}_1 = \lambda_0(y - \hat{y}_1)$$

$$u = u^*(t) - A\left[k_2(\hat{y}_2 - \dot{y}^*(t)) + k_1(y - y^*(t)) + k_0\int_0^t (y - y^*(\sigma))d\sigma + z_1\right]$$

$$(4.224)$$

The closed-loop system tracking error behavior $e = y - y^*(t)$ is, approximately, governed by

$$e^{(3)} + k_2\ddot{e} + k_1\dot{e} + k_0 e = 0 \qquad (4.225)$$

For the simulations, we have chosen $A = 1$, $B = 0.5$, $C = 1$. The reference trajectory was set to be a smooth rest-to-rest maneuver, synthesized by means of a Bézier polynomial. Fig. 4.36 depicts the performance of the proposed GPI observer based controller.

REFERENCES

[1] M. Fliess, R. Marquez, E. Delaleau, H. Sira Ramírez, Correcteurs proportionnels-intégraux généralisés, ESAIM: Control, Optimisation and Calculus of Variations 7 (2002) 23–41.
[2] R. Madoński, P. Herman, Survey on methods of increasing the efficiency of extended state disturbance observers, ISA Transactions 56 (2015) 18–27.
[3] Y. Huang, W. Xue, Z. Gao, H. Sira-Ramírez, D. Wu, M. Sun, Active disturbance rejection control: methodology, practice and analysis, in: 2014 33rd Chinese Control Conference, IEEE, 2014, pp. 6643–6648.
[4] R.H. Middleton, G.C. Goodwin, Digital Control and Estimation: A Unified Approach, Prentice Hall Professional Technical Reference, ISBN 0132116650, 1990.
[5] A. Feuer, G.C. Goodwin, Sampling in Digital Signal Processing and Control, Springer Science & Business Media, 1996.
[6] N. Martínez-Fonseca, L.A. Castañeda, A. Uranga, A. Luviano-Juárez, I. Chairez, Robust disturbance rejection control of a biped robotic system using high-order extended state observer, ISA Transactions 62 (2016) 276–286.
[7] A. Poznyak, Advanced Mathematical Tools for Control Engineers: vol. 1: Deterministic Systems, vol. 1, Elsevier, 2008.
[8] H.K. Khalil, Nonlinear Systems, Prentice Hall, ISBN 9780130673893, 2002.
[9] M.W. Spong, S. Hutchinson, M. Vidyasagar, Robot Modeling and Control, vol. 3, Wiley, New York, 2006.
[10] V.I. Utkin, Sliding Modes and Their Application in Variable Structure Systems, MIR, Moscow, 1978.
[11] L.B. Freidovich, H.K. Khalil, Performance recovery of feedback-linearization-based designs, IEEE Transactions on Automatic Control 53 (10) (2008) 2324–2334.
[12] M.A. Chadwick, V. Kadirkamanathan, S.A. Billings, Analysis of fast-sampled non-linear systems: generalised frequency response functions for δ-operator models, Signal Processing 86 (2006) 3246–3257.
[13] G.C. Goodwin, R. Middleton, High-speed digital signal processing and control, Proceedings of the IEEE 80 (2) (1992) 240–259.
[14] A. Tesfaye, M. Tomizuka, Zeros of discretized continuous systems expressed in the Euler operator — an asymptotic analysis, IEEE Transactions on Automatic Control 40 (4) (1995) 743–747.
[15] K.J. Astrom, C.C. Hang, B.C. Lim, A new Smith predictor for controlling a process with an integrator and long dead-time, IEEE Transactions on Automatic Control 39 (2) (1994) 343–345.
[16] M. Fliess, C. Join, Model-free control, International Journal of Control 86 (12) (2013) 2228–2252.
[17] M. Fliess, C. Join, H. Sira-Ramirez, Non-linear estimation is easy, International Journal of Modelling, Identification and Control 4 (1) (2008) 12–27.
[18] H. Sira-Ramírez, C. García-Rodríguez, J. Cortés-Romero, A. Luviano-Juárez, Algebraic Identification and Estimation Methods in Feedback Control Systems, John Wiley & Sons, 2014.
[19] P.J. Torvik, R.L. Bagley, On the appearance of the fractional derivative in the behavior of real materials, Journal of Applied Mechanics 51 (2) (1984) 294–298.
[20] I. Podlubny, Fractional Differential Equations: An Introduction to Fractional Derivatives, Fractional Differential Equations, to Methods of Their Solution and Some of Their Applications, Mathematics in Science and Engineering, vol. 198, Academic Press, 1998.
[21] C. Trinks, P. Ruge, Treatment of dynamic systems with fractional derivatives without evaluating memory-integrals, Computational Mechanics 29 (6) (2002) 471–476.

Chapter 5

Case Studies

Chapter Points

- In this chapter, we present several applications of Active Disturbance Rejection Control for the trajectory tracking and regulation of a number of physically oriented examples. In some cases we present realistic simulation results.
- The presented experimental results on several complex nonlinear systems confirm the effectiveness of Active Disturbance Rejection Control and its dual variation, Flat filtering control, for dealing with systems exhibiting significant exogenous uncertainties and endogenous uncertainties in the form of unmodeled nonlinearities.
- The obtained results are shown to also extend to a class of control-input delayed systems.

5.1 INTRODUCTION

This chapter is devoted to present several applications examples of ADRC schemes to the control of physical systems. The examples are all constituted by nonlinear dynamical systems of single-variable or multivariable type. We pay specific attention to modeling issues to benefit the application-oriented reader.

The first three examples presented in this chapter deal with simulations only, whereas the rest of the examples presented, six in total, illustrate actual laboratory implementation of the ADRC schemes. Section 5.2 presents a simulation example concerning a nonlinear two-degree-of-freedom robot manipulator on which a trajectory tracking task is imposed. A flat filtering approach is used for the tracking of a set of rest-to-rest output reference trajectories. Only the angular positions of the robot waste and arm are assumed to be measured. In this first example, Coriolis terms are neglected, whereas the gravitational terms are cancelled thanks to the measurement of the generalized position coordinates. Section 5.3 presents simulation results on a fully actuated three-degree-of-freedom omnidirectional robot. In this instance, an observer-based ADRC scheme is implemented. For a laboratory implementation of the GPI observer-based ADRC controller, the reader is referred to [1].

Section 5.4 deals with an unfrequent combination of a single-link robotic manipulator and a nonlinear multivariable synchronous motor of the permanent

Active Disturbance Rejection Control of Dynamic Systems. http://dx.doi.org/10.1016/B978-0-12-849868-2.00005-8
Copyright © 2017 Elsevier Inc. All rights reserved.
173

magnet synchronous type, acting as a drive for the single-link system. The flatness of the system is exploited to independently regulate the two flat outputs. The flat outputs are one of the armature circuit currents, the quadrature current, and the motor shaft angular position. Section 5.5 shifts attention to actual laboratory experiments and begins by considering a nonlinear pendulum system in all detail. A high-dimensional GPI observer is used as the Extended State Observer. The accuracy of the disturbance estimation also allows us to carry out an independent off-line identification procedure of the ignored linearities and nonlinearities affecting the system behavior. These identification efforts are used, via their corresponding parameters, in the observer-based ADRC implementation. Section 5.6 presents a nontraditional control example constituted by the Thomson jumping ring. Experimental results show that the observer-based ADRC controller is capable of making the system track a rather wild position reference trajectory for the levitated ring. The reader is referred to [2] for an ADRC synchronization task on a set of two Thomson's jumping ring prototypes. Section 5.7 centers around the observer-based multivariable ADRC control of an actual Delta robot in a trajectory tracking task. Section 5.8 is concerned with the ADRC control of a nonlinear input-delayed system constituted by a flywheel pendulum type of system. In this example a generalization of ADRC is proposed to include a Smith Predictor type of control option. The use of high-dimensional Extended State Observers allows us to compute enough time derivatives of the disturbance input. These derivatives are used in a disturbance signal predictor, based on the first few terms of a Taylor series expansion of the future disturbance around the present. The successful implementation, however, implied that only relatively small delays of perfectly known nature resulted in a feasible rest-to-rest trajectory tracking control.

5.2 A TWO-DEGREE-OF-FREEDOM ROBOTIC ARM

Consider the mathematical model of a two-degree-of-freedom robotic manipulator. This anthropomorphic arm, provided with a waste and shoulder, is capable of angular displacements, denoted by θ, around a vertical axis orthogonal to the horizontal plane. The angular motions of the shoulder, achieved by rotations around an axis perpendicular to the vertical axes, are denoted by ϕ. We consider two control input torques τ_1 and τ_2 acting, respectively, on the axis describing θ and ϕ. This manipulator is depicted in Fig. 5.1.

5.2.1 Derivation of the Model via Euler–Lagrange Equations

We use spherical coordinates to define the position of the end effector, idealized to be a point on the surface of the sphere acting as the configuration space. The

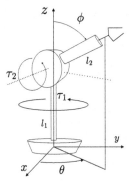

FIGURE 5.1 Two-degree-of-freedom robotic arm.

kinetic energy of the system is given by

$$K = \frac{1}{2}m_2 l_2^2 (\dot{\phi}^2 + \sin^2 \dot{\phi}^2) \qquad (5.1)$$

The potential energy is

$$V = m_2 g (l_1 + l_2 \cos \phi) - \tau_1 \theta - \tau_2 \phi \qquad (5.2)$$

The Lagrangian of the system is then obtained as

$$\mathcal{L} = \frac{1}{2}m_2 l_2^2 (\dot{\phi}^2 + \sin^2 \dot{\phi}^2) - m_2 g (l_1 + l_2 \cos \phi) + \tau_1 \theta + \tau_2 \phi \qquad (5.3)$$

The mathematical model of the system is then derived from the Euler–Lagrange equations, obtaining the following differential equations in matrix form:

$$\begin{bmatrix} m_2 l_2^2 \sin^2 \phi & 0 \\ 0 & m_2 l_2^2 \end{bmatrix} \begin{bmatrix} \ddot{\theta} \\ \ddot{\phi} \end{bmatrix} + \begin{bmatrix} m_2 l_2^2 \dot{\phi} \sin \phi \cos \phi & m_2 l_2^2 \dot{\theta} \sin \phi \cos \phi \\ -m_2 l_2^2 \dot{\theta} \sin \phi \cos \phi & 0 \end{bmatrix} \begin{bmatrix} \dot{\theta} \\ \dot{\phi} \end{bmatrix}$$

$$+ \begin{bmatrix} 0 \\ -m_2 g l_2 \sin \phi \end{bmatrix} = \begin{bmatrix} \tau_1 \\ \tau_2 \end{bmatrix} \qquad (5.4)$$

5.2.2 Normalization

A normalization of the differential equations is achieved via the following amplitude and time scaling operations:

$$t' = t\sqrt{\frac{g}{l_2}}, \quad u_i = \frac{\tau_i}{m_2 g l_2}, i = 1, 2$$

We obtain the following set of normalized equations for the system:

$$\sin^2 \phi \ddot{\theta} + 2\dot{\phi}\dot{\theta} \sin\phi \cos\phi = u_1 \tag{5.5}$$

$$\ddot{\phi} - \dot{\theta}^2 \sin\phi \cos\phi - \sin\phi = u_2 \tag{5.6}$$

A state-dependent locally invertible input coordinate transformation of the form

$$u_1 = \sin^2 \phi v_1 + 2\dot{\phi}\dot{\theta} \sin\phi \cos\phi \tag{5.7}$$

$$u_2 = v_2 - \dot{\theta}^2 \sin\phi \cos\phi - \sin\phi \tag{5.8}$$

with singularities located at $\phi = 0$ and $\phi = \pi$ yields the following multivariable linear system:

$$\ddot{\theta} = v_1 \tag{5.9}$$

$$\ddot{\phi} = v_2 \tag{5.10}$$

5.2.3 A Flat-Filtering-Based Controller

A set of linear flat filtering controllers for the transformed system is readily obtained as

$$v_1 = \ddot{\theta}^*(t) - \left[\frac{k_2 s^2 + k_1 s + k_0}{s(s+k_3)}\right](\theta - \theta^*(t)) \tag{5.11}$$

$$v_2 = \ddot{\phi}^*(t) - \left[\frac{\gamma_2 s^2 + \gamma_1 s + \gamma_0}{s(s+\gamma_3)}\right](\phi - \phi^*(t)) \tag{5.12}$$

Fig. 5.2 shows a block diagram depicting the closed-loop controlled subsystem for the θ variable. The reader is referred to Appendix B for an explanation of this block diagram realization of a closed-loop system using flat filtering.

The corresponding controllers in original variables u_1 and u_2 are simply

$$u_1 = \sin^2 \phi \left\{\ddot{\theta}^*(t) - \left[\frac{k_2 s^2 + k_1 s + k_0}{s(s+k_3)}\right](\theta - \theta^*(t))\right\} \tag{5.13}$$

$$u_2 = -\sin\phi + \left\{\ddot{\phi}^*(t) - \left[\frac{\gamma_2 s^2 + \gamma_1 s + \gamma_0}{s(s+\gamma_3)}\right](\phi - \phi^*(t))\right\} \tag{5.14}$$

where $k_2, \ldots, k_0$ and $\gamma_2, \ldots, \gamma_0$ correspond to a Hurwitz polynomial of fourth order of the form $(s^2 + 2\xi\omega_n s + \omega_n^2)^2$.

The controllers are found to be robust with respect to the neglected nonlinearities, thanks to the extra integrator located in the denominator of the transfer function.

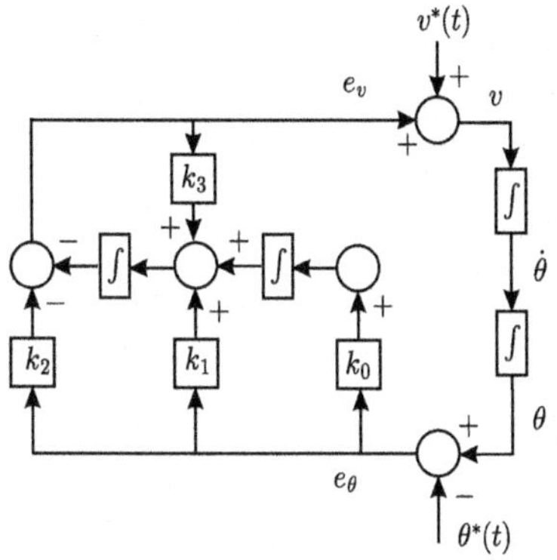

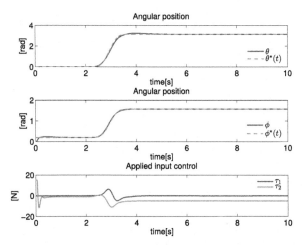

FIGURE 5.2 Block diagram of the GPI control for the robotic arm.

FIGURE 5.3 Performance of the flat filtering controllers on a two-degree-of-freedom robotic system.[1]

5.2.3.1 Simulations

For simulations, we have set the following parameters:

$$m_1 = 0.5 \text{ [kg]}, \quad L_1 = 1 \text{ [m]}, \quad m_2 = 0.5 \text{ [kg]}, \quad L_2 = 1 \text{ [m]}, \quad g = 9.8 \text{ [m/s}^2]$$

A trajectory tracking task, rest-to-rest trajectory for θ and ϕ during the time interval from $t = 2$ s to $t = 4$ s, is shown in Fig. 5.3. The values of ξ and ω_n for constants in u_1 and u_2 were set to be $\xi = 1$, $\omega_n = 6$ and $\xi = 1$, $\omega_n = 10$, respectively.

---

1. For interpretation of the references to color in this and the following figures, the reader is referred to the web version of this chapter.

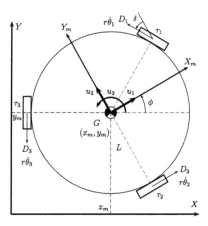

FIGURE 5.4 Schematic of the omnidirectional robot.

5.3 AN OMNIDIRECTIONAL ROBOT

In this section, we address the problem of trajectory tracking control for a three-wheel omnidirectional robot using an observer-based active disturbance rejection control strategy.

5.3.1 Mathematical Model

In this subsection, the dynamic equations of the omnidirectional robot are obtained via the Newton Second Law of Motion. This solution is based on the results obtained in [3,4].

Fig. 5.4 shows a top view of omnidirectional robot configuration, where $X_m - Y_m$ is the mobile frame set in the mass center of the robot G. In this case, the axis X_m is aligned with respect to the axis to the 3D wheel (this assumption is made without loss of generality since we can choose any one of the wheels as a reference). The other wheels are displaced 30° to each side of the Y_m axis. The fixed frame $X - Y$ provides the position with respect to an arbitrary point in the laboratory environment. We address this coordinate system as the global position coordinate frame.

Other parameters required to describe the model are the orientation angle ϕ, the total mass M_p, the moment of inertia I_p, the distance L from the center to each wheel, the wheel common radius r, and the angular velocities θ_i with $i = 1, 2, 3$. Additionally, we require the moment of inertia I_{ri} and the friction force D_i of the ith wheel.

The point S in the $X_m - Y_m$ frame is defined as

$$S = \begin{bmatrix} x, y \end{bmatrix}^T \tag{5.15}$$

Using Newton's Second Law

$$M\ddot{S} = F \tag{5.16}$$

where $F = \begin{bmatrix} F_x, F_y \end{bmatrix}^T$ is the vector of generalized forces in global coordinates, applied in the center of mass of the robot, and M is a positive definite symmetric matrix $(\text{diag}\left(M_p, M_p\right) > 0)$.

The rotation of the robot with respect to the global frame is characterized by ϕ, and the rotational transformation from global to body coordinate frame is expressed as

$$R = \begin{bmatrix} \cos(\phi) & -\sin(\phi) \\ \sin(\phi) & \cos(\phi) \end{bmatrix} \tag{5.17}$$

Using this transformation, the following relation is obtained in the body coordinates frame:

$$M\left(\dot{R}\dot{S}_m + R\ddot{S}_m\right) = RF_m \tag{5.18}$$

where $S_m = [x_m, y_m]^T$ and $F_m = [f_x, f_y]^T$ represent, respectively, the position vector of the center of mass and the vector of applied forces to the center of mass in body coordinates. The last equation can be simplified as follows:

$$M\left\{\begin{bmatrix} 0 & -\dot{\phi} \\ \dot{\phi} & 0 \end{bmatrix}\begin{bmatrix} \dot{x}_m \\ \dot{y}_m \end{bmatrix} + \begin{bmatrix} \ddot{x}_m \\ \ddot{y}_m \end{bmatrix}\right\} = \begin{bmatrix} f_x \\ f_y \end{bmatrix} \tag{5.19}$$

The angular moment of the mobile robot is computed as

$$I_p\ddot{\phi} = M_I \tag{5.20}$$

where I_p is the moment of inertia, and M_I is the moment of the force around the center of mass center.

From (5.19) and (5.20) the dynamics with respect to the body frame is described by

$$\begin{bmatrix} M_p & 0 & 0 \\ 0 & M_p & 0 \\ 0 & 0 & I_p \end{bmatrix}\begin{bmatrix} \ddot{x}_m \\ \ddot{y}_m \\ \ddot{\phi} \end{bmatrix} + \begin{bmatrix} 0 & -M_p & 0 \\ M_p & 0 & 0 \\ 0 & 0 & 0 \end{bmatrix}\begin{bmatrix} \dot{x}_m \\ \dot{y}_m \\ \dot{\phi} \end{bmatrix} = \begin{bmatrix} f_x \\ f_y \\ M_I \end{bmatrix} \tag{5.21}$$

The terms f_x, f_y, and M_I are computed as follows:

$$\begin{bmatrix} f_x \\ f_y \\ M_I \end{bmatrix} = \begin{bmatrix} -\sin(\delta) & -\sin(\delta) & 1 \\ \cos(\delta) & -\cos(\delta) & 0 \\ L & L & L \end{bmatrix}\begin{bmatrix} D_1 \\ D_2 \\ M_I \end{bmatrix} \tag{5.22}$$

where D_1, D_2, and D_3 are the forces introduced by each wheel, L is the distance from wheels to the center of mass, and $\delta = \pi/6$ is the angle between the vertical axis and the direction of the D vectors.

Assume that the angular moments of inertia of the wheels I_r are all the same, say, $I_{r_i} = I_r$. The wheel dynamics is described by

$$I_r \ddot{\theta}_i = \tau_i - r D_i \qquad (5.23)$$

where τ_i is the applied input torque. From the inverse kinematic model of the omnidirectional mobile robot, with respect to the mobile frame, we obtain

$$\begin{bmatrix} r\dot{\theta}_1 \\ r\dot{\theta}_2 \\ r\dot{\theta}_3 \end{bmatrix} = \begin{bmatrix} -\sin(\delta) & \cos(\delta) & L \\ -\sin(\delta) & -\cos(\delta) & L \\ 1 & 0 & L \end{bmatrix} \begin{bmatrix} \dot{x}_m \\ \dot{y}_m \\ \dot{\phi} \end{bmatrix} \qquad (5.24)$$

Differentiating Eq. (5.24) and multiplying by $1/r$, we get the resulting equation

$$\begin{bmatrix} \ddot{\theta}_1 \\ \ddot{\theta}_3 \\ \ddot{\theta}_3 \end{bmatrix} = \frac{1}{r} \begin{bmatrix} -\sin(\delta) & \cos(\delta) & L \\ -\sin(\delta) & -\cos(\delta) & L \\ 1 & 0 & L \end{bmatrix} \begin{bmatrix} \ddot{x}_m \\ \ddot{y}_m \\ \ddot{\phi} \end{bmatrix} \qquad (5.25)$$

Expressing (5.23) in the matrix form, substituting (5.25) and (5.22), and using the numerical value of δ, the wheels dynamics is represented by

$$\begin{bmatrix} f_x \\ f_y \\ M_I \end{bmatrix} = \frac{1}{r} \begin{bmatrix} -\dfrac{1}{2} & -\dfrac{1}{2} & 1 \\ \dfrac{\sqrt{3}}{2} & -\dfrac{\sqrt{3}}{2} & 0 \\ L & L & L \end{bmatrix} \begin{bmatrix} \tau_1 \\ \tau_2 \\ \tau_3 \end{bmatrix} - \frac{I_r}{r^2} \begin{bmatrix} \dfrac{3}{2} & 0 & -2L \\ 0 & \dfrac{3}{2} & 0 \\ -2L & 0 & 3L^2 \end{bmatrix} \begin{bmatrix} \ddot{x}_m \\ \ddot{y}_m \\ \ddot{\phi} \end{bmatrix} \qquad (5.26)$$

Finally, substituting (5.26) into (5.21), the dynamics of the omnidirectional mobile robot is described as

$$\begin{bmatrix} \dfrac{3I_r}{2r^2} + M_p & 0 & 0 \\ 0 & \dfrac{3I_r}{2r^2} + M_p & 0 \\ 0 & 0 & I_p + \dfrac{3L^2 I_r}{r^2} \end{bmatrix} \begin{bmatrix} \ddot{x}_m \\ \ddot{y}_m \\ \ddot{\phi} \end{bmatrix} + \dot{\phi} \begin{bmatrix} 0 & -M_p & 0 \\ M_p & 0 & 0 \\ 0 & 0 & 0 \end{bmatrix} \begin{bmatrix} \dot{x}_m \\ \dot{y}_m \\ \dot{\phi} \end{bmatrix}$$

$$= \frac{1}{r} \begin{bmatrix} -\dfrac{1}{2} & -\dfrac{1}{2} & 1 \\ \dfrac{\sqrt{3}}{2} & -\dfrac{\sqrt{3}}{2} & 0 \\ \dfrac{L}{L} & \dfrac{L}{L} & \dfrac{L}{L} \end{bmatrix} \begin{bmatrix} \tau_1 \\ \tau_2 \\ \tau_3 \end{bmatrix} \tag{5.27}$$

To express (5.27) in terms of the global (inertial) frame, it is necessary to transform this equation once again. The transformation consists in a planar rotation based on the orientation angle ϕ:

$$T(\phi) = \begin{bmatrix} \cos(\phi) & \sin(\phi) & 0 \\ -\sin(\phi) & \cos(\phi) & 0 \\ 0 & 0 & 1 \end{bmatrix} \tag{5.28}$$

Thus, the velocity and acceleration are described as

$$\begin{bmatrix} \dot{x}_m \\ \dot{y}_m \\ \dot{\phi} \end{bmatrix} = \begin{bmatrix} \cos(\phi) & \sin(\phi) & 0 \\ -\sin(\phi) & \cos(\phi) & 0 \\ 0 & 0 & 1 \end{bmatrix} \begin{bmatrix} \dot{x} \\ \dot{y} \\ \dot{\phi} \end{bmatrix} \tag{5.29}$$

$$\begin{bmatrix} \ddot{x}_m \\ \ddot{y}_m \\ \ddot{\phi} \end{bmatrix} = \begin{bmatrix} \cos(\phi) & \sin(\phi) & 0 \\ -\sin(\phi) & \cos(\phi) & 0 \\ 0 & 0 & 1 \end{bmatrix} \begin{bmatrix} \ddot{x} \\ \ddot{y} \\ \ddot{\phi} \end{bmatrix} + \begin{bmatrix} -\sin(\phi)\dot{\phi} & \cos(\phi)\dot{\phi} & 0 \\ -\cos(\phi)\dot{\phi} & -\sin(\phi)\dot{\phi} & 0 \\ 0 & 0 & 0 \end{bmatrix} \begin{bmatrix} \dot{x} \\ \dot{y} \\ \dot{\phi} \end{bmatrix} \tag{5.30}$$

Substituting (5.29) and (5.30) into (5.27) and then simplifying, we express the model in terms of the global frame of reference as

$$\begin{bmatrix} \dfrac{3I_r}{2r^2} + M_p & 0 & 0 \\ 0 & \dfrac{3I_r}{2r^2} + M_p & 0 \\ 0 & 0 & \dfrac{3L^2 I_r}{r^2} + I_p \end{bmatrix} \begin{bmatrix} \ddot{x} \\ \ddot{y} \\ \ddot{\phi} \end{bmatrix} + \dot{\phi} \begin{bmatrix} 0 & \dfrac{3I_r}{2r^2} & 0 \\ -\dfrac{3I_r}{2r^2} & 0 & 0 \\ 0 & 0 & 0 \end{bmatrix} \begin{bmatrix} \dot{x} \\ \dot{y} \\ \dot{\phi} \end{bmatrix}$$

$$= \frac{1}{r} \begin{bmatrix} -\sin(\delta + \phi) & -\sin(\delta - \phi) & \cos(\phi) \\ \cos(\delta + \phi) & -\cos(\delta - \phi) & \sin(\phi) \\ L & L & L \end{bmatrix} \begin{bmatrix} \tau_1 \\ \tau_2 \\ \tau_3 \end{bmatrix} \tag{5.31}$$

This system can be expressed in the form

$$D\ddot{q} + C(\dot{q})\dot{q} = B\tau \tag{5.32}$$

The control system can be developed on the basis of Extended State Observers of the GPI type. Using the fact that the omnidirectional robot has a positive definite inertia matrix, it is possible to express system (5.32) in the simplified form

$$\ddot{q} = D^{-1}B\tau + \xi(t) \tag{5.33}$$

where $\xi(t)$ contains both neglected nonlinearities (Coriolis effects) and external disturbance inputs. Proposing an approximation by means of a first-degree polynomial (a second-order chain of integrator disturbance model approximation) for each element of $\xi(t)$, we propose the following ADRC controller:

$$
\begin{aligned}
\tau &= B^{-1}D\left[\ddot{q}^* - K_D(\hat{q}_2 - \dot{q}^*) - K_P(q - q^*) - \hat{\rho}_0\right]\\
\dot{\hat{q}}_1 &= \hat{q}_2 + \lambda_3(q - \hat{q}_1)\\
\dot{\hat{q}}_2 &= D^{-1}B\tau + \hat{\rho}_0 + \lambda_2(q - \hat{q}_1)\\
\dot{\hat{\rho}}_0 &= \hat{\rho}_1 + \lambda_1(q - \hat{q}_1)\\
\dot{\hat{\rho}}_1 &= \lambda_0(q - \hat{q}_1)\\
\hat{\rho}_0 &= \hat{\xi}
\end{aligned}
\tag{5.34}
$$

where the matrices and the observer gains are $K_D = \mathrm{diag}(K_{d1}, K_{d2}, K_{d3})$ and $K_P = \mathrm{diag}(K_{p1}, K_{p2}, K_{p3})$, $K_D, K_P > 0$, and $\lambda_{i1}, \lambda_{i2}, \lambda_{i3}$, $i = 0, 1, 2, 3$, $\lambda_{i,j} \in \mathbb{R}$, are selected such that the associated poles of the dominant linear dynamics have its roots located deep in the left side of the complex plane, guaranteeing the stability of the tracking errors.

5.3.2 Simulation Results

To validate the proposed controllers, some numerical simulations were carried out. The system was simulated in Matlab Simulink. The parameters of the omnidirectional robot were as follows: $I_r = 0.6$ [kgm^2], $M_p = 7.25$ [kg], $r = 0.05$ [m], $L = 0.175$ [m], $I_p = 0.20$ [kgm^2], $\delta = \pi/6$. The controller gains were set such that each linear dominant tracking error dynamics matches the following stable dynamics in the Laplace operator: $s^2 + 2\zeta\omega_n s + \omega_n^2$ with $\zeta = 2$ and $\omega_n = 5$. The observer injection gains were set to match the following linear dominant injection error dynamics in the Laplace transform notation: $\left(s^2 + 2\zeta_1\omega_{n1}s + \omega_{n1}^2\right)\left(s^2 + 2\zeta_2\omega_{n2}s + \omega_{n2}^2\right)$ with $\zeta_1 = 1$, $\zeta_2 = 1.5$, $\omega_{n1} = 15$, $\omega_{n2} = 20$.

Fig. 5.5 depicts the tracking results for a reference trajectory consisting of a circle of radius 0.75 [m], centered in the origin. In this image, the red line represents the actual robot position, and the blue dashed line denotes the reference trajectory. Notice that the error convergence has no peaking effects as

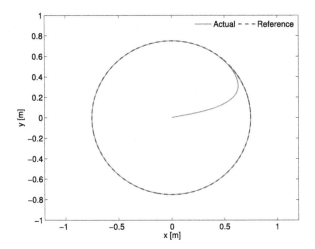

FIGURE 5.5 Trajectory tracking of the omnidirectional robot for a circular reference.

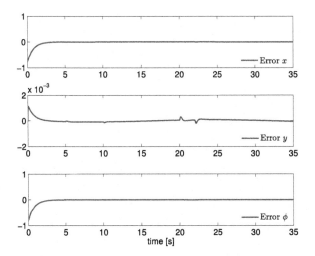

FIGURE 5.6 Tracking error signals.

seen in the tracking error dynamics (Fig. 5.6) despite the assumption of the lack of information of the dynamic modeling and the effects of the nonlinearities on the manipulator behavior. To show the disturbance rejection, two external forces were applied. The first one at $t = 5$ [s] consisted on a unit step disturbance input applied on the x axis of the vehicle. This force was applied during 5 seconds. The other torque disturbance input of the step kind, with a magnitude of 2 [Nm], was applied during 2.5 s at $t = 20$ [s]. As shown in Fig. 5.5, these disturbances did not affect the general tracking performance. The disturbance estimations and

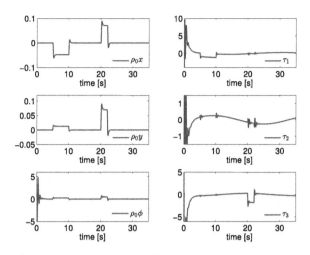

FIGURE 5.7 Disturbance estimates and control inputs.

the torque inputs are shown in Fig. 5.7. Notice the effect of the disturbance estimations in the system control inputs. After the reference is reached, the robot position remains in a vicinity of the reference trajectory, as expected.

5.4 A SINGLE-LINK MANIPULATOR DRIVEN BY A SYNCHRONOUS MOTOR

Consider a one-degree-of-freedom manipulator. The system consists of a point mass m, a link of negligible mass, and the moment of inertia. The length of the link is L_1. This manipulator is controlled by a pair τ applied to the pivot axis. We assume that only the angular position of the pendulum is measured. The corresponding angular velocity is not measured. All the single-link manipulator parameters are assumed to be known.

5.4.1 Derivation of the Model via Euler–Lagrange Equations

From Fig. 5.8 we have the Cartesian coordinate locations of m_1 given by

$$x_{m1} = L_1 \sin(\theta); \quad y_{m1} = -L_1 \cos(\theta) \tag{5.35}$$

The velocities associated with the mass are obtained by straightforward time differentiation of their respective coordinate positions:

$$\dot{x}_{m1} = -L_1 \dot{\theta} \cos(\theta), \quad \dot{y}_{m1} = -L_1 \dot{\theta} \sin(\theta) \tag{5.36}$$

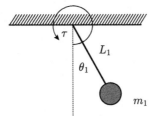

FIGURE 5.8 Pendulum system.

The kinetic energy is obtained as follows:

$$K_{m1} = \frac{1}{2}m_1(L_1^2\dot{\theta}^2) \tag{5.37}$$

The potential energy associated with the bob mass position and the work of the external input force is expressed as

$$V_{m1} = -m_1 g L_1 \cos(\theta) - \tau\theta \tag{5.38}$$

The Lagrangian of the system is computed as

$$\mathcal{L} = K_{m1} + V_{m1} = \frac{1}{2}m_1(L_1^2\dot{\theta}^2) - m_1 g L_1 \cos(\theta) - \tau\theta \tag{5.39}$$

Using the Euler–Lagrange equations of the system are given by

$$\frac{d}{dt}\left(\frac{\partial\mathcal{L}}{\partial\dot{\theta}}\right) - \frac{\partial\mathcal{L}}{\partial\theta} = 0$$

we obtain

$$m_1 L_1^2\ddot{\theta} + m_1 g L_1 \sin(\theta) = \tau \tag{5.40}$$

Introducing an auxiliary control v for a state-dependent input coordinate transformation yields

$$\tau = m_1 g L_1 \sin(\theta) + m_1 L_1^2 v$$

where the model of the system has been reduced to

$$\ddot{\theta} = v \tag{5.41}$$

We propose the following classical controller with feed-forward compensation for the simplified system:

$$v = \ddot{\theta}^*(t) - \left[\frac{bs + c}{s + a}\right](\theta - \theta^*(t)) \tag{5.42}$$

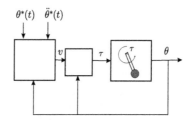

FIGURE 5.9   Control scheme for the single-link manipulator.

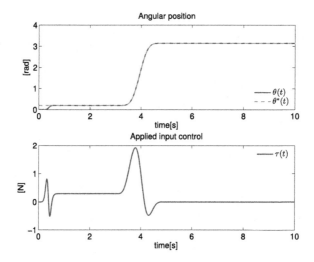

FIGURE 5.10   Trajectory tracking performance of the controller with gravity compensation.

The resulting nonlinear control is just

$$\tau = m_1 g L_1 \sin(\theta) + m_1 L_1^2 \left[\ddot{\theta}^*(t) - \left[\frac{bs + c}{s + a} \right] (\theta - \theta^*(t)) \right] \qquad (5.43)$$

Such a controller allows for the accurate tracking of the angular position. The values of a, b, and c correspond with a Hurwitz polynomial in the form $(s^2 + 2\xi\omega_n s + \omega_n^2)(s + p)$, where

$$a = 2\omega_n \xi + p, \quad b = 2 p\omega_n \xi + \omega_n^2, \quad c = p\omega_n^2 \qquad (5.44)$$

Fig. 5.9 shows a block diagram of the single-link manipulator with a flat filtering approach. On the other hand, in Fig. 5.10, we show a simulation of the trajectory tracking for the angular position in the manipulator, where the values for simulation were set to be

$$m_1 = 0.3 \text{ [kg]}, \quad L_1 = 0.5 \text{ [m]}, \quad g = 9.8 \text{ [m/s}^2\text{]}, \quad \xi = 1, \quad \omega_n = 30, \quad p = 30 \qquad (5.45)$$

5.4.2 Problem Formulation

A Permanent Magnet Synchronous Motor (PMSM) is coupled to a single-link manipulator. It is desired to control the angular position θ of the manipulator so that a given smooth output reference trajectory $\theta^*(t)$ is accurately tracked. The control input to the composite systems is the applied armature voltage V_d applied to the electrical part of the motor.

5.4.3 Including the Model of a PMSM Motor

The control input torque for the single-link manipulator system can be provided via a PMSM motor. A mathematical model of a PMSM, described by the following multivariable nonlinear system, written in a d–q frame, is given by Chiasson [5]:

$$L_s \frac{di_d}{dt} = -R_s i_d + n_p \dot{\theta}_m L_s i_q + v_d$$

$$L_s \frac{di_q}{dt} = -R_s i_q - (n_p L_s i_d + K_m) \dot{\theta}_m + v_q$$

$$J \frac{d^2\theta_m}{dt^2} = K_m i_q - B\dot{\theta}_m - \tau_L \tag{5.46}$$

where i_d and i_q are the direct and quadrature currents, respectively, v_d and v_q are the input controls, θ_m is the motor shaft angular position, L_s and R_s denote, respectively, the stator resistance and the stator inductance, and J and B represent the moment of inertia and the rotational friction coefficient. The parameter K_m is the back-emf (back electro-motive force) term obtained through $K_m = n_p \lambda_m$, where n_p is the number of pole pairs, λ_m is the flux linkage of the stator winding due the flux produced by the rotor magnets, and τ_L is the torque demanded by the mechanical load.

To connect both systems, the pendulum and the DC motor dynamics, we set

$$\tau_L = \tau, \quad \theta_m = \theta \tag{5.47}$$

The coupled system given by the models (5.40) and (5.46) is described by

$$L_s \frac{di_d}{dt} = -R_s i_d + n_p \dot{\theta} L_s i_q + v_d$$

$$L_s \frac{di_q}{dt} = -R_s i_q - (n_p L_s i_d + K_m)\dot{\theta} + v_q$$

$$(J + mL_1^2)\frac{d^2\theta}{dt^2} = K_m i_q - B\dot{\theta} - m_1 g L_1 \sin(\theta) \tag{5.48}$$

By (5.48) the quadrature current i_q is

$$i_q = \frac{(J + m_1 L_1^2)\ddot{\theta} + B\dot{\theta} + m_1 g L_1 \sin(\theta_1)}{K_m} \tag{5.49}$$

Substituting i_q from (5.49) into (5.48), we obtain the input–output representation of the coupled system:

$$L_s \frac{di_d}{dt} + R_s i_d - \frac{n_p L_s}{K_m}\left[(J + m_1 L_1^2)\ddot{\theta} + B\dot{\theta} + m_1 g L_1 \sin(\theta)\right]\dot{\theta} = v_d$$

$$\frac{L_s(J + m_1 L_1^2)}{K_m}\theta^{(3)} + \left[\frac{B L_s}{K_m} + \frac{R_s(J + m_1 L_1^2)}{K_m}\right]\ddot{\theta}$$

$$+ \left[\frac{R_s B}{K_m} + \frac{L_s m_1 g L_1 \cos(\theta)}{K_m} + n_p L_s i_d + K_m\right]\dot{\theta} + \frac{R_s m_1 g L_1 \sin(\theta)}{K_m} = v_q \tag{5.50}$$

5.4.3.1 A Flat-Filtering Approach

To control the angular position, it seems necessary to measure the angular velocity $\dot{\theta}$ and the angular acceleration $\ddot{\theta}$. In practice, this results inconvenient due the noise associated with physical measurements of the time derivatives of given signals. To avoid the measurements of the angular velocity $\dot{\theta}$ and of the angular acceleration $\ddot{\theta}$, we propose the following control with nominal compensation of variables:

$$v_d = R_s i_d - L_s \left[\frac{k_1 s + k_0}{s + k_2}\right] i_d$$

$$v_q = \frac{L_s(J + m_1 L_1^2)}{K_m}[\theta^*(t)]^{(3)} + \frac{R_s}{K_m} m_1 g L_1 \sin(\theta)$$

$$- \frac{L_s(J + m_1 L_1^2)}{K_m}\left[\frac{\gamma_2 s^2 + \gamma_1 s + \gamma_0}{s^2 + \gamma_4 s + \gamma_3}\right](\theta - \theta^*(t)) \tag{5.51}$$

Such a controller drives the current i_d to zero and the angular position to track the desired angular reference trajectory.

In closed loop, the resulting systems are decoupled perturbed linear systems with characteristic polynomials of third and fifth orders, respectively. We equate these polynomials, term by term, with desired Hurwitz polynomials of the forms $(s^2 + 2\xi_1\omega_1 + \omega_1^2)(s + p_1)$ and $(s^2 + 2\xi_2\omega_2 + \omega_2^2)^2(s + p_2)$. This yields the following equalities to determine the controller parameters:

$$k_2 = 2\omega_1\xi_1 + p_1$$

$$k_1 = 2p_1\omega_1\xi_1 + \omega_1^2$$

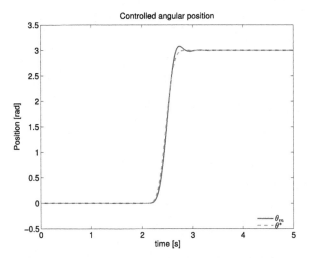

FIGURE 5.11 Trajectory tracking performance of the proposed controllers for the single-link manipulator driven by a PMSM motor, including nominal compensation.

$$k_0 = p_1\omega_1^2$$
$$\gamma_4 = 4\omega_2\xi_2 + p_2$$
$$\gamma_3 = 4\omega_2^2\xi_2^2 + 4p_2\omega_2\xi_2 + 2\omega_2^2$$
$$\gamma_2 = 4p_2\omega_2^2\xi_2^2 + 4\omega_2^3\xi_2 + 2p_2\omega_2^2$$
$$\gamma_1 = 4p_2\omega_2^3\xi_2 + \omega_2^4$$
$$\gamma_0 = p_2\omega_2^4 \tag{5.52}$$

5.4.3.2 Simulation Results

For computer simulations, we used the following set of parameter values:

$$m_1 = 0.3 \text{ [kg]}, \quad L_1 = 0.5 \text{ [m]}, \quad J = 0.182e^{-3} \text{ [kgm}^2\text{]},$$
$$L_s = 6.365e{-3} \text{ [H]}, \quad R_s = 1.6 \text{ [}\Omega\text{]}, \quad g = 9.8 \text{ [m/s}^2\text{]},$$
$$B = 8.7081e{-05} \text{ [Nms]}, \quad K_m = 0.468 \text{ [V/rad/s]}, \quad n_p = 2$$
$$\xi_1 = 1, \quad \omega_1 = 50, \quad p_1 = 50, \quad \xi_2 = 1, \quad \omega_2 = 100, \quad p_2 = 100$$

In Fig. 5.11 the trajectory tracking of a rest-to-rest maneuver is depicted for the angular position of the manipulator. The angular trajectory rises from 0 to 3 radians, during a time interval of 1 second, $t = [2, 3]$ [s]. The controllers include nominal precompensation terms related to the desired reference trajectory. The direct and quadrature current and the respectively control inputs are shown in Fig. 5.12.

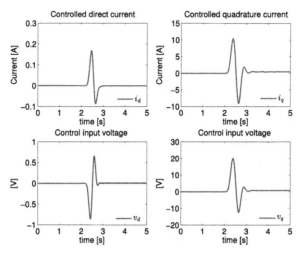

FIGURE 5.12 Direct-quadrature current and control variables with nominal compensation for trajectory tracking task.

5.4.3.3   An Active Disturbance Rejection Controller-Based Approach

Eq. (5.50) is of the form

$$\theta^{(3)} = \left[\frac{K_m}{L_s(J + m_1 L_1^2)} \right] v_q - \xi(\theta, \dot{\theta}, \ddot{\theta}, i_d) \tag{5.53}$$

where

$$\xi(\theta, \dot{\theta}, \ddot{\theta}, i_d) = \frac{K_m}{L_s(J + m_1 L_1^2)} \left\{ \left[\frac{BL_s}{K_m} + \frac{R_s(J + m_1 L_1^2)}{K_m} \right] \ddot{\theta} \right.$$
$$+ \left[\frac{R_s B}{K_m} + \frac{L_s m_1 g L_1 \cos(\theta)}{K_m} + n_p L_s i_d + K_m \right] \dot{\theta} + \left. \frac{R_s m_1 g L_1 \sin(\theta)}{K_m} \right\} \tag{5.54}$$

As customary in the ADRC approach, we can rewrite the above model in a simplified manner as follows:

$$\theta^{(3)} = \left[\frac{K_m}{L_s(J + m_1 L_1^2)} \right] v_q + \xi(t) \tag{5.55}$$

The expression $\xi(t)$ includes all the neglected nonlinearities and is taken to represent the *total of disturbance*. For control design purposes, we consider the simplified version of the system associated with the previous nonlinear system.

The combined PMSM-link manipulator model is differentially flat (see [6,7]).

Consider the following third-order model of the angular position of the motor shaft:

$$\frac{d^3\theta_m}{dt^3} = v + \xi(t) \tag{5.56}$$

where the variable v is an auxiliary control input. The controller design route is now clear and it follows the GPI observer-based Active Disturbance Rejection Controller methodology. The feedback controller is given by

$$v = [\theta^*(t)]^{(3)} - k_2(\theta_2 - \ddot{\theta}^*(t)) - k_1(\theta_1 - \dot{\theta}^*(t)) - k_0(\theta - \theta^*(t)) - z_1(t) \tag{5.57}$$

where

$$
\begin{aligned}
\dot{\theta}_0 &= \theta_1 + \gamma_3(\theta - \theta_0) \\
\dot{\theta}_1 &= \theta_2 + \gamma_2(\theta - \theta_0) \\
\dot{\theta}_2 &= v + z_1 + \gamma_1(\theta - \theta_0) \\
\dot{z}_1 &= \gamma_0(\theta - \theta_0)
\end{aligned}
\tag{5.58}
$$

This controller globally uniformly asymptotically stabilizes the tracking error $e = \theta_r - \theta_m^*(t)$ and its time derivatives $e^{(j)} = \theta_r^{(j)} - [\theta_m^*(t)]^{(j)}$, $j = 1, 2, 3$, to a small as desired vicinity of the origin of the tracking error phase space $(e, \dot{e}, \ddot{e})$, provided that the sets of real coefficients $\gamma_0, \ldots, \gamma_3$ and k_0, k_1, k_2 are chosen such that the roots of the polynomials in the complex variable s

$$
\begin{aligned}
po(s) &= s^4 + \gamma_3 s^3 + \gamma_2 s^2 + \gamma_1 s + \gamma_0 \\
pc(s) &= s^3 + k_2 s^2 + k_1 s + k_0
\end{aligned}
\tag{5.59}
$$

are located sufficiently far to the left of the imaginary axis in the complex plane s. The values of $\gamma_3, \ldots, \gamma_0$ and $k_2, \ldots, k_0$ correspond with Hurwitz polynomials of fourth and third orders, respectively; these Hurwitz polynomials are of the forms $(s^2 + \xi_1\omega_1 s + \omega_1^2)^2$ and $(s^2 + 2\xi_2\omega_2 s + \omega_2^2)(s + p_2)$, where

$$k_2 = 2\omega_2\xi_2 + p_2; \quad k_1 = 2p_2\omega_2\xi_2 + \omega_2^2; \quad k_0 = p_2\omega_2^2$$

$$\gamma_3 = 4\xi_1\omega_1; \quad \gamma_2 = 4\xi_1^2\omega_1^2 + 2\omega_1^2; \quad \gamma_1 = 4\xi_1\omega_1^3; \quad \gamma_0 = \omega_1^4;$$

Figs. 5.13 and 5.14 show the performance of the trajectory tracking controller for the pendulum angular position. The ADRC approach substantially improved the performance when compared with the flat filtering approach due to the use of state observers that allow better disturbance cancellation in the task of stabilization of the tracking error to zero.

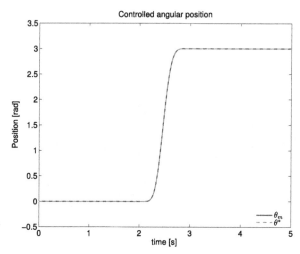

FIGURE 5.13 Trajectory tracking performance of the ADRC controller for the angular position of the single-link manipulator driven by a PMSM motor.

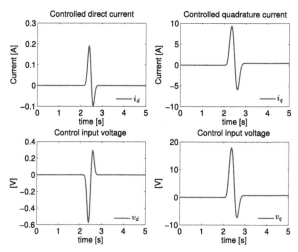

FIGURE 5.14 Trajectory tracking performance of the ADRC controller for direct-quadrature current and control variables.

5.5 NONLINEAR PENDULUM SYSTEM

5.5.1 Experimental Results for a Nonlinear Pendulum System

Consider a pendulum system consisting of a solid metal bar of length L, mass m_c, and an extra mass m_a attached to its tip (see Fig. 5.15). The pendulum is actuated by a dc motor through a gear box. The angular displacement of the motor shaft is denoted by θ_m, and the pendulum angular position is de-

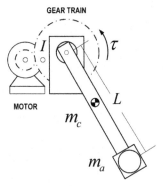

FIGURE 5.15 Schematics of the experimental prototype.

noted by θ. The model of the nonlinear system from the motor side, where all measurements are carried out, is given by

$$\left[(\frac{m_c}{3} + m_a)L^2 + I\right]\ddot{\theta} + (\frac{m_c}{2} + m_a)gL\sin(\theta) = \tau \qquad (5.60)$$

where I includes the reflected moment of inertia through the gears and that of the rotor of the DC motor. The parameter N stands for the gear ratio. The torque τ is related, via $\tau = \frac{Nk_\tau}{R_a}V(t) - k_\tau k_v N^2 \dot{\theta}$, to the voltage applied to the motor, where k_τ is the motor torque constant, and R_a is the armature circuit resistance.

Terms such as viscous friction, Coulomb friction (typically acting through a sign function of the angular velocity), dead zone phenomena present in the gears, air resistance, etc. are not specifically modeled in (5.60). These torques, however, were found to significantly affect the motion of the pendulum. All these perturbation inputs and the position-dependent gravitational nonlinearity of the pendulum are assumed to be unknown and, according to our proposal, considered to be a bounded perturbation input signal, lumped into a single additive time-varying function $\tilde{\xi}(t)$. The simplified model of the pendulum system, in terms of the pendulum angle, is given by

$$\ddot{\theta} = \kappa V(t) + \tilde{\xi}(t), \quad \kappa = \frac{Nk_\tau}{R_a\left[(M + 3m)L^2 + I\right]} \qquad (5.61)$$

The constant control input gain κ is assumed to be known. However, the linear-observer–linear-controller scheme to be used proves to be quite robust with respect to significant mismatches in variations of this control input gain κ (see the next subsection). It is desired to track a given angular position reference trajectory $\theta^*(t)$ by means of the proposed linear observer-based linear controller scheme, in spite of the lack of knowledge of various system state-dependent nonlinearities, whose effects are jointly modeled as an unknown, but bounded, perturbation input signal $\tilde{\xi}(t)$.

The simplified open-loop output reference trajectory tracking error dynamics $e = \theta - \theta^*(t)$ is given by

$$\ddot{e} = \kappa V(t) + \xi(t) \tag{5.62}$$

where $\xi(t)$ comprises the previously described perturbation $\tilde{\xi}(t)$, which also contains the influence of the (unknown) nominal control input $\tau^*(t)$, on the output tracking error dynamics, and some other unforeseen effects of disturbances. Specifically,

$$\xi(t) = \tilde{\xi}(t) - \kappa \tau^*(t) \tag{5.63}$$

A PD controller, with estimated angular velocity tracking error feedback (obtained from the GPI observer), is given by

$$V(t) = \frac{1}{\kappa}\left[-2\zeta_c \omega_{nc}\hat{\dot{e}} - \omega_{nc}^2 e - \hat{\xi}(t)\right] \tag{5.64}$$

with $\hat{\dot{e}} = \hat{e}_2$ and $\hat{\xi}(t) = \hat{z}_1(t)$. These last two variables are online generated by the GPI observer. The observer is given by

$$
\begin{aligned}
\dot{\hat{e}}_1 &= \hat{e}_2 - \lambda_7(e - \hat{e}_1) \\
\dot{\hat{e}}_2 &= \kappa\tau + \hat{z}_1 - \lambda_6(e - \hat{e}_1) \\
\dot{\hat{z}}_1 &= \hat{z}_2 - \lambda_5(e - \hat{e}_1) \\
\dot{\hat{z}}_2 &= \hat{z}_3 - \lambda_4(e - \hat{e}_1) \\
&\vdots \\
\dot{\hat{z}}_6 &= \lambda_0(e - \hat{e}_1)
\end{aligned}
\tag{5.65}
$$

We set $m = 6$ for the polynomial internal model of the unknown disturbance input, $\xi(t)$. The observer reconstruction error of the output tracking error $\tilde{e} = e - \hat{e}_1$ satisfies, approximately, the following linear dynamics:

$$\tilde{e}^{(8)} + \lambda_7\tilde{e}^{(7)} + \cdots + \lambda_1\dot{\tilde{e}} + \lambda_0\tilde{e} \approx 0 \tag{5.66}$$

The design coefficients $\{\lambda_7, \ldots, \lambda_0\}$ were chosen so that the predominantly linear response of the reference tracking error observer reconstruction error $\tilde{e}$ exhibited the following desired characteristic polynomial: $p(s) = (s^2 + 2\zeta_o\omega_{no}s + \omega_{no}^2)^2$ with $\zeta_o, \omega_{no} > 0$.

The prototype of the pendulum system is shown in Fig. 5.16. It has the following nominal parameter values: $m_c = 0.268$ [kg], $m_a = 0.1$ [kg], $L = 0.60$ [m], $I = 0.0103$ [kgm^2], $N = 16$, $R_a = 2.983$ [Ω], $k_\tau = 0.0724$ [Nm/A],

FIGURE 5.16 Experimental prototype of pendulum system.

$k_v = 0.0687$ [Nms/rad]. To be able to generate comparisons of the estimated perturbation with the best possible perturbation identification, we found, via a careful model validation through experimentation, the following disturbance parameters for the viscous friction (B), the Coulomb friction magnitude (F_c), and the dead zone angular range:

$$B = 0.4 \text{ [Nms/rad]}, \quad F_c = 0.3 \text{ [Nm]}, \quad \text{Dead zone} = 0.3 \text{ [rad]}$$

We set the observer design parameters to be $\zeta_o = 3$ and $p_o = 12$. The controller design parameters were specified as $\zeta_c = 1$ and $\omega_{nc} = 2$. A diagram of the experimental platform of pendulum is shown in Fig. 5.17. The incremental encoder used to measure the pendulum angular position is a series CP 350-1000-LD-1/4 encoder of 1000 pulses per revolution. The acquisition card was a Sensoray Model 626 with 6 counters, 4 D/A outputs, and 16 A/D inputs in a 48 digital bidirectional I/O pins. The card acts as an interface for the Matlab-Simulink Real-time Windows Target environment. The sampling time was set to 0.001 [s].

Fig. 5.18 depicts the quality of the controller performance in a trajectory tracking task of a sinusoidal signal for the angular position coordinate θ. The output reference trajectory $\theta^*(t)$ was specified as

$$\theta^*(t) = \frac{\pi}{2} \sin(\omega t), \text{ where } \omega = \frac{2\pi}{5} \text{ [rad/s]} \tag{5.67}$$

The output reference trajectory tracking error, $e(t) = \theta(t) - \theta^*(t)$, is depicted in Fig. 5.19. The tracking error is seen to be uniformly bounded, in steady state, approximately within the interval $[-0.01, 0.01]$ [rad].

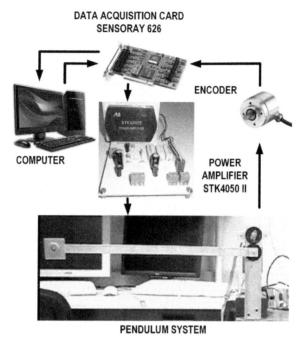

FIGURE 5.17 Experimental control implementation system.

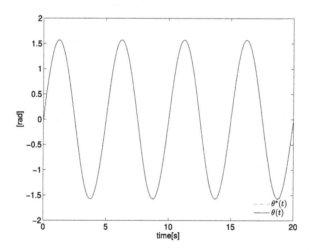

FIGURE 5.18 Reference trajectory tracking performance.

Fig. 5.20 exhibits the applied input voltage as estimated from the DC motor current measurements. Figs. 5.21–5.22 depict, in a step-by-step fashion, the comparison of the observer-estimated disturbance input signal $\hat{\xi}(t)$ and the

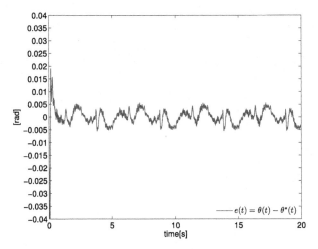

FIGURE 5.19 Performance of GPI observer-based controller in a trajectory tracking task.

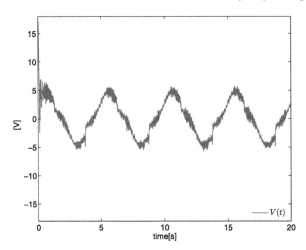

FIGURE 5.20 Control input torque for output reference trajectory tracking task.

identified disturbance input $\tilde{\xi}(t)$ affecting the system motions. The linear GPI observer-based estimation follows quite closely the computed perturbation input $\xi(t)$, based on an experimental identification of the acting perturbation nonlinearities.

5.6   THE THOMSON RING

5.6.1   The Nonlinear Thomson Ring Model

Consider the Thomson jumping ring system, which is shown in Fig. 5.23. The variable y denotes the distance of the ring measured from the end of the core.

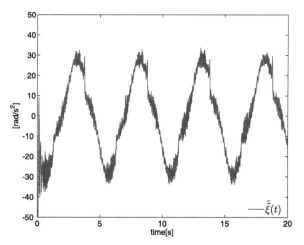

FIGURE 5.21 Online, GPI observer-based estimated nonlinear state-dependent perturbation input signal $\hat{\tilde{\xi}}(t)$.

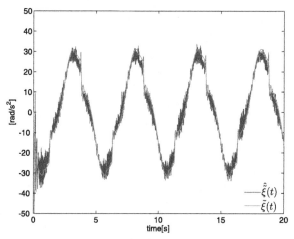

FIGURE 5.22 Gravitational + viscous + Coulomb + "backlash" perturbation input signal $\tilde{\xi}(t)$ compared with experimentally identified perturbation input $\hat{\tilde{\xi}}(t)$.

The other circuit dimensional parameters are indicated in the figure. The ring is free to move up and down in the direction of the core with zero friction between itself and the core made from a solid piece of ferrous metal. The ring is made from a nonmagnetic electrical conductor (aluminium).

A sinusoidal input voltage V_c generates an input current I_c on the coil around the core. As a result, a levitating force is obtained, whose average value may effectively counteract the effect of gravity. It can be demonstrated (see [8] and [9]

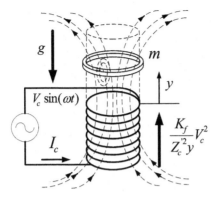

FIGURE 5.23 Schematic diagram of the Thomson ring.

for details) that the dynamics (controlled) of the ring is governed by the following second-order nonlinear differential equation:

$$m\frac{d^2y}{dt} = -mg + \frac{K_f}{Z_c'^2 y}V_c^2 \tag{5.68}$$

where m is the mass of the ring, g is the acceleration of gravity, Z_c' is an impedance magnitude related to the ring, and V_c is the amplitude of a sinusoidally modulated input voltage applied to the circuit, which acts as the control input to the system. The strictly positive parameter K_f exhibits a complex dependence on the magnetic field, the dimensions of the circuit, and the position of the ring from the end of the coil. In this work, its value is taken to be a constant precisely determined at a nominal equilibrium condition. In actual operation, especially under time-varying trajectory tracking conditions, the control input gain of the system exhibits a noticeable variation that cannot be easily measured. The control scheme to be proposed will be remarkably robust with respect to such unknown variations.

5.6.2 Flatness and the Linearized Model of the Thomson Jumping Ring System

System (5.68) is flat since the control input V_c is differentially parameterized by y. Indeed,

$$V_c = \sqrt{\frac{m Z_c'^2 y}{K}(\ddot{y} + g)} \tag{5.69}$$

Relation (5.69) clearly demands that downward accelerated motions beyond "$1G$" are not meaningfully modeled by (5.68). Negative values of the input V_c

produce the same effect as its absolute values. Hence, the assumption $V_c \geq 0$ is inherent to the problem. Note that free fall corresponds to zero applied input voltage V_c. Naturally, the ring velocity dy/dt is a differential function of y. Given a nominal reference trajectory $y^*(t)$ for the flat output, the corresponding nominal input $V_c^*(t)$ is given by

$$V_c^* = \sqrt{\frac{m Z_c'^2 y^*}{K_f} (\ddot{y}^* + g)} \tag{5.70}$$

that is,

$$m \frac{d^2 y^*}{dt} = -mg + \frac{K_f}{Z_c'^2 y^*} [V_c^*(t)]^2 \tag{5.71}$$

A constant equilibrium point for the ring position is obtained from (5.68) as

$$y = \bar{y}, \quad \dot{y} = \ddot{y} = 0, \quad \bar{V}_c = \sqrt{\frac{m Z_c'^2 \bar{y}}{K_f}} \tag{5.72}$$

The tangent linearization of (5.68) around the equilibrium point (5.72) is given by

$$m \ddot{y}_\delta = \left(\frac{2 K_f \bar{V}_c}{Z_c'^2 \bar{y}} \right) V_{c\delta} - \left(\frac{K_f \bar{V}_c^2}{Z_c'^2 \bar{y}^2} \right) y_\delta \tag{5.73}$$

with $y_\delta = y - \bar{y}$ and $V_{c\delta} = V_c - \bar{V}_c$ being the incremental variables of the linearized model. Clearly, $\dot{y}_\delta = \dot{y}$ and $\ddot{y}_\delta = \ddot{y}$.

5.6.3 Problem Formulation

Traditionally, the control objective consists in driving the incremental output y_δ and its time derivative $\dot{y}_\delta$ to zero, so that the ring position is regulated toward its nominal equilibrium value (5.72) and achieves a stable controlled rest condition. Our main departure consists in using the locally valid tangent linearization model (5.73) to design an output feedback control law geared to effectively achieve arbitrary position reference trajectory tracking beyond the region of validity of the incremental model. The problem is formulated as follows:

Given an arbitrary smooth position reference trajectory $y^*(t)$ for the ring position y, device a linear feedback control law so that the incremental variable $y_\delta = y - \bar{y}$, approximately governed by (5.73), tracks the desired incremental reference position $y_\delta^* = y^*(t) - \bar{y}$, irrespectively of the effects of gravity and of the exogenous perturbation inputs possibly affecting the ring position.

5.6.4 The Global Ultralocal Model of the Reference Trajectory Tracking Error Dynamics

From the definitions of the incremental position and of the desired incremental reference position given above, the reference trajectory tracking error $e_{y\delta} = y_\delta - y^*_\delta(t)$ precisely coincides with the tracking error $y - y^*(t)$. Moreover, $\ddot{e}_{y\delta} = \ddot{y}_\delta - \ddot{y}^*_\delta = \ddot{y}_\delta - \ddot{y}^*(t)$. Notice that, according to (5.73) and (5.71), the tracking error $e_{y\delta}$ satisfies

$$\ddot{e}_{y\delta} = \left(\frac{2K_f \bar{V}_c}{m Z'^2_c \bar{y}}\right) V_{c\delta} - \left(\frac{K_f \bar{V}^2_c}{m Z'^2_c \bar{y}^2}\right) y_\delta + \left(g - \frac{K_f}{Z'^2_c y^*}[V^*_c(t)]^2\right) \qquad (5.74)$$

We consider the following linear global ultralocal model of the tracking error dynamics (see Fliess and Join [10] for the use of the local fixed-order version of ultralocal models). In this case, the ultralocal model of the tracking error dynamics retains only the constant linearization input gain and the order of the system, that is,

$$\ddot{e}_{y\delta} = \left(\frac{2K_f \bar{V}_c}{m Z'^2_c \bar{y}}\right) V_{c\delta} + \xi(t) \qquad (5.75)$$

where the time-varying function $\xi(t)$ represents an unstructured version of the state-dependent disturbance input. This quantity is to be treated as an unknown time-varying signal. More precisely,

$$\xi(t) = g - \left(\frac{K_f \bar{V}^2_c}{m Z'^2_c \bar{y}^2}\right) y_\delta(t) - \frac{K_f}{Z'^2_c y^*}[V^*_c(t)]^2 \qquad (5.76)$$

We point out to the fact that the unknown perturbation $\xi(t)$ in the model (5.75) is observable from the measurable position coordinate in the sense of Diop and Fliess [11]. Indeed, from Eq. (5.75) we have

$$\xi(t) = \ddot{e}_{y\delta} - \left(\frac{2K_f \bar{V}_c}{m Z'^2_c \bar{y}}\right) V_{c\delta} = \ddot{y} - \ddot{y}^*(t) - \left(\frac{2K_f \bar{V}_c}{m Z'^2_c \bar{y}}\right) V_{c\delta} \qquad (5.77)$$

In other words, $\xi(t)$ can be written in terms of the output y, a finite number of its time derivatives, the known quantity $y^*(t)$ and the control input $V_{c\delta}$. Let p be a given integer (typically, p is a low-order integer, that is, $3 \le p \le 6$ in an "almost everywhere" sense where needed). It is assumed that the disturbance input $\xi(t)$ and its time derivatives $\xi^{(k)}(t)$ for $k = 1, 2, \ldots, p$ are uniformly absolutely bounded. This condition guarantees the existence of solutions for $e_{y\delta}$ (see Gliklikh [12]) in (5.74) and (5.75).

5.6.5 A GPI Observer-Based ADRC for the Jumping Ring

Let p be a given integer. We have the following result.

Theorem 1. *Let $\{\gamma_1, \gamma_2\}$ and $\{\lambda_{p+1}, \ldots, \lambda_0\}$ be sets of real numbers, chosen so that the polynomials, in the complex variable s,*

$$p_c(s) = s^2 + \gamma_1 s + \gamma_0 \tag{5.78}$$

$$p_o(s) = s^{(p+2)} + \lambda_{(p+1)} s^{(p+1)} + \cdots + \lambda_1 s + \lambda_0 \tag{5.79}$$

are Hurwitz polynomials that exhibit their roots sufficiently far from the imaginary axis in the complex plane. Then, the linear feedback controller

$$V_c = \bar{V}_c - \left(\frac{m Z_c'^2 \bar{y}}{2 K_f \bar{V}_c}\right) \left[\gamma_1 \hat{e}_{2\delta} + \gamma_0 e_\delta + \hat{\xi}(t)\right] \tag{5.80}$$

drives the output reference trajectory tracking error $e_{y\delta}$ and its time derivative $\dot{e}_{y\delta}$ to be ultimately uniformly bounded around a neighborhood of zero, provided that the quantities $\hat{e}_{2y\delta}$ and $\hat{\xi}(t) = \hat{z}_1$ are online generated by the following linear GPI observer:

$$\dot{\hat{e}}_{1y\delta} = \hat{e}_{2y\delta} + \lambda_{(p+1)}(e_{y\delta} - \hat{e}_{1y\delta})$$

$$\dot{\hat{e}}_{2y\delta} = \left(\frac{2 K_f \bar{V}_c}{m Z_c'^2 \bar{y}}\right) V_{c\delta} + \hat{z}_1 + \lambda_p(e_{y\delta} - \hat{e}_{1y\delta})$$

$$\dot{\hat{z}}_1 = \hat{z}_2 + \lambda_5(e_{y\delta} - \hat{e}_{1y\delta})$$

$$\dot{\hat{z}}_2 = \hat{z}_3 + \lambda_{(p-2)}(e_{y\delta} - \hat{e}_{1y\delta}) \tag{5.81}$$

$$\vdots$$

$$\dot{\hat{z}}_{(p-1)} = \hat{z}_p + \lambda_1(e_{y\delta} - \hat{e}_{1y\delta})$$

$$\dot{\hat{z}}_p = \lambda_0(e_{y\delta} - \hat{e}_{1y\delta})$$

The observation error $\tilde{e} = e_{y\delta} - e_{1y\delta}$ and its time derivative $\tilde{e}^{(j)}$, $j = 1, 2, \ldots, p + 1$, ultimately uniformly converge toward a small as desired neighborhood of the origin of the observation error phase space. Moreover, the larger, in absolute value, the negative real part of the dominant roots of $p_o(s)$, the smaller the radius of the bounding ball of such a vicinity. As a consequence, the disturbance estimation error $\xi(t) - \hat{z}_1(t) = \xi(t) - \hat{\xi}(t)$ is ultimately uniformly bounded by a small as desired neighborhood of zero.

Proof. The proof is based on the fact that the estimation error $\tilde{e} = e_{y\delta} - e_{1y\delta}$ satisfies the following perturbed linear differential equation (see [13]):

$$\tilde{e}^{(p+2)} + \lambda_{(p+1)} \tilde{e}^{(p+1)} + \cdots + \lambda_1 \tilde{e} + \lambda_0 = \xi^{(p)}(t) \tag{5.82}$$

Since $\xi^{(p)}(t)$ is assumed to be uniformly absolutely bounded, there exists coefficients λ_k, $k = 0, \ldots, p + 1$, such that $\tilde{e}$ and its time derivatives converge to a small vicinity of the origin, provided that the roots of the associated characteristic polynomial, in the complex variable s,

$$s^{(p+2)} + \lambda_{(p+1)}s^{(p+1)} + \cdots + \lambda_1 s + \lambda_0 \tag{5.83}$$

are all located deep into the left half of the complex plane. The further away from the imaginary axis of the complex plane are these roots located, the smaller the neighborhood of the origin, in the estimation error phase space, where the estimation error $\tilde{e}$ will remain ultimately uniformly bounded. Since $\ddot{\tilde{e}}$ and $\tilde{e}$ are forced to converge to a small as desired vicinity of the origin, the underlying disturbance estimation error $\xi(t) - \hat{z}_1$ is ultimately uniformly absolutely bounded by a small neighborhood of zero. This fact clearly depicts the self-updating character of the implicit time polynomial model, adopted in $\hat{z}_1$, for the lumped disturbance $\xi(t)$. The output tracking error $e_{y\delta} = y - y^{*(t)}$ evolves according to the following linear perturbed dynamics:

$$\ddot{e}_{y\delta} + \gamma_1 \dot{e}_{y\delta} + \gamma_0 e_{y\delta} = \left[\xi(t) - \hat{\xi}(t) \right] \tag{5.84}$$

Since $\xi(t) - \hat{\xi}(t)$ and $\dot{e}_{y\delta}$ are ultimately convergent to a small as desired vicinity of zero, the right hand side of (5.84) is, therefore, uniformly absolutely bounded, and the previous result applies. Choose the controller coefficients $\{\gamma_0, \gamma_1\}$ so that the associated closed-loop characteristic polynomial $p_c(s)$ exhibits its roots moderately far from the imaginary axis in the left half portion of the complex plane. It follows that the tracking error $e_{y\delta}$ and its first-order time derivative are guaranteed to asymptotically converge toward a vicinity of the origin of the tracking error phase space. Their trajectories remain ultimately absolutely uniformly bounded. $\qquad\square$

5.6.6 Experimental Results

Fig. 5.24 shows a diagram of the experimental platform used for the Thomson Jumping Ring. The dimensions of the Thomson ring considered in the prototype are given in Fig. 5.25. The main coil has a primary winding of 1200 turns of AWG copper wire of 0.574 [mm] of diameter. The wire is wrapped around a stainless steel core. The ring to be levitated is an aluminum ring of mass $m = 1.44 \times 10^{-3}$[kg]. The nominal value of the parameter is $K_f = 1.162 \times 10^{-3}$[Nm$\Omega$/V]. The nominal control input gain is 2.243×10^{-4}[m/s^2 V], and $Z'_c = 44.75$ [Ω]. The position of the ring to the top of the coil was measured by

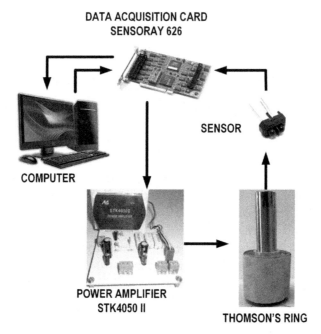

FIGURE 5.24 Experimental system block diagram.

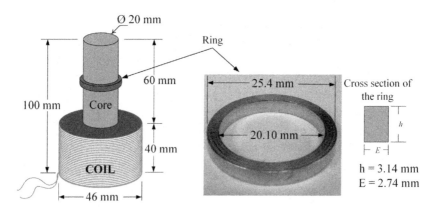

FIGURE 5.25 Prototype main dimensions.

means of an array of infrared sensors, coupled with a noise filtering stage. All data were channeled to a PC by means of a data acquisition board (Sensoray: Model 626). The proposed control strategy was implemented on the PC making use of a Matlab-Simulink platform. The obtained control signals were injected to the prototype (see Fig. 5.26) by considering a previous conditioning stage constituted by a power amplifier (Sanyo: Model STK4050II). The parameters

FIGURE 5.26   Experimental Thomson's ring platform.

associated with the model of the circuit were obtained following the procedures described in [8].

Terms such as friction, air resistance, parasitic effects, etc. are not specifically modeled in (5.68). All these perturbation inputs and the position-depending variation of the gain parameter K and gravitational acceleration force g are present in the plant. The constant control input gain $2K_f \bar{V}_c / m Z_c'^2 \bar{y}$ is assumed to be known with the parameter K calculated on the basis of the nominal equilibrium point in accordance with the formulas in [8] and [9]. The value of the gain was obtained setting the nominal equilibrium height to be $\bar{y} = 20$ [mm] and using the corresponding constant voltage amplitude $\bar{V}_c = 56$ [V]. The following position reference trajectory was implemented in the software of the platform with $\omega = 0.628$ [rad/s]:

$$y^*(t) = 15 + 5\sin(2\omega t)\cos(\omega t) + 2\sin(15\omega t) + 2\cos(20\omega t) \qquad (5.85)$$

The second-order closed-loop dominant characteristic polynomial $p_c(s)$ was chosen to be of the form $p_c(s) = s^2 + 2\zeta_c\omega_{nc}s + \omega_{nc}^2$ with, $w_{nc} = 30$ and $\zeta_c = 3$. The linear GPI observer was implemented by letting $p = 6$, so that the dominant characteristic polynomial for the estimation error was set to be of the form $p_o(s) = (s^2 + 2\zeta_o\omega_{no}s + \omega_{no}^2)^4$ with $w_{no} = 20$ and $\zeta_o = 2$.

The proposed linear-observer linear-controller scheme proves to be quite robust with respect to significant mismatches due to the variations of the control input gain (see [13]) since it can be inferred from the high-quality features of the obtained trajectory tracking. The tracking performance is depicted in Fig. 5.27, where the desired reference trajectory is superimposed to the actual ring position

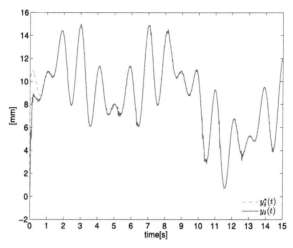

FIGURE 5.27 Ring position reference trajectory tracking with observer-based linear ADRC controller.

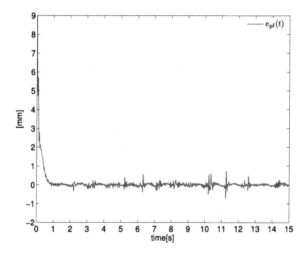

FIGURE 5.28 Reference trajectory tracking error.

trajectory. The corresponding output trajectory tracking error $e_\delta = Y - Y^*(t)$ is shown in Fig. 5.28. As it can be seen, the trajectory tracking error is uniformly absolutely bounded by a small neighbourhood of zero. Fig. 5.29 shows the resulting control voltage where the amplitude of control signal is modulated by a sinusoidal signal with frequency $f = 60$ [Hz]. Fig. 5.30 depicts a zoom of the input voltage showing the sinusoidal character of the carrier signal. Fig. 5.31 shows the GPI observer-based estimated state-dependent time-varying pertur-

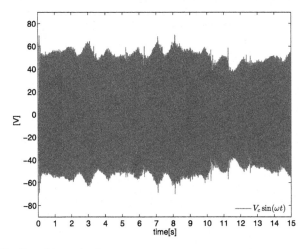

FIGURE 5.29 Control input signal.

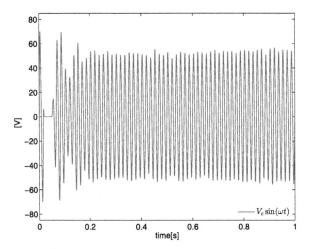

FIGURE 5.30 Zoom of control input signal.

bation input signal. Note that the lumped perturbation estimation (Fig. 5.31) determines the form of the control inputs, which tends to cancel out the additive disturbance inputs.

It should be mentioned that the ring position measurement has a high sensitivity with respect to ambient light, so that they require an initial calibration at each experiment. Also, along the experiments, we can find peaks in the estimated disturbance due to friction between core and ring, which are not included explicitly on the model.

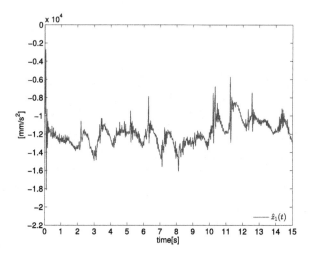

FIGURE 5.31 State-dependent online estimated input signal.

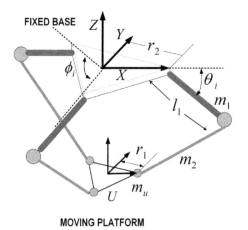

MOVING PLATFORM

FIGURE 5.32 Schematics of the Delta Robot.

5.7 TRAJECTORY TRACKING CONTROL OF A DELTA ROBOT

Fig. 5.32 depicts the Delta Robot, which consists in a three-degree-of-freedom parallel manipulator. The dynamical model is given as follows [14]:

$$D(\Theta)\ddot{\Theta} + G(\Theta) + W(\Theta, U)\lambda = \tau \qquad (5.86)$$

where $\Theta = \begin{bmatrix} \theta_1, & \theta_2, & \theta_3 \end{bmatrix}^T$ is the actuated articular position vector, $U = \begin{bmatrix} u_x, & u_y, & u_z \end{bmatrix}^T$ represents the Cartesian positions, $\lambda = \begin{bmatrix} \lambda_1, & \lambda_2, & \lambda_3 \end{bmatrix}^T$ denotes the Lagrange multipliers, $\tau = \begin{bmatrix} \tau_1, & \tau_2, & \tau_3 \end{bmatrix}^T$ is the vector of the external torques and forces, $D(\Theta)$ denotes the generalized inertia matrix, $G(\Theta)$ contains the gravitational terms, where g is the gravitational acceleration term, and $W(\Theta, U)$ is the restriction vector, which indicates the coupling restrictions between the mobile platform, the parallel links, and the Delta Robot arms, whose length is given by l_1, l_2; r_1 and r_2 are the distances from the geometric center of the platforms to any of their vertex. In this case, the coupling dynamics is computed by the Lagrange multipliers. $D(\Theta), G(\Theta)$, and $W(\Theta, U)$ are given as follows:

$$D(\Theta) = \begin{bmatrix} (\frac{1}{3}m_1 + m_2)l_1^2 & 0 & 0 \\ 0 & (\frac{1}{3}m_1 + m_2)l_1^2 & 0 \\ 0 & 0 & (\frac{1}{3}m_1 + m_2)l_1^2 \end{bmatrix} \quad (5.87)$$

$$G(\Theta) = \begin{bmatrix} (\frac{1}{2}m_1 + m_2)gl_1 \cos(\theta_1) \\ (\frac{1}{2}m_1 + m_2)gl_1 \cos(\theta_2) \\ (\frac{1}{2}m_1 + m_2)gl_1 \cos(\theta_3) \end{bmatrix} \quad (5.88)$$

$$W(\Theta, U) = (-1)$$
$$\cdot \begin{bmatrix} 2l_1 \left[(u_x \cos(\phi_1) + u_y \sin(\phi_1) + r_2 - r_1) \sin(\theta_1) - u_z \cos(\theta_1) \right] \\ 2l_1 \left[(u_x \cos(\phi_2) + u_y \sin(\phi_2) + r_2 - r_1) \sin(\theta_2) - u_z \cos(\theta_2) \right] \\ 2l_1 \left[(u_x \cos(\phi_3) + u_y \sin(\phi_3) + r_2 - r_1) \sin(\theta_3) - u_z \cos(\theta_3) \right] \end{bmatrix} \quad (5.89)$$

The terms λ_i, $i = 1, 2, 3$, are obtained by the following system of equations:

$$2 \sum_{i=1}^{3} \lambda_i (u_x + r_2 \cos(\phi_i) - r_1 \cos(\phi_i) - l_1 \cos(\phi_i) \cos(\theta_i))$$
$$= (m_u + 3m_2)\ddot{u}_x - f_{ux}$$

$$2 \sum_{i=1}^{3} \lambda_i (u_y + r_2 \sin(\phi_i) - r_1 \sin(\phi_i) - l_1 \sin(\phi_i) \cos(\theta_i)) \quad (5.90)$$
$$= (m_u + 3m_2)\ddot{u}_y - f_{uy}$$

$$2 \sum_{i=1}^{3} \lambda_i (u_z - l_1 \sin(\theta_i)) = (m_u + 3m_2)\ddot{u}_z + (m_u + 3m_2)g - f_{uz}$$

where f_{ux}, f_{uy}, and f_{uz} are the external forces for the mobile platform in the x, y, and z axes, m_1 is the input link mass, m_2 denotes the parallel bars link mass, and m_u is the mobile platform mass.

5.7.1 GPI Control Design

Consider a vector of desired articular positions $\Theta^* = \begin{bmatrix} \theta_1^* & \theta_2^* & \theta_3^* \end{bmatrix}$. Multiplying (5.86) by $D(\Theta)^{-1}$, we have

$$\ddot{\Theta} = D^{-1}(\Theta)\tau - M^{-1}(\Theta)[G(\Theta) + W(\Theta, U)\lambda] \qquad (5.91)$$

Let $\tilde{\xi}$ denote the uncertain terms.

$$\tilde{\xi} = -D^{-1}(\Theta)[G(\Theta) + W(\Theta, U)\lambda] \qquad (5.92)$$

On the other hand, consider that each actuator is affected by the train gear disturbances represented by additive disturbance external inputs $\eta(t)$. The lumped disturbance input vector is given by

$$\xi(t) = \tilde{\xi} + \eta(t) \qquad (5.93)$$

Thus, the dynamics governing the disturbed system is viewed, for observer design purposes, as

$$\ddot{\Theta} = D^{-1}(\Theta)\tau + \xi(t) \qquad (5.94)$$

where the inverse of $D(\Theta)$ is obtained as

$$D^{-1}(\Theta) = \begin{bmatrix} \frac{1}{(\frac{1}{3}m_1+m_2)l_1^2} & 0 & 0 \\ 0 & \frac{1}{(\frac{1}{3}m_1+m_2)l_1^2} & 0 \\ 0 & 0 & \frac{1}{(\frac{1}{3}m_1+m_2)l_1^2} \end{bmatrix} \qquad (5.95)$$

The relation between the motor torque and the input voltage is computed as

$$\tau = \begin{bmatrix} (k_\tau N/R_a)V_{\theta 1}(t) \\ (K_\tau N/R_a)V_{\theta 2}(t) \\ (K_\tau N/R_a)V_{\theta 3}(t) \end{bmatrix} \qquad (5.96)$$

where k_τ is the torque constant, N is the gear relation, R_a is the armature resistance, and $V_{\theta i}$ is the voltage input. It is assumed that all the motors have identical features. Using Eqs. (5.95) and (5.96) in (5.94) leads to

$$\ddot{\Theta} = \frac{1}{(\frac{1}{3}m_1 + m_2)l_1^2} \begin{bmatrix} 1 & 0 & 0 \\ 0 & 1 & 0 \\ 0 & 0 & 1 \end{bmatrix} \begin{bmatrix} (k_\tau N/R_a)V_{\theta 1}(t) \\ (k_\tau N/R_a)V_{\theta 2}(t) \\ (k_\tau N/R_a)V_{\theta 3}(t) \end{bmatrix} + \xi(t) \qquad (5.97)$$

Eq. (5.97) consists of a set of three decoupled perturbed linear systems. It is, then possible to define three independent control inputs of the form (5.98)

to solve the robust trajectory tracking problem for the three angular coordinates
$e_{\theta i} = \theta_i(t) - \theta_i^*(t)$, $i = 1, 2, 3$.

It is assumed that $\xi(t)$ is unknown but uniformly absolutely bounded. Thus, it is necessary to estimate the disturbance input in an online fashion by means of an observer. We then cancel the actual disturbance in the feedback control law.

Let $\theta_{0i} = \int_0^t \theta_i(\tau)d\tau$, $i = 1, 2, 3$. We propose the following extended-order GPI observer-based control [15]:

$$V_i = \frac{R_a(\frac{1}{3}m_1 + m_2)l_1^2}{k_\tau N} \left[\ddot{\theta}_i^*(t) - \kappa_0(\hat{\theta}_{1i} - \theta_i^*) - \kappa_1(\hat{\theta}_{2i} - \dot{\theta}_i^*) - \hat{\xi}_i(t) \right]$$

$$\dot{\hat{\theta}}_{0i} = \hat{\theta}_{1i} + \lambda_{7\theta i}(\theta_{0i} - \hat{\theta}_{0i})$$

$$\dot{\hat{\theta}}_{1i} = \hat{\theta}_{2i} + \lambda_{6\theta i}(\theta_{0i} - \hat{\theta}_{0i})$$

$$\dot{\hat{\theta}}_{2i} = \frac{k_\tau N}{R_a(\frac{1}{3}m_1 + m_2)l_1^2} V_i \hat{z}_{1\theta i} + \lambda_{5\theta i}(\theta_{0i} - \hat{\theta}_{0i})$$

$$\dot{\hat{z}}_{1\theta i} = \hat{z}_{2\theta i} + \lambda_{4\theta i}(\theta_{0i} - \hat{\theta}_{0i})$$

$$\dot{\hat{z}}_{2\theta i} = \hat{z}_{3\theta i} + \lambda_{3\theta i}(\theta_{0i} - \hat{\theta}_{0i})$$

$$\dot{\hat{z}}_{3\theta i} = \hat{z}_{4\theta i} + \lambda_{2\theta i}(\theta_{0i} - \hat{\theta}_{0i})$$

$$\dot{\hat{z}}_{4\theta i} = \hat{z}_{5\theta i} + \lambda_{1\theta i}(\theta_{0i} - \hat{\theta}_{0i})$$

$$\dot{\hat{z}}_{5\theta i} = \lambda_{0\theta i}(\theta_{0i} - \hat{\theta}_{0i})$$

$$\dot{\hat{\xi}}_i = z_{1\theta i} \tag{5.98}$$

The linear dominant part of each injection error dynamics is defined by the following characteristic polynomials expressed in terms of the Laplace operator s:

$$p_{oi}(s) = s^8 + \lambda_{7\theta i}s^7 + \lambda_{6\theta i}s^6 + \lambda_{5\theta i}s^5 + \lambda_{4\theta i}s^4 + \lambda_{3\theta i}s^3$$
$$+ \lambda_{2\theta i}s^2 + \lambda_{1\theta i}s + \lambda_{0\theta i} \tag{5.99}$$

The observer gain parameters $\lambda_{j\theta i}$ for $i = 1, 2, 3$ and $j = 1, 2, \ldots, 7$ are chosen so that each characteristic polynomial of the injection dominant dynamics has its roots in the left half of complex plane, sufficiently far from the imaginary axis. To achieve the last objective, we propose Hurwitz polynomials of the form

$$p_{oi}(s) = (s^2 + 2\zeta_{\theta i}\omega_{\theta i}s + \omega_{\theta i}^2)^4 \tag{5.100}$$

as dominant characteristic polynomials of the closed-loop dynamics. Finally, by choosing the controller gain parameters κ_{0i}, κ_{1i} such that the associated characteristic polynomial of the closed-loop system is of the form

$$s^2 + 2\zeta_{ci}\omega_{nci}s + \omega_{nci}^2 \tag{5.101}$$

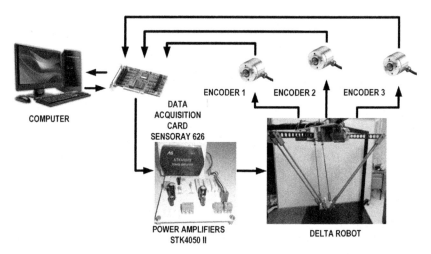

FIGURE 5.33 Block diagram of delta robot controller.

the tracking error and its time derivatives are then commanded to exponentially converge to a small vicinity of the origin of the tracking error phase space.

5.7.2 Experimental Results

Fig. 5.33 shows a schematic diagram of the experimental platform. The position of each arm was obtained by means of incremental encoders of 2000 counts per revolution to measure the angular position of each gear box shaft. The position data were sent to the main controller by means of a data acquisition board Sensoray Model 626. The control strategy was implemented in the Matlab-Simulink platform, and the devised control signals were transferred to the actuators through three power amplifiers Sanyo: Model STK4050II. The delta robot actuators were three dc geared motors NISCA: Model NC5475. The motor parameters are then following: A torque constant $k_\tau = 0.0724$ [Nm/A], the armature resistance is $R_a = 2.983$ [Ω], and the gear relation is $N = 16$. Finally, the sampling time was set to be 0.001 [s].

The delta robot system parameters are $l_1 = 0.25$ [m], $l_2 = 0.50$ [m], $r_2 = 0.045$ [m], $r_1 = 0.1$ [m], $m_1 = 0.168$ [kg], $m_2 = 0.3$ [kg], and $m_u = 0.215$ [kg] respectively. To illustrate the tracking results, a reference trajectory in the Cartesian space x–y–z was proposed. In this case, the trajectory consisted in a circle centered at the origin of the x–y plane, with radius 200 [mm]. The initial conditions, at $t = 0$ [s], for the delta robot configuration were set as $x^*(0) = 0$ [mm], $y^*(0) = 0$ [mm], and $z^*(0) = -396.2$ [mm]; then, at $t = 3$ [s], the end effector of the robot was moved to the position $x^* = 200$ [mm], $y^* = 0$ [mm], $z^* = -396.2$ [mm]. Then, at $t = 4.5$ [s], the reference trajectory is set to be

FIGURE 5.34 Delta robot system prototype.

$x^*(t) = 200 \sin(\frac{2\pi}{5}(t-4.5))$ [mm], $y^*(t) = 200 \cos(\frac{2\pi}{5}(t-4.5))$ [mm], $z^*(t) = -396.2$ [mm]. This circle is traced in 5 seconds, and, finally, at $t = 9.5$ [s], the robot turns out to its initial configuration. The inverse kinematics was used to find the desired joint angles θ_1^*, θ_2^*, and θ_3^* [14]. The initial conditions for the joint variables in the robot were $[\theta_1(0) = 0$, $\theta_2(0) = 0$, and $\theta_3(0) = 0$. The observer gain parameters were set to be $\zeta_{o1} = \zeta_{o2} = \zeta_{o3} = 5$, $\omega_{o1} = \omega_{o3} = 20$, $\omega_{o2} = 22$. The controller design parameters were specified to be $\zeta_{c1} = \zeta_{c2} = 2$, $\zeta_{c3} = 2.5$, $\omega_{nc1} = 14$, $\omega_{nc2} = \omega_{nc3} = 15$.

Fig. 5.34 shows the experimental platform delta robot, and the quality of the trajectory tracking performance in the x–y–z plane, obtained on the basis of the proposed linear output feedback observer–controller scheme, is depicted in Fig. 5.35, whereas Fig. 5.36 shows the tracking results in each actuated angular coordinate. The tracking error is detailed in Fig. 5.37, where it is shown that the angular error is restricted to a vicinity of radius 0.02 [rad] approximately, or 5 [mm] in the Cartesian space for the proposed design parameters. The produced control voltages are provided in Fig. 5.38. Notice that the lumped disturbance estimations (Fig. 5.39) determine the form of the control inputs, which tends to cancel out the additive disturbance inputs.

5.8 A TIME-DELAYED FLYWHEEL SYSTEM

5.8.1 Smith Predictor GPI Controller

Consider the following disturbed flywheel system with input time delay:

$$\ddot{q}(t) = M^{-1}(q)u(t - T) + \psi(t, q, \dot{q}) \tag{5.102}$$

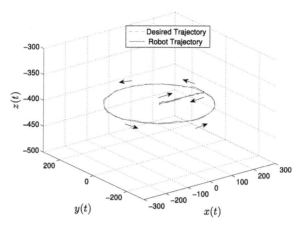

FIGURE 5.35 x, y, z-directions reference trajectory tracking with observer-based linear controller.

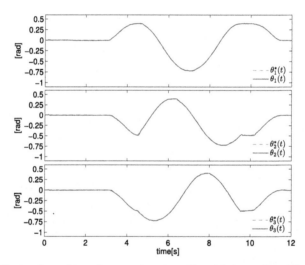

FIGURE 5.36 Angular position reference trajectory tracking with observer-based linear controller.

where q, $\dot{q}$ are the states of the system, $M(q)$ is a positive definite symmetric matrix, u is the control input, $\psi(\cdot)$ denotes a lumped disturbance input, and T is a time delay. From (5.102), the "forward system" is defined as follows:

$$\ddot{q}_f(t) = M^{-1}(q)_f u(t) + \overline{\psi}(t + T, q_f, \dot{q}_f) \qquad (5.103)$$

where $\overline{\psi}(t + T, q_f, \dot{q}_f)$ is the *predicted disturbance input* to be reproduced, in an approximate form, using the estimated states of the original system. Taking

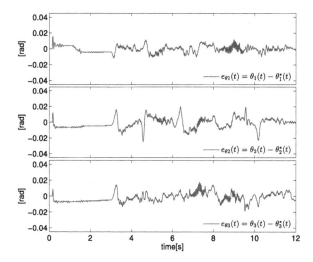

FIGURE 5.37 Output reference trajectory tracking error evolution.

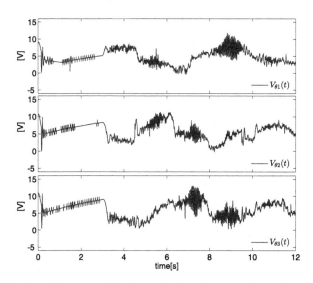

FIGURE 5.38 Control input torque for output reference trajectory tracking task.

a Taylor series expansion, we obtain the following disturbance input predictor:

$$\overline{\psi}(t+T) = \overline{\psi}(t) + \dot{\overline{\psi}}(t)T + \frac{1}{2!}\ddot{\overline{\psi}}(t)T^2 + \cdots \qquad (5.104)$$

Using the input disturbance estimator and a truncated version of (5.104), we obtain an approximate disturbance input predictor as follows:

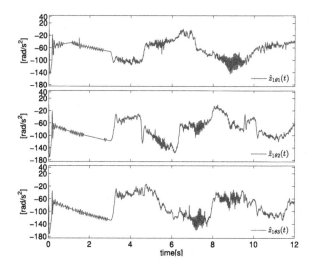

FIGURE 5.39 Disturbance estimate translation dynamics.

$$\widehat{\overline{\psi}}(t+T) = \hat{z}_1(t) + \hat{z}_2(t)T + \ldots + \frac{1}{(p-1)!}\hat{z}_p T^{p-1} \qquad (5.105)$$

As it occurs in polynomial series approximation, increasing the value of p allows a better approximation; however, the numerical complexity of the observer increases.

We have the following control law for the forward system using the disturbance predictor estimation:

$$u(t) = M(q)\Big[-\widehat{\overline{\psi}}(t+T) + \ddot{q}^*(t+T)$$
$$- \big(\kappa_1[\dot{e}_f(t) + \dot{e}_{rr}(t)] + \kappa_0[e_f(t) + e_s(t)]\big)\Big] \qquad (5.106)$$
$$e_f(t) = \hat{q}(t) - q_f^*(t)$$
$$e_s(t) = q(t) - q_f(t-T)$$

where κ_1, κ_0 are diagonal positive definite matrices with the control gains of a classical multivariable PD controller such that the linear dominant dynamics is Hurwitz; $\hat{q}^{(j)}(t)$, $j = 0, 1$, are supplied by the GPI observer, and $q_f^{(j)}(t)$ are, through algebraic manipulations, available for measurement. The terms $e_f^{(j)}$ are introduced to handle possible errors in the disturbance prediction as a part of the Smith Predictor methodology. These terms use the difference between the plant output and the time-delayed forward output to compensate possible differences between the delayed system and the delayed forward model.

Fig. 5.40 shows a schematic of the control design.

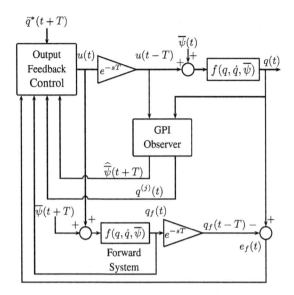

FIGURE 5.40 Control scheme for ADRC-Smith Predictor.

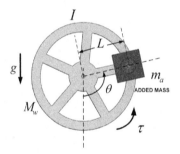

FIGURE 5.41 Schematics of flywheel system.

5.8.2 Control of a Perturbed Flywheel with Delayed Input

Consider a perturbed flywheel system (see Fig. 5.41) with delayed input. The flywheel consists of an aluminium wheel of radius L, mass M_w, and an extra perturbation mass m_a attached on the circumference of the wheel. The system is actuated by a DC motor, and the angular displacement of the rotor shaft is denoted by ϕ. The model of the nonlinear system is given by

$$(I_m + I + m_a L^2)\ddot{\phi} + m_a g L \sin \phi = \tau(t - T) \qquad (5.107)$$

where T is the known fixed time delay, I is the inertia of the flywheel, and I_m is the inertia of motor shaft-rotor. Terms such as the viscous friction, Coulomb friction (typically acting through a *sign* function of the angular velocity), air resistance, etc. are not specifically modeled in (5.107). These torques, however, were found to significantly act on the motion of the flywheel. All these disturbances are assumed to be unknown, and they are given by the disturbance term $\eta(t, \phi, \dot{\phi})$, which, together with the position-dependent gravitational nonlinearity of the extra mass on flywheel, are, according to our proposal, considered to form a bounded generalized disturbance input signal, lumped into a single additive time-varying function $\tilde{\psi}(t)$.

$$\tilde{\psi}(t) = (1/(I_m + I + m_a L^2)) (m_a g L \sin \theta + \eta(t)) \tag{5.108}$$

The simplified model of the flywheel system, in terms of the flywheel angle, is given by

$$\ddot{\phi}(t) = (1/(I_m + I + m_a L^2)) \tau(t - T) + \tilde{\psi}(t) \tag{5.109}$$

From the DC motor dynamics, the relation between the torque and the voltage input is given by

$$\tau(t - T) = \frac{K_\tau N}{R_a} V(t - T) - K_\tau K_v N^2 \dot{\theta} \tag{5.110}$$

where R_a [Ω] is the armature resistance, K_τ [N $\cdot$ m/A] denotes the torque constant, and K_v [N $\cdot$ m $\cdot$ s/rad] is the back electromotive force constant. Using (5.110) in (5.109), we get

$$\ddot{\phi}(t) = M^{-1} V(t - T) + \overline{\psi}(t) \tag{5.111}$$

where

$$M^{-1} = \frac{K_\tau}{R_a (I_m + I + m_a L^2)}, \quad \overline{\psi}(t) = \tilde{\psi} - K_\tau K_v N^2 \dot{\phi} \tag{5.112}$$

According to the problem formulation, given a reference angular position trajectory, say $\theta^*(t)$, it is desired to devise a Smith Predictor-based GPI observer-based active disturbance rejection controller to obtain a robust tracking of the reference function in spite of the unmodeled dynamics, nonlinear effects, and external disturbances, all of them lumped in the disturbance input $\overline{\psi}$.

We propose the following GPI observer with forward input and approximation parameter $p = 6$:

$$\dot{\widehat{\phi}}_0 = \hat{\phi}_1 + \lambda_8 (\phi_0 - \hat{\phi}_0)$$
$$\dot{\widehat{\phi}}_1 = \hat{\phi}_2 + \lambda_7 (\phi_0 - \hat{\phi}_0)$$
$$\dot{\widehat{\phi}}_2 = M^{-1} V(t - T) + \hat{z}_1 + \lambda_6 (\phi_0 - \hat{\phi}_0)$$
$$\dot{\widehat{z}}_1 = \hat{z}_2 + \lambda_5 (\phi_0 - \hat{\phi}_0)$$

$$\vdots$$

$$\hat{\hat{z}}_5 = \hat{z}_6 + \lambda_1(\phi_0 - \hat{\phi}_0)$$
$$\hat{\hat{z}}_6 = \lambda_0(\phi_0 - \hat{\phi}_0)$$
$$\theta_0 = \int \phi(t)dt$$
$$\hat{\hat{\psi}} = \hat{z}_1 \tag{5.113}$$

which leads to a characteristic polynomial for the integral error in the linear dominant part:

$$P(s) = s^9 + \lambda_8 s^8 + \lambda_7 s^7 + \ldots + \lambda_2 s^2 + \lambda_1 s + \lambda_0 \tag{5.114}$$

The observer gain parameters λ_j for $j = 0, 1, 2, \cdots, 8$ are chosen using the following procedure. Consider a desired characteristic polynomial $p(s)$ of the form

$$a_n s^n + a_{n-1} s^{n-1} + \cdots + a_2 s^2 + a_1 s + a_0, \qquad a_i > 0 \tag{5.115}$$

and let α_i be the characteristic ratios of $p(s)$. It is said that $p(s)$ is Hurwitz when the following two conditions hold:

A) $\alpha_1 > 2$;

B) $\alpha_k = \dfrac{\sin\left(\frac{k\pi}{n}\right) + \sin\left(\frac{\pi}{n}\right)}{2\sin\left(\frac{k\pi}{n}\right)} \alpha_1$

for $k = 2, 3, \ldots, n - 1$. The construction of the all-pole stable characteristic polynomial involves only α_1, which we require to be larger than 2. Thus, this result allows us to characterize the reference all-pole systems by adjusting a single parameter α_1 to achieve the desired damping. Since the *generalized time constant* can be chosen independently of α_i, the coefficients of $p(s)$ are calculated using the procedure given in [16]: For arbitrary a_0 and $\tau > 0$,

$$a_1 = \tau a_0$$

$$a_i = \frac{\tau^i a_0}{\alpha_{i-1} \alpha_{i-2}^2 \alpha_{i-3}^3 \cdots \alpha_1^{i-1}}$$

for $i = 2, 3, \ldots, n$

$$\lambda_j = \left(\frac{a_j}{a_n}\right)$$

for $j = 0, 1, , 3, \ldots, 8$

which helps to find a fast tracking response avoiding overshooting effects.

The GPI observer provides an online approximation of the disturbance input $\overline{\psi}$ by means of the observer variable $\hat{z}_1$. Besides, the observer also obtains a finite number of time derivatives of the disturbance input given by $\hat{z}_2, \ldots, \hat{z}_6$. Using the Taylor series expansion, the prediction of the disturbance input may be calculated as follows:

$$\widehat{\overline{\psi}}(t + T) \approx \hat{z}_1(t) + \hat{z}_2(t)T + \frac{1}{2!}\hat{z}_3(t)T^2 + \frac{1}{3!}\hat{z}_4(t)T^3$$
$$+ \frac{1}{4!}\hat{z}_5(t)T^4 + \frac{1}{5!}\hat{z}_4(t)T^5 \tag{5.116}$$

Using the Smith predictor methodology, the forward system with prediction period T for the perturbed linear system (5.111) is

$$\ddot{\phi}_f = M^{-1}V(t) + \hat{z}_1(t + T) \tag{5.117}$$

The trajectory tracking error of the forward plant is defined as follows:

$$e_{\phi f} = \phi_f(t) - \phi^*(t + T) \tag{5.118}$$

and the error between the delayed forward system and the actual plant is defined as

$$e_{\phi s} = \phi(t) - \phi_f(t - T) \tag{5.119}$$

where $e_{\phi s}$ represents the discrepancies between the forward system and the actual system, to be used as a compensation term. We propose a PD with a disturbance compensation term controller for the forward system

$$V(t) = -M\left[\hat{z}_1(t + T) + 2\zeta_c \omega_{nc}(\dot{e}_{\phi s} + \dot{e}_{\phi f}) + \omega_{nc}^2(e_{\phi s} + e_{\phi f}) - \ddot{\phi}_f^*(t)\right] \tag{5.120}$$

with $\zeta_c, \omega_{nc} > 0$.

5.8.3 Experimental Results

Some experiments were carried out on the test bed system to show the effectiveness of the control strategy. Fig. 5.42 shows a schematic of the control system connection. The angular position of the motor shaft was obtained by means of an incremental encoder of 1000 counts per revolution. The position data was sent to the main controller by means of a data acquisition card Sensoray Model 626. The controller was implemented in the Matlab-Simulink platform, and the control signals were transferred to the actuator through a power amplifier Sanyo: Model STK4050II. The actuator consisted in a DC motor NISCA:

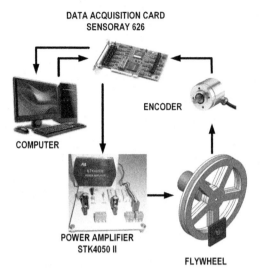

FIGURE 5.42 Flywheel system prototype.

FIGURE 5.43 Experimental system block diagram.

Model NC5475. The motor parameters were the following: The motor inertia $I_m = 4.0101 \times 10^{-5}$ [kgm^2], a torque constant $K_\tau = 0.0724$ [Nm/A], the armature resistance was $R_a = 2.983$ [Ω], and the back electromotive force constant, $K_v = 0.0687$ [Nms/rad], respectively. The inertia of the flywheel was $I = 9.611 \times 10^{-3}$ [kgm^2], the flywheel mass $M_w = 1.315$ [kg], and the added mass $m_a = 0.1$ [kg] (see Fig. 5.43). The flywheel radius parameter was $L = 0.1$ [m], and the input delay time T was chosen to be 0.07 [s]. Finally, the sampling time was set to be 0.0005 [s].

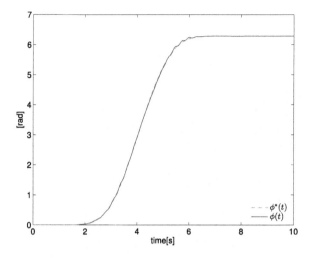

FIGURE 5.44 Tracking behavior position reference trajectory.

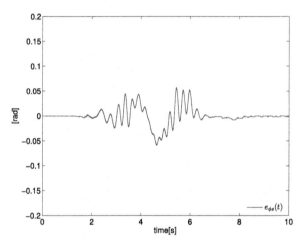

FIGURE 5.45 Reference trajectory tracking error.

The initial conditions for the system were $\phi(0) = 0$. The observer gain parameters were set as follows: $n = 9$, $\tau = 6$, $a_0 = 15^{10}$, $\alpha_1 = 3.5$. The controller design parameters were specified to be $\zeta_c = 3$ and $\omega_{nc} = 5$.

The tracking results using the Smith predictor-based control in a rest-to-rest path are depicted in Fig. 5.44, where the presented results, for a constant time delay T, were satisfactory since the error signal remained bounded as stated in the problem approach, with an absolute peak error less than 0.2 [rad] (see Fig. 5.45). Fig. 5.46 shows the control inputs for the tracking process, and, finally, Fig. 5.47 depicts the disturbance input estimation and its prediction.

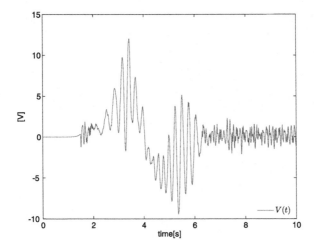

FIGURE 5.46 Voltage control input.

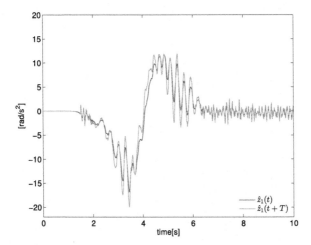

FIGURE 5.47 Disturbance input predictor.

5.9 CONTROL OF ROBOT MANIPULATORS WITH DELAYED INPUTS

Motivated by the previous example, consider the following fully actuated model of a controlled manipulator system, including a delayed control input:

$$M(\mathbf{q})\ddot{\mathbf{q}} + C(\mathbf{q}, \dot{\mathbf{q}})\dot{\mathbf{q}} + \mathbf{g}(\mathbf{q}) = \tau(t - T) \qquad (5.121)$$

with $\mathbf{q} \in \mathbb{R}^n$.

We consider, as usual, a simplified system with time-varying (state-dependent) control input gain,

$$\ddot{\mathbf{q}} = M^{-1}(\mathbf{q})\tau(t - T) + \tilde{\varphi}(t) \qquad (5.122)$$

with $\tilde{\varphi}(t)$ a sufficiently smooth unknown but bounded signal devoid of high-frequency components.

It is desired to have the output vector $\mathbf{q}$ of generalized joints positions track a given desired trajectory $\mathbf{q}_d(t)$. Let $\tau_d(t - T)$ be the delayed version of the nominal control input $\tau_d(t)$. This signal, which we do not need to synthesize in our approach, may be nominally computed off-line as follows:

$$\tau_d(t) = M(\mathbf{q}_d(t + T))\ddot{\mathbf{q}}_d(t + T) \qquad (5.123)$$

where we do not complement the computation with the "nominal predicted perturbation input," given that it represents an unknown quantity. Thus, our nominal control input $\tau_d(t)$ and its delayed version $\tau_d(t - T)$ will always be in significant error with respect to its possible actual value. We henceforth deal with the tracking error directly and avoid explicitly involving such a nominal control input $\tau_d(t)$ and the "nominal perturbation input" $\tilde{\varphi}_d(t)$. We proceed as follows: We consider the following perturbed tracking error system dynamics with $\mathbf{e} = \mathbf{q} - \mathbf{q}_d(t)$:

$$
\begin{aligned}
\ddot{\mathbf{q}} - \ddot{\mathbf{q}}_d(t) &= M^{-1}(\mathbf{q})\tau(t - T) + \tilde{\varphi}(t) \\
&\quad - M^{-1}(\mathbf{q}_d(t))\tau_d(t - T) - \tilde{\varphi}_d(t) \qquad (5.124) \\
&= M^{-1}(\mathbf{q})\tau(t - T) + \varphi(t)
\end{aligned}
$$

i.e.,

$$\ddot{\mathbf{e}} = M^{-1}(\mathbf{q})\tau(t - T) + \varphi(t) \qquad (5.125)$$

where now

$$\varphi(t) = \tilde{\varphi}(t) - M^{-1}(\mathbf{q}_d(t))\tau_d(t - T) + \tilde{\varphi}(t) - \tilde{\varphi}_d(t) \qquad (5.126)$$

The Smith predictor methodology postulates a *forward system* representing an updated version of the plant and conformed, in our perturbed case, as follows:

$$\ddot{\mathbf{e}}_f = M^{-1}(\mathbf{q}_f)\tau(t) + \varphi(t + T) \qquad (5.127)$$

To conform this forward system, we must synthesize a *prediction estimate* $\hat{\varphi}(t + T)$ of the unknown advanced perturbation input $\varphi(t + T)$. For this, we use a GPI observer on the delayed tracking error system itself.

Let $\mathbf{e}_1 = \mathbf{e}$, $\mathbf{e}_2 = \dot{\mathbf{e}}$. We have that the system dynamics evolves according to

$$\dot{\mathbf{e}}_1 = \mathbf{e}_2 \tag{5.128}$$
$$\dot{\mathbf{e}}_2 = M^{-1}(\mathbf{q}_1)\tau(t - T) + \varphi(t)$$

Note that $\mathbf{e}_f(t) = \mathbf{q}_f(t) - \mathbf{q}_d(t + T)$ and $\ddot{\mathbf{q}}_f = M^{-1}(\mathbf{q}_f)\tau(t) + \tilde{\varphi}(t + T)$ in accordance with the predictive nature of the forward system.

Suppose that we model the uncertain perturbation input $\varphi(t)$ as an n-dimensional vector of $(r - 1)$th-degree arbitrary time polynomial inputs so that

$$\frac{d^r}{dt^r}\varphi(t) \approx 0 \tag{5.129}$$

or, else, such an rth time derivative of its entries represent completely negligible quantities. Letting $\mathbf{z}_1 = \varphi(t)$, we have the following local, but self-updating, dynamical model of the perturbation input

$$\frac{d^r}{dt^r}\mathbf{z}_1 = \frac{d^r}{dt^r}\varphi(t) \approx 0 \tag{5.130}$$

We use the following approximate model for the perturbed delayed system:

$$\begin{aligned} \dot{\mathbf{e}}_1 &= \mathbf{e}_2 \\ \dot{\mathbf{e}}_2 &= M^{-1}(\mathbf{q}_1)\tau(t - T) + \mathbf{z}_1 \\ \dot{\mathbf{z}}_1 &= \mathbf{z}_2 \\ &\vdots \\ \dot{\mathbf{z}}_{r-1} &= \mathbf{z}_r \\ \dot{\mathbf{z}}_r &= \mathbf{0} \end{aligned} \tag{5.131}$$

A GPI observer for the delayed perturbed system is readily proposed as

$$\begin{aligned} \dot{\hat{\mathbf{e}}}_1 &= \hat{\mathbf{e}}_2 + \Lambda_{r+1}(\mathbf{e}_1 - \hat{\mathbf{e}}_1) \\ \dot{\hat{\mathbf{e}}}_2 &= M^{-1}(\mathbf{q}_1)\tau(t - T) + \mathbf{z}_1 + \Lambda_r(\mathbf{e}_1 - \hat{\mathbf{e}}_1) \\ \dot{\hat{\mathbf{z}}}_1 &= \hat{\mathbf{z}}_2 + \Lambda_{r-1}(\mathbf{e}_1 - \hat{\mathbf{e}}_1) \\ \dot{\hat{\mathbf{z}}}_2 &= \hat{\mathbf{z}}_3 + \Lambda_{r-2}(\mathbf{e}_1 - \hat{\mathbf{e}}_1) \\ &\vdots \\ \dot{\hat{\mathbf{z}}}_{r-1} &= \hat{\mathbf{z}}_r + \Lambda_1(\mathbf{e}_1 - \hat{\mathbf{e}}_1) \\ \dot{\hat{\mathbf{z}}}_r &= \Lambda_0(\mathbf{e}_1 - \hat{\mathbf{e}}_1) \end{aligned} \tag{5.132}$$

Let $\tilde{\mathbf{e}}_i = \mathbf{e}_i - \hat{\mathbf{e}}_i$, $i = 1, 2$, and $\tilde{z}_j = \mathbf{z}_j - \hat{\mathbf{z}}_j$, $j = 1, 2, \ldots, r$. The reconstruction error dynamics is governed by

$$
\begin{aligned}
\dot{\tilde{\mathbf{e}}}_1 &= \tilde{\mathbf{e}}_2 - \Lambda_{r+1}\tilde{\mathbf{e}}_1 \\
\dot{\tilde{\mathbf{e}}}_2 &= \tilde{\mathbf{z}}_1 - \Lambda_r\tilde{\mathbf{e}}_1 \\
\dot{\tilde{\mathbf{z}}}_1 &= \tilde{\mathbf{z}}_2 - \Lambda_{r-1}\tilde{\mathbf{e}}_1 \\
&\;\;\vdots \\
\dot{\tilde{\mathbf{z}}}_{r-1} &= \tilde{\mathbf{z}}_r - \Lambda_1\tilde{\mathbf{e}}_1 \\
\dot{\tilde{\mathbf{z}}}_r &= -\Lambda_0\tilde{\mathbf{e}}_1
\end{aligned}
\tag{5.133}
$$

i.e., we have that the reconstruction error dynamics satisfies

$$
\tilde{\mathbf{e}}_1^{(r+2)} + \Lambda_{r+1}\tilde{\mathbf{e}}_1^{(r+1)} + \cdots + \Lambda_1\dot{\tilde{\mathbf{e}}}_1 + \Lambda_0\tilde{\mathbf{e}}_1 \approx \mathbf{0} \tag{5.134}
$$

Large values in the observer gains contribute to make the effect of the neglected right side of the perturbed reconstruction error dynamics be negligible and, thus, have the reconstruction error approximately behave as the following linear dynamics:

$$
\tilde{\mathbf{e}}_1^{(r+2)} + \Lambda_{r+1}\tilde{\mathbf{e}}_1^{(r+1)} + \cdots + \Lambda_1\dot{\tilde{\mathbf{e}}}_1 + \Lambda_0\tilde{\mathbf{e}}_1 = \mathbf{0} \tag{5.135}
$$

which we take for valid.

In fact, the reconstruction error dynamics evolves according to

$$
\tilde{\mathbf{e}}_1^{(r+2)} + \Lambda_{r+1}\tilde{\mathbf{e}}_1^{(r+1)} + \cdots + \Lambda_1\dot{\tilde{\mathbf{e}}}_1 + \Lambda_0\tilde{\mathbf{e}}_1 = \varphi^{(r)}(t) \tag{5.136}
$$

Our assumption $\varphi^{(r)}(t) \approx \mathbf{0}$ further justifies our procedure.

As a consequence, choosing the entries of the (preferably diagonal positive) gain matrices Λ_j, $j = 1, \ldots, r+2$, so that the closed-loop eigenvalues of the reconstruction error dynamics lie deep in the left half of the complex plane, the reconstruction error $\mathbf{e}_1$ and its various time derivatives rapidly converge toward a small neighborhood of zero. In turn, the variables $\mathbf{z}_j$, $j = 1, \ldots, r$, approximately converge toward the time derivatives of the perturbation input $\varphi(t)$ in the following way:

$$
\mathbf{z}_j \to \varphi^{(j-1)}(t), \quad j = 1, 2, \ldots \tag{5.137}
$$

Using the Taylor series development formula for the prediction of a vector time signal $\varphi(t)$, we may propose the following perturbation input predictor:

$$
\varphi(t + T) = \varphi(t) + \dot{\varphi}(t)T + \frac{1}{2!}\ddot{\varphi}(t)T^2 + \cdots \tag{5.138}
$$

Then consider the natural approximation to the predicted perturbation signal $\varphi(t * t)$ with available time derivatives at time t of the signal $\varphi(t)$ as the following Taylor polynomial in the time interval value T:

$$\hat{\varphi}(t + T) = \mathbf{z}_1 + \mathbf{z}_2 T + \frac{1}{2!}\mathbf{z}_3 T^2 + \cdots + \frac{1}{[\frac{r}{2}]!}\mathbf{z}_{[\frac{r}{2}+1]}T^{[\frac{r}{2}]} \qquad (5.139)$$

where we have taken only the integer part of $\frac{r}{2}$, denoted here by $[\frac{r}{2}]$, as the most reasonable degree for the approximating Taylor polynomial of the predicted value of the vector of signals $\varphi(t)$.

The forward system is, hence, completely determined, although in an approximate manner, by

$$\ddot{\mathbf{e}}_f = M^{-1}(\mathbf{q_f})\tau(t) + \hat{\varphi}(t + T) \qquad (5.140)$$

According to the Smith predictor methodology, we may propose a variety of simple linear controllers that either take or, alternatively, do not take into account the actual perturbed plant output vector $\mathbf{q}$. One such possibility, which completely overrides the perturbed plant, is the following proportional derivative controller:

$$\tau(t) = M(\mathbf{q}_f)\left[-\hat{\varphi}(t + T) - K_1\dot{\mathbf{e}}_f - K_0\mathbf{e}_f\right] \qquad (5.141)$$

with K_1 and K_0 being suitable positive definite symmetric matrices (preferably diagonal matrices) that provoke an asymptotically exponentially stable character to the origin of the tracking error $\mathbf{e}$ as a solution of the homogeneous system

$$\ddot{\mathbf{e}}_f + K_1\dot{\mathbf{e}}_f + K_0\mathbf{e}_f = \mathbf{0} \qquad (5.142)$$

Naturally, the synthesized control input $\tau(t)$ is shared by both the forward plant and the controlled perturbed delayed robot manipulator dynamics.

The second possibility is represented by considering the synthesis of the error between the plant output $\mathbf{q}(t)$ and the delayed output of the forward system $\mathbf{q}_f(t - T)$. Then let

$$\mathbf{err}(t) = \mathbf{q}(t) - \mathbf{q}_f(t - T) \qquad (5.143)$$

This error, depicting discrepancies between the actual plant output and the delayed output of the forward system, is used as an additional perturbation input to the feedback forward system output. The controller is thus synthesized as

$$\tau(t) = -\hat{\varphi}(t + T) - K_1\frac{d}{dt}\left(\mathbf{e}_f + \mathbf{err}(t)\right) - K_0\left(\mathbf{e}_f + \mathbf{err}(t)\right) \qquad (5.144)$$

Since this error is defined as $\mathbf{err}(t) = \mathbf{q}(t) - \mathbf{q}_f(t - T)$, its time derivative may be computed using suitable signals from the previously proposed observer and signals readily available from the forward plant emulation, i.e.,

$$\frac{d}{dt}(\mathbf{err}(t)) = \hat{\mathbf{q}}_2(t) - \dot{\mathbf{q}}_f(t - T) = \hat{\mathbf{e}}_2 + \dot{\mathbf{q}}_d(t) - \dot{\mathbf{q}}_f(t - T) \qquad (5.145)$$

If the forward system is perfect, then the scheme coincides with the traditional plant overriding Smith predictor controller; otherwise, modeling errors in the forward system can be suitably compensated via this scheme.

Further interesting modifications of the Smith predictor scheme, in particular, those using the Mikusinsky transform, proposed by Marquez, Fliess, and Mounier [17] can also be considered with great advantages and simplicity.

5.9.1 Time-Delayed Delta Robot

Consider the Delta-type robot [14], which consists of a three-degree-of-freedom (DOF) parallel robot. Fig. 5.32 depicts the Delta Robot. The dynamic model with delayed input is given as follows:

$$M(\Theta)\ddot{\Theta}(t) + G(\Theta, t) - R(\Theta, P(t))\lambda = \tau(t - T) \qquad (5.146)$$

where we suppose that $G(\Theta)$, $R(\Theta, P)$ are unmodeled dynamics and that each actuator is affected by the train gear disturbances ($\eta(t) = [\eta_1\ \eta_2\ \eta_3]^T$). Lumping the last terms leads to the following disturbance vector:

$$\psi(t) = \frac{1}{(\frac{1}{3}m_a + m_b)a^2}[-G(\Theta) + R(\Theta, P)] + \eta(t) \qquad (5.147)$$

Thus, the dynamics governing the tracking error of the simplified disturbed system is

$$\ddot{\Theta} = M^{-1}(\Theta)\tau(t - T) + \psi(t) \qquad (5.148)$$

The relation between the motor torque and the voltage input is

$$\tau(t - T) = \begin{bmatrix} (K_1 N/R_a)V_1(t - T) \\ (K_1 N/R_a)V_2(t - T) \\ (K_1 N/R_a)V_3(t - T) \end{bmatrix} \qquad (5.149)$$

Using (5.149) and (5.148), we have:

$$\ddot{\Theta} = \frac{1}{(\frac{1}{3}m_a + m_b)a^2}\begin{bmatrix} 1 & 0 & 0 \\ 0 & 1 & 0 \\ 0 & 0 & 1 \end{bmatrix}\begin{bmatrix} (K_1 N/R_a)V_1(t - T) \\ (K_1 N/R_a)V_2(t - T) \\ (K_1 N/R_a)V_3(t - T) \end{bmatrix} + \psi(t) \quad (5.150)$$

Eq. (5.150) consists of a set of three decoupled disturbed systems of the form (5.122). Then it is possible to define three independent control inputs of the form (5.144) to solve the robust trajectory tracking problem.

According to the procedure proposed for robot manipulators with delayed inputs, the forward disturbed system is proposed as

$$
\ddot{\Theta}_f = \frac{1}{(\frac{1}{3}m_a + m_b)a^2}
\begin{bmatrix} 1 & 0 & 0 \\ 0 & 1 & 0 \\ 0 & 0 & 1 \end{bmatrix}
\begin{bmatrix} (K_1 N/R_a)V_1(t) \\ (K_1 N/R_a)V_2(t) \\ (K_1 N/R_a)V_3(t) \end{bmatrix}
+ \psi(t+T) \quad (5.151)
$$

where $\psi(t+T)$ is the vector of disturbed input signals to be estimated in an online fashion using the truncated power series expansions in combination with the GPI observers. It is assumed that ψ is unknown but uniformly absolutely bounded.

Let us define $\hat{e}_{10}$, $\hat{e}_{20}$, and $\hat{e}_{\theta 30}$ the estimates of the integral tracking errors $e_{10} = \int_0^t (\theta_1(\tau) - \theta_1^*(\tau))d\tau$, $e_{20} = \int_0^t (\theta_2(\tau) - \theta_2^*(\tau))d\tau$, and $e_{30} = \int_0^t (\theta_3(\tau) - \theta_3^*(\tau))d\tau$. Now, consider the estimates $\hat{e}_{11}$, $\hat{e}_{21}$, and $\hat{e}_{\theta 31}$ of the tracking errors $e_{11} = \theta_1(t) - \theta_1^*(t)$, $e_{\theta 21} = \theta_2(t) - \theta_2^*(t)$, and $e_{31} = \theta_3(t) - \theta_3^*(t)$. In a similar fashion, let $\hat{e}_{12}$, $\hat{e}_{22}$, and $\hat{e}_{\theta 32}$ be the estimates of the velocity tracking error states given by $e_{12} = \dot{\theta}_1(t) - \dot{\theta}_1^*(t)$, $e_{\theta 22} = \dot{\theta}_2(t) - \dot{\theta}_2^*(t)$, and $e_{32} = \dot{\theta}_3(t) - \dot{\theta}_3^*(t)$, respectively.

The reconstruction errors associated with the tracking errors are defined as follows: $\tilde{e}_{10} = e_{10} - \widehat{e}_{10}$, $\tilde{e}_{20} = e_{20} - \widehat{e}_{20}$, $\tilde{e}_{30} = e_{30} - \widehat{e}_{30}$, $\tilde{e}_{11} = e_{11} - \widehat{e}_{11}$, $\tilde{e}_{21} = e_{21} - \widehat{e}_{21}$, and $\tilde{e}_{31} = e_{31} - \widehat{e}_{31}$.

Thus, the set of GPI observers for the reference tracking errors in the input delayed Delta robot are proposed as

$$
\begin{aligned}
\dot{\hat{e}}_{j0} &= \hat{e}_{j1} + \lambda_{(p+2)j}\tilde{e}_{j0} \\
\dot{\hat{e}}_{j1} &= \hat{e}_{j2} + \lambda_{(p+1)j}\tilde{e}_{j0} \\
\dot{\hat{e}}_{j2} &= (K_j N/R_a)V_j(t-T) + \hat{z}_{1j} + \lambda_{pj}\tilde{e}_{j0} \qquad (5.152) \\
\dot{\hat{z}}_{1j} &= \hat{z}_{2j} + \lambda_{(p-1)j}\tilde{e}_{j0} \\
\dot{\hat{z}}_{2j} &= \hat{z}_{3j} + \lambda_{(p-2)j}\tilde{e}_{j0} \\
&\vdots \\
\dot{\hat{z}}_{pj} &= \lambda_{0j}\tilde{e}_{j0} \\
j &= 1, 2, 3
\end{aligned}
$$

Consider the approximation parameter $p = 5$. The linear dominant part of the each injection error dynamics is defined by the following characteristic polynomials expressed in terms of the Laplace operator s:

$$p_{oj}(s) = s^8 + \lambda_{7\theta i} s^7 + \lambda_{6\theta i} s^6 + \lambda_{5\theta i} s^5 + \lambda_{4\theta i} s^4$$
$$+ \lambda_{3\theta i} s^3 + \lambda_{2\theta i} s^2 + \lambda_{1\theta i} s + \lambda_{0\theta i} \qquad (5.153)$$

The observer gain parameters $\lambda_{j\theta i}$ for $i = 1, 2, 3$ and $j = 1, 2 \ldots, 7$ are chosen in such a way that each characteristic polynomial of the injection dominant dynamics has its roots in the left half of complex plane, sufficiently far of the imaginary axis. To achieve the last objective, Hurwitz polynomials of the form:

$$p_{oi}(s) = (s^2 + 2\zeta_{\theta i}\omega_{\theta i} s + \omega_{\theta i}^2)^4 \qquad (5.154)$$

are proposed as dominant characteristic polynomials of the closed-loop dynamics.

Using the truncated Taylor series expansion to predict the lumped disturbance functions $\psi_j(t + T)$, we propose the following estimator:

$$\hat{\psi}_j(t + T) = \hat{z}_{1j} + \hat{z}_{2j} T + \hat{z}_{3j} \frac{T^2}{2!} + \hat{z}_{4j} \frac{T^3}{3!} + \hat{z}_{5j} \frac{T^4}{4!} \qquad (5.155)$$

Now, let us define the errors associated with the prediction process and the Smith predictor control design:

$$e_{rr} = \begin{bmatrix} \theta_1 - \theta_{1f}(t - T) \\ \theta_2 - \theta_{2f}(t - T) \\ \theta_3 - \theta_{3f}(t - T) \end{bmatrix} e_f = \begin{bmatrix} \theta_{1f}(t) - \theta_1^*(t + T) \\ \theta_{2f}(t) - \theta_2^*(t + T) \\ \theta_{3f}(t) - \theta_3^*(t + T) \end{bmatrix} \qquad (5.156)$$

Finally, the output feedback controller is given by

$$\tau_j(t) = \frac{R_a(\frac{1}{3}m_a + m_b)a^2}{K_j N} \left[-\hat{\psi}_j(t + T) - \kappa_{1j}\frac{d}{dt}(e_{fj} + e_{rrj})\kappa_{0j}(e_{fj} + e_{rrj}) \right] \qquad (5.157)$$

The controller includes a compensation for the disturbance prediction functions $\psi_j(t + T)$, $j = 1, 2, 3$. The cancelation is carried out through the disturbance observer extended states and the use of the tracking velocities $\hat{\theta}_{1j}$. The controller gain parameters κ_{0i}, κ_{1i} are chosen such that the associated dominant characteristic polynomials of the closed-loop systems

$$p_j(s) = s^2 + \kappa_{1j} s + \kappa_{0j} \qquad (5.158)$$

locate their roots deep into the left half of the complex plane. In particular, to emulate stable responses of second-order systems, we can propose a location of

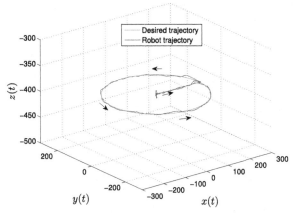

FIGURE 5.48 x, y, z-directions reference trajectory tracking with observer-based linear controller.

the form

$$s^2 + 2\zeta_{cj}\omega_{ncj}s + \omega_{ncj}^2 \tag{5.159}$$

with $\zeta_{cj}, \omega_{ncj} > 0$.

5.9.2 Experimental Results

The controller was applied to achieve a reference trajectory in the Cartesian space x–y–z. To start the main trajectory, the robot tracked a line between the initial point $(0, 0, -400)$ [mm] and the point $(200, 0, -400)$ [mm] in the x–y–z space. Then, the trajectory was a circle centered at the origin of the x–y plane with radius 200 [mm] for $z = 400$ [mm]. The inverse kinematics was used to find the joint angles θ_{11}, θ_{12}, and θ_{13}. The initial conditions for the joint variables in the robot were $\theta_{11}(0) = 0$, $\theta_{12}(0) = 0$, and $\theta_{13}(0) = 0$. The observer gain parameters were set to be

$$\begin{bmatrix} \zeta_{o1} \\ \zeta_{o2} \\ \zeta_{o3} \end{bmatrix} = \begin{bmatrix} 5 \\ 5 \\ 5 \end{bmatrix} \qquad \begin{bmatrix} \omega_{o1} \\ \omega_{o2} \\ \omega_{o3} \end{bmatrix} = \begin{bmatrix} 20 \\ 22 \\ 20 \end{bmatrix}$$

The controller design parameters were specified to be

$$\begin{bmatrix} \zeta_{c1} \\ \zeta_{c2} \\ \zeta_{c3} \end{bmatrix} = \begin{bmatrix} 2 \\ 2 \\ 2 \end{bmatrix} \qquad \begin{bmatrix} \omega_{nc1} \\ \omega_{nc2} \\ \omega_{nc3} \end{bmatrix} = \begin{bmatrix} 13 \\ 13 \\ 13 \end{bmatrix}$$

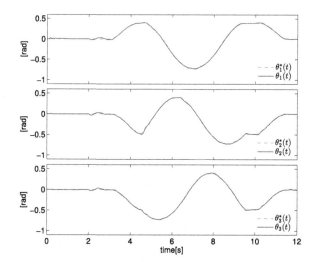

FIGURE 5.49 Tracking behavior for the actuated joints.

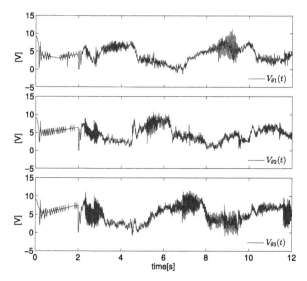

FIGURE 5.50 Voltage control inputs.

Fig. 5.25 shows the experimental delta robot test bed. The tracking results in the $x-y-z$ space, obtained by the proposed Smith predictor-based GPI output feedback controller, are shown in Fig. 5.48. Fig. 5.49 depicts the behavior of the controller in each actuated joint. Fig. 5.50 shows the control inputs (in voltage) for the tracking process. The last results show that the approximation is acceptable in spite of the input time delay, and, finally, Figs. 5.51 and 5.52 depict the disturbance input predictors.

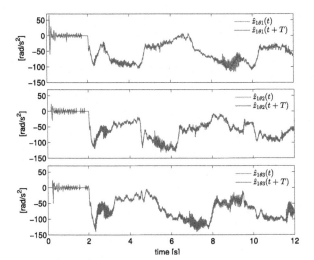

FIGURE 5.51   Disturbance input predictor.

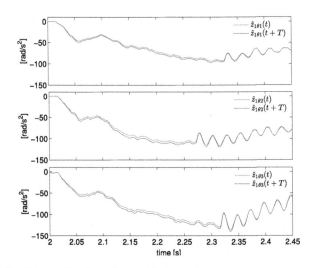

FIGURE 5.52   Zoom on disturbance input predictor.

REFERENCES

[1] H. Sira-Ramírez, C. López-Uribe, M. Velasco-Villa, Linear observer-based active disturbance rejection control of the omnidirectional mobile robot, Asian Journal of Control 15 (1) (2013) 51–63.

[2] M. Ramirez-Neria, J.L. Garcia-Antonio, H. Sira-Ramirez, M. Velasco-Villa, R. Castro-Linares, An active disturbance rejection control of leader–follower Thomson's jumping rings, Control Theory & Applications 30 (12) (2013) 1564–1572.

[3] J.A. Vázquez-Santacruz, M. Velasco-Villa, Path-tracking dynamic model based control of an omnidirectional mobile robot, Proceedings of the IFAC World Conference 41 (2) (2008) 5365–5370.

[4] M. Velasco-Villa, H. Rodríguez-Cortés, H. Sira-Ramirez, I. Estrada-Sanchez, J.A. Vázquez, Dynamic trajectory-tracking control of an omnidirectional mobile robot based on passive approach, in: Advances in Robot Manipulators, INTECH, 2010, pp. 299–314.

[5] J. Chiasson, Modeling and High Performance Control of Electric Machines, IEEE Press Series on Power Engineering, vol. 26, John Wiley & Sons, 2005.

[6] H. Sira-Ramírez, J. Linares-Flores, A. Luviano-Juarez, J. Cortés-Romero, Ultramodelos Globales y el Control por Rechazo Activo de Perturbaciones en Sistemas No lineales Diferencialmente Planos, Revista Iberoamericana de Automática e Informática Industrial RIAI 12 (2) (2015) 133–144.

[7] J. Linares-Flores, C. García-Rodríguez, H. Sira-Ramírez, O.D. Ramírez-Cárdenas, Robust backstepping tracking controller for low-speed PMSM positioning system: design, analysis, and implementation, IEEE Transactions on Industrial Informatics 11 (5) (2015) 1130–1141.

[8] N. Barry, R. Casey, Elihu Thomson's jumping ring in a levited closed-loop control experiment, IEEE Transactions on Education 42 (1) (1999) 72–80.

[9] J.L. García-Antonio, R. Castro-Linares, M. Velasco-Villa, Synchronization control of a magnetic levitation system powered by alternating current, in: 9th IEEE International Conference on Control & Automation, Santiago, Chile, 2011, pp. 794–799.

[10] M. Fliess, C. Join, Model-free control, International Journal of Control 86 (12) (2013) 2228–2252.

[11] S. Diop, M. Fliess, Nonlinear observability, identifiability, and persistent trajectories, in: Proceedings of the 30th IEEE Conference on Decision and Control, IEEE, 1991, pp. 714–719.

[12] Y.E. Gliklikh, Necessary and sufficient conditions for global-in-time existence of solutions of ordinary, stochastic and parabolic differential equations, Abstract and Applied Analysis 6 (1) (2006) 1–17.

[13] H. Sira-Ramírez, M. Ramírez-Neria, A. Rodríguez-Angeles, On the linear control of nonlinear mechanical systems, in: 49th IEEE Conference on Decision and Control, Atlanta, USA, 2010, pp. 1999–2004.

[14] L.W. Tsai, Robot Analysis: The Mechanics of Serial and Parallel Manipulators, Wiley–Interscience, 1999.

[15] D.L. Martinez-Vazquez, A. Rodriguez-Angeles, H. Sira-Ramirez, Robust GPI observer under noisy measurements, in: 6th International Conference on Electrical Engineering, Computing Science and Automatic Control, 2009.

[16] Y.C. Kim, L.H. Keel, S.P. Bhattacharyya, Transient response control via characteristic ratio assignment, IEEE Transactions on Automatic Control 48 (1) (2003) 2238–2244.

[17] R. Marquez, M. Fliess, H. Mounier, A non-conventional robust PI-controller for the Smith predictor, in: Decision and Control, 2001. Proceedings of the 40th IEEE Conference on, vol. 3, IEEE, 2001, pp. 2259–2260.

Chapter 6

The Challenging Case
of Underactuated Systems

Chapter Points

- Difficult to control under-actuated, nonflat systems, which exhibit a controllable linearization around a certain equilibrium point are considered to illustrate the effectiveness of the ADRC schemes synthesized on the basis of a tangent linearization of the plant around a given equilibrium point. The control tasks entitle, however, arbitrary output reference trajectory tracking, taking the system state away from the equilibrium point thus overcoming a traditional obstacle to linearization-based control of nonlinear systems.

- Trajectory planning is carried out for the incremental flat output variables, demanding significant excursions from the initial equilibrium point. The observer-based ADRC scheme estimates the neglected nonlinearities, excited by the features of the planned reference trajectory, and cancels them forcing the controlled plant dynamics to behave as a linear system around the prescribed trajectory.

- ADRC in any of its two forms, extended observer based or flat filter control, will efficiently handle the neglected nonlinearities of the linearized system in the trajectory tracking task while taking the system operation far from the vicinity of the initial equilibrium point where the linearized model is assumed to be valid.

- An important property of controllable tangent linearization models of underactuated systems is that even-order time derivatives of the incremental flat outputs are functions of the measurable incremental position variables alone. Similarly, odd-time derivatives of the flat output are functions of the velocity variables alone. This property is exploited to substantially simplify the observer-based ADRC scheme design, thus obtaining a robust controlled performance that only requires position measurements, linear controllers, and rather unrestricted reference trajectories.

6.1 INTRODUCTION

In this chapter, we explore the relevance of ADRC schemes in a particular but highly visible class of nondifferentially flat systems, namely, the class of underactuated nonflat systems with controllable tangent linearizations (see [1–3]). The lack of a flat output in the full nonlinear coordinates of the system intro-

Active Disturbance Rejection Control of Dynamic Systems. http://dx.doi.org/10.1016/B978-0-12-849868-2.00006-X
Copyright © 2017 Elsevier Inc. All rights reserved.   **235**

duces noncontrollable *defect* variables and makes it impossible to plan trajectories without solving differential equations [4,5]. Tangent linearization of such systems around an equilibrium point turns out to be usually controllable (i.e., flat), thus allowing us to formulate the trajectory tracking or stabilization problem in a direct manner via a suitable planning of the incremental flat output variables. ADRC will force the controlled system to act as a linear system along the prescribed trajectory. Observer-based ADRC will provide an estimate of the neglected nonlinear terms, which can be directly canceled by turning the control problem into a linear problem, regardless of the excursions demanded from the equilibrium point. GPI control will dually attenuate the nonlinear disturbances in a radical manner while controlling the remaining linear dynamics (see Appendix B).

Underactuated systems have been the object of sustained research efforts for several decades now (see, e.g., [6], [7], [8], [1], [9], [10], [11], [12], among others). Interesting contributions have been given from the geometric viewpoint described in configuration spaces and corresponding Lagrangian formulations (see the authoritative contributions [13], [2], [14]). From the flatness viewpoint, several interesting schemes have been proposed entitling: exploiting the Liouvillian character of the nonflat systems exhibiting an integrable defect (see [15], [16], [17], [4]) and high-frequency control inducing flatness in an average sense. The area of robust control of perturbed nonflat systems has been very little explored ([18], [19], [20]). Here, we illustrate how ADRC schemes in combination with flatness are especially suitable for successfully controlling underactuated nonflat systems with controllable linearizations while being possibly subject to unforeseen exogenous disturbances, to unmodeled dynamics, and to the effects of ignored nonlinearities due to the underlying tangent linearization scheme.

In this chapter, we present several examples of underactuated nonflat systems whose tangent linearization model is controllable. The flat output of the linearized model is shown to systematically exhibit a remarkable property that enormously facilitates the extended observer and linear feedback controller design for the linearized input-to-flat output dynamics. Even-order time derivatives of the flat output can be invariably expressed in terms of linear combinations of the measurable position variables alone (see Appendix A). Even-order time derivatives of the flat output become, before hitting the control inputs, functions of the measurable generalized position variables alone. On the other hand, odd-order time derivatives of the flat output are, in turn, functions of the original incremental generalized velocities alone. These facts make it unnecessary to devote efforts to estimate, via ESO, all the phase variables of the linearized input–output system. In fact, only incremental velocities, or systems momenta, are required to complete the feedback loop. Thus, in observer-based ADRC

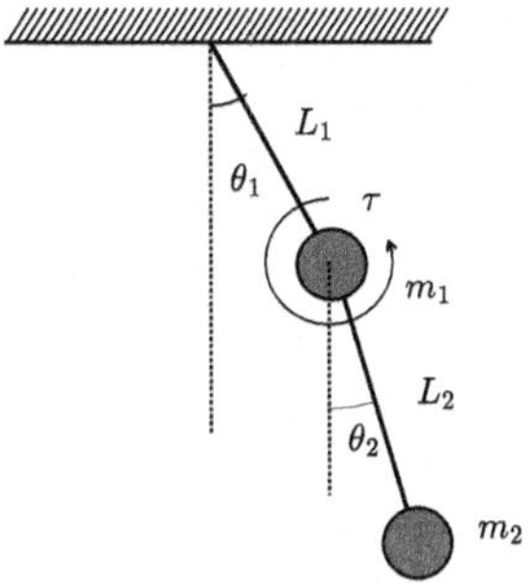

FIGURE 6.1   The Acrobot.

schemes, the observer design task is decomposed into several low-order observers that take linear combinations of position variables as inputs and produce
estimates of specific velocities in the system. As it is well known by now, in the
case of flat-filtering-based control, only output variables need to be measured.
However, flat filtering can also take advantage of the natural cascaded structure
of the controllable linearized underactuated systems.

Some illustrative design examples presented in this chapter are evaluated
using digital computer simulations. Experimental results are also included regarding ADRC control of challenging underactuated systems such as the Furuta
pendulum and the ball-and-beam system.

6.2   CONTROLLING THE PENDUBOT VIA TANGENT LINEARIZATION

The pendubot and its counterpart, the acrobot, constitute underactuated systems
that are not feedback linearizable and, hence, nondifferentially flat (for further
information; see [11], [21], [22], [23], and references therein). These systems
are easy to build but quite difficult to control. In this section, we propose a
straightforward output feedback controller based on the incremental flat output
of the linearized model of the system. Linearization is carried out around a natural equilibrium point of the system. We proceed to stabilize the system around
such an equilibrium point using an observer-based ADRC controller.

Fig. 6.1 shows a schematic diagram of the acrobot. The main assumptions
are:

1. The masses are concentrated at the end of the links;
2. The links have negligible mass, and all the moments of inertia are due to the
   bob of the penduli;
3. The masses and the link lengths are different.

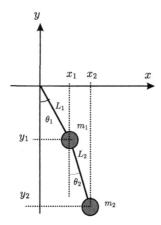

FIGURE 6.2   Generalized coordinates for the acrobot system.

6.2.1   Derivation of the Model via Euler–Lagrange Equations

The two masses need four coordinates to be located on the plane. However, there are two restrictions regarding the fixed distances from the mass centers to their respective pivots. The configuration space must then have dimension 2. The two shown angles θ_1, θ_2 constitute a possible choice for the generalized coordinates describing the system.

The Cartesian coordinate locations of the masses m_1 and m_2 are given by (see Fig. 6.2)

$$x_1 = L_1 \sin\theta_1, \quad y_1 = -L_1 \cos\theta_1$$
$$x_2 = L_1 \sin\theta_1 + L_2 \sin\theta_2, \quad y_2 = -L_1 \cos\theta_1 - L_2 \cos\theta_2 \tag{6.1}$$

The corresponding velocities are then given by

$$\dot{x}_1 = L_1 \dot{\theta}_1 \cos\theta_1, \quad \dot{y}_1 = L_1 \dot{\theta}_1 \sin\theta_1$$
$$\dot{x}_2 = L_1 \dot{\theta}_1 \cos\theta_1 + L_2 \dot{\theta}_2 \cos\theta_2, \quad \dot{y}_2 = L_1 \dot{\theta}_1 \sin\theta_1 + L_2 \dot{\theta}_2 \sin\theta_2 \tag{6.2}$$

The kinetic energies associated with the two masses are, respectively,

$$K_1 = \frac{1}{2}m_1 \left(\dot{x}_1^2 + \dot{y}_1^2 \right) = \frac{1}{2}m_1 L_1^2 \dot{\theta}_1^2$$
$$K_2 = \frac{1}{2}m_2 \left(\dot{x}_2^2 + \dot{y}_2^2 \right) = \frac{1}{2}m_2 \left[L_1^2 \dot{\theta}_1^2 + L_2^2 \dot{\theta}_2^2 + 2L_1 L_2 \dot{\theta}_1 \dot{\theta}_2 \cos(\theta_1 - \theta_2) \right]$$
$$\tag{6.3}$$

The potential energies, including the external control action τ, are obtained as

$$U_1 = mgL_1(1 - \cos\theta_1)$$
$$U_2 = m_2 g \left[L_1 + L_2 - L_1 \cos\theta_1 - L_2 \cos\theta_2 \right] - \tau\theta_2 \tag{6.4}$$

The system Lagrangian eliminating constant terms is then expressed as

$$\mathcal{L} = K_1 + K_2 - (U_1 + U_2)$$
$$= \frac{1}{2}(m_1 + m_2)L_1^2\dot{\theta}_1^2 + \frac{1}{2}m_2L_2^2\dot{\theta}_2 + m_2L_1L_2\dot{\theta}_1\dot{\theta}_2\cos(\theta_1 - \theta_2)$$
$$+ (m_1 + m_2)gL_1\cos\theta_1 + m_2gL_2\cos\theta_2 + \tau\theta_2 \tag{6.5}$$

The Euler–Lagrange equations of the system are, in this case, of the form

$$\frac{d}{dt}\left(\frac{\partial\mathcal{L}}{\partial\dot{\theta}_1}\right) - \frac{\partial\mathcal{L}}{\partial\theta_1} = 0, \quad \frac{d}{dt}\left(\frac{\partial\mathcal{L}}{\partial\dot{\theta}_2}\right) - \frac{\partial\mathcal{L}}{\partial\theta_2} = 0 \tag{6.6}$$

Using these formulas, we obtain the implicit model of the system as

$$(m_1 + m_2)L_1^2\ddot{\theta}_1 + m_2L_1L_2\ddot{\theta}_2\cos(\theta_1 - \theta_2) + m_2L_1L_2\dot{\theta}_2^2\sin(\theta_1 - \theta_2)$$
$$+ (m_1 + m_2)gL_1\sin\theta_1 = 0$$
$$m_2L_2^2\ddot{\theta}_2 + m_2L_1L_2\ddot{\theta}_1\cos(\theta_1 - \theta_2) - m_2L_1L_2\dot{\theta}_1^2\sin(\theta_1 - \theta_2)$$
$$+ m_2gL_2\sin\theta_2 = \tau \tag{6.7}$$

or, in matrix form,

$$M(q)\ddot{q} + C(q, \dot{q})\dot{q} + G(q) = F \tag{6.8}$$

which reads as follows:

$$\begin{bmatrix} (m_1 + m_2)L_1^2 & m_2L_1L_2\cos(\theta_1 - \theta_2) \\ m_2L_1L_2\cos(\theta_1 - \theta_2) & m_2L_2^2 \end{bmatrix}\begin{bmatrix} \ddot{\theta}_1 \\ \ddot{\theta}_2 \end{bmatrix}$$
$$+ \begin{bmatrix} 0 & m_2L_1L_2\dot{\theta}_2\sin(\theta_1 - \theta_2) \\ -m_2L_1L_2\dot{\theta}_1\sin(\theta_1 - \theta_2) & 0 \end{bmatrix}\begin{bmatrix} \dot{\theta}_1 \\ \dot{\theta}_2 \end{bmatrix}$$
$$+ \begin{bmatrix} (m_1 + m_2)gL_1\sin\theta_1 \\ m_2gL_2\sin\theta_2 \end{bmatrix} = \begin{bmatrix} 0 \\ \tau \end{bmatrix} \tag{6.9}$$

6.2.2 Normalization of the Model

Generally speaking, normalization of a system clears the underlying features of the mathematical structure of physical systems and, definitely, helps to ease the simulation tasks. We may alternatively first normalize the Lagrangian or, as we are doing here, proceed to normalize the set of the obtained system equations. Define the following variables and dimensionless parameters:

$$t' = t\sqrt{\frac{g}{L_2}}, \quad u = \frac{\tau}{m_2gL_2}, \quad \epsilon = \frac{1}{1 + (m_2/m_1)}, \quad \rho = \frac{L_1}{L_2} \tag{6.10}$$

We then obtain the following normalized system:

$$\rho\ddot{\theta}_1 + \epsilon\ddot{\theta}_2\cos(\theta_1 - \theta_2) + \epsilon\dot{\theta}_2^2\sin(\theta_1 - \theta_2) + \sin\theta_1 = 0$$
$$\ddot{\theta}_2 + \rho\ddot{\theta}_1\cos(\theta_1 - \theta_2) - \rho\dot{\theta}_1^2\sin(\theta_1 - \theta_2) + \sin\theta_2 = u \qquad (6.11)$$

where we have used, in an abuse of notation, the dots to indicate time derivatives with respect to the normalized time scale t'.

6.2.3 Problem Formulation

Given an arbitrary initial position for the configuration variables θ_1, θ_2, it is desired to bring the pendubot system to the stable equilibrium $\theta_1 = 0$, $\theta_2 = 0$ based only on position measurements.

6.2.4 Tangent Linearization

Linearization around the equilibrium point

$$\bar{\theta}_1 = 0, \quad \bar{\theta}_2 = 0, \quad \bar{\dot{\theta}}_1 = 0, \quad \bar{\dot{\theta}}_2 = 0 \text{ with } \bar{u} = 0 \qquad (6.12)$$

yields the following incremental dynamics, linearized dynamics, or tangent linearization model[1]:

$$\begin{bmatrix} \rho & \epsilon \\ \rho & 1 \end{bmatrix}\begin{bmatrix} \ddot{\theta}_1 \\ \ddot{\theta}_2 \end{bmatrix} + \begin{bmatrix} \theta_1 \\ \theta_2 \end{bmatrix} = \begin{bmatrix} 0 \\ u \end{bmatrix} \qquad (6.13)$$

The following virtual angular displacement constitutes a flat output of the linearized system:

$$\varphi = \rho\theta_1 + \epsilon\theta_2 \qquad (6.14)$$

The flat output φ yields the following string of relations from where the observability property of the incremental flat output is easily confirmed:

$$\varphi = \rho\theta_1 + \epsilon\theta_2, \quad \dot{\varphi} = \rho\dot{\theta}_1 + \epsilon\dot{\theta}_2, \quad \ddot{\varphi} = \rho\ddot{\theta}_1 + \epsilon\ddot{\theta}_2 = -\theta_1, \quad \varphi^{(3)} = -\dot{\theta}_1 \qquad (6.15)$$

Notice that the incremental flat output acceleration $\ddot{\varphi}$ is a function of the position variables alone. In fact, it coincides with the negative of the (incremental) angular position of the first pendulum. Similarly, the odd time derivatives of the incremental flat output depends only upon velocities of the incremental generalized coordinates of the system.

---

1. Due to the nature of the equilibrium point, located at the origin of the state space, the incremental variables coincide with the actual variables.

The incremental flat output produces the following differential parameterization of the incremental state variables of the system:

$$\theta_1 = -\ddot{\varphi}, \quad \dot{\theta}_1 = -\varphi^{(3)}, \quad \theta_2 = \frac{1}{\epsilon}[\varphi + \rho\ddot{\varphi}], \quad \dot{\theta}_2 = \frac{1}{\epsilon}\left[\dot{\varphi} + \rho\varphi^{(3)}\right] \quad (6.16)$$

Finally, the incremental control input u is expressible in terms of the flat output as

$$u = \rho\left(\frac{1-\epsilon}{\epsilon}\right)\varphi^{(4)} + \frac{1}{\epsilon}(1+\rho)\ddot{\varphi} + \left(\frac{1}{\epsilon}\right)\varphi \quad (6.17)$$

The input-to-flat output dynamics is then a fourth-order dynamics, confirming that the variable φ is indeed the flat output with which there is no zero dynamics associated.

If the variable φ is driven to zero along with all its time derivatives (the first, second, and third time derivatives), then θ_1 and $\dot{\theta}_1$ are driven to zero. In turn, θ_2 is also driven to zero. The angular velocity $\dot{\theta}_2$ is also driven to zero, and under such circumstances, u also converges to zero, reaching the desired equilibrium input.

The system to be controlled is thus represented by the following fourth-order linear time-invariant system:

$$\begin{aligned}
\varphi^{(4)} &= \left[\frac{\epsilon}{\rho(1-\epsilon)}\right]u - \left[\frac{1+\rho}{\rho(1-\epsilon)}\right]\ddot{\varphi} - \left[\frac{1}{\rho(1-\epsilon)}\right]\varphi \\
&= \left(\frac{\epsilon}{\rho(1-\epsilon)}\right)\left[u + \frac{\theta_1}{\epsilon} - \theta_2\right] \quad (6.18)
\end{aligned}$$

6.2.5 A Flatness-Based Linear Controller

First, for comparison purposes, we design a linear full-state feedback controller before proceeding with an ESO observer-based ADRC feedback scheme.

A linear incremental state feedback controller stabilizes, exponentially asymptotically to zero, the phase variables associated with the flat output φ, namely, the variables, $(\varphi, \dot{\varphi}, \ddot{\varphi}, \varphi^{(3)})$.

Let v be an auxiliary control input. The partial state feedback control law

$$\begin{aligned}
u &= \frac{\rho(1-\epsilon)}{\epsilon}\left\{-\left[\frac{1+\rho}{\rho(1-\epsilon)}\right]\theta_1 + \frac{1}{\rho(1-\epsilon)}(\rho\theta_1 + \epsilon\theta_2) + v\right\} \\
&= \frac{1}{\epsilon}\left[-\theta_1 + \epsilon\theta_2 + \frac{\rho(1-\epsilon)}{\epsilon}v\right] \quad (6.19)
\end{aligned}$$

reduces the problem to controlling a pure integration system represented by the equation $\varphi^{(4)} = v$. A traditional prescription for v that requires positions and velocities of the incremental generalized coordinates is given by

$$v = -\gamma_3 \varphi^{(3)} - \gamma_2 \ddot{\varphi} - \gamma_1 \dot{\varphi} - \gamma_0 \varphi$$
$$= \gamma_3 \dot{\theta}_1 + \gamma_2 \theta_1 - \gamma_1(\rho\dot{\theta}_1 + \epsilon\dot{\theta}_2) - \gamma_0(\rho\theta_1 + \epsilon\theta_2) \qquad (6.20)$$

The above traditional incremental state feedback controller drives the incremental flat output to zero via the appropriate choice of the feedback gains $\gamma_3, \cdots, \gamma_0$.

The incremental closed-loop system responses are governed by the location of the roots of the characteristic polynomial

$$p(s) = s^4 + \gamma_3 s^3 + \gamma_2 s^2 + \gamma_1 s + \gamma_0 \qquad (6.21)$$

6.2.6 Simulations

We set the following parameter values:

$$m_1 = 0.5 \text{ [kg]}, \quad m_2 = 1 \text{ [kg]}, \quad L_1 = 1 \text{ [m]}, \quad L_2 = 0.75 \text{ [m]}, \quad g = 9.8 \text{ [m/s}^2]. \qquad (6.22)$$

The initial angular displacement of mass m_1 was taken to be $\theta_1(0) = 0.5$ [rad] with zero angular velocity, whereas $\theta_2(0) = 0$ [rad] and $\dot{\theta}_2 = 0$ [rad/s].

The closed-loop poles of the linear controller were chosen as the roots of the following desired characteristic polynomial:

$$p_d(s) = (s^2 + 2\xi\omega_n s + \omega_n^2)^2$$
$$= s^4 + 4\xi\omega_n s^3 + (2\omega_n^2 + 4\xi^2\omega_n^2)s^2 + 4\xi\omega_n^3 s + \omega_n^4 \qquad (6.23)$$

We obtain a controller design via a term-by-term identification of the coefficients of the closed-loop system characteristic polynomial $p(s)$ with those of the desired characteristic polynomial $p_d(s)$. Fig. 6.3 shows the performance of the feedback control scheme in stabilization task and Fig. 6.4 shows the transient behavior of the pendubot.

6.2.7 An ADRC Controller-Based on the Tangent Linearization Model

Clearly, the flat output φ is measurable since the configuration variables θ_1, θ_2 are assumed to be measurable. This is translated, in particular, to the fact that the acceleration of the (incremental) flat output $\ddot{\varphi}$ is also measurable. This means that we can decompose the Extended State Observer design in two complementary parts.

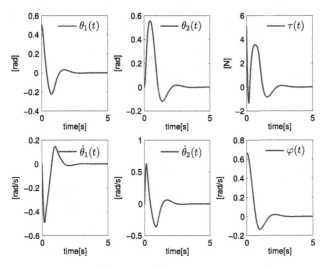

FIGURE 6.3 Response of nonlinear pendubot system to linear state feedback control.

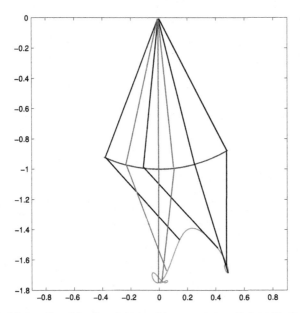

FIGURE 6.4 Motion of pendubot from initial position toward controlled stabilization.

1. The flat output acceleration variable $\ddot{\varphi}$ may be considered to be a known measured auxiliary input represented by $-\theta_1$. Such an input is driving the double integration subsystem yielding the flat output velocity $\dot{\varphi}$ and the flat output φ itself.

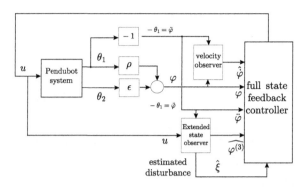

FIGURE 6.5 Stabilizing ADRC scheme for the pendubot.

2. A second-order system considers the fourth-order time derivative of the flat output as the second-order time derivative of the flat output acceleration. This yields a relation with the system input u and the generalized incremental position variables θ_1 and θ_2. This last second-order subsystem accepts, through the input channel, all the effects of the neglected nonlinearities and all possible lumped exogenous perturbation inputs affecting the system.

A set of two observers can then be devised for the above-described two constitutive subsystems. The block diagram shown in Fig. 6.5 clarifies the procedure.

The input–output representation of the pendubot system given in (6.19) may be represented as

$$\varphi^{(4)} = \frac{\epsilon}{\rho(1 - \epsilon)}u + \xi(t) \tag{6.24}$$

where $\xi(t)$ represents the effects of the neglected nonlinear terms in the linearization process.

Consider the following relations involving the even-order time derivatives of the incremental flat output:

$$\ddot{\varphi} = -\theta_1, \quad \varphi^{(4)} = \frac{\epsilon}{\rho(1 - \epsilon)}\left[u + \frac{\theta_1}{\epsilon} - \theta_2\right] \tag{6.25}$$

These relations represent dynamical subsystems with inputs given by known linear combinations of either the generalized position variables or the position variables and the control input.

For the first subsystem, the corresponding observer, which does not need to be an ESO observer, is synthesized on the basis of the subsystem auxiliary input $-\theta_1$ and the measured flat output φ identified with y_0. As a result, the first observer produces the estimate of the unmeasured flat output velocity $\dot{\varphi}$, here denoted by $\hat{y}_1$:

$$\frac{d}{dt}\hat{y}_0 = \hat{y}_1 + \lambda_1(y_0 - \hat{y}_0)$$

$$\frac{d}{dt}\hat{y}_1 = -\theta_1 + \lambda_0(y_0 - \hat{y}_0) \qquad (6.26)$$

For the second subsystem, the corresponding observer is synthesized on the basis of the above input–output relation taking the flat output acceleration as an auxiliary output signal. As a result, this second observer produces the estimate of the unmeasured flat output third-order time derivative $\varphi^{(3)}$, here denoted by $\hat{y}_3$:

$$\frac{d}{dt}\hat{y}_2 = \hat{y}_3 + \lambda_4(-\theta_1 - \hat{y}_2)$$

$$\frac{d}{dt}\hat{y}_3 = \hat{z} + \frac{\epsilon}{\rho(1-\epsilon)}\left[u + \frac{\theta_1}{\epsilon} - \theta_2\right] + \lambda_3(-\theta_1 - \hat{y}_2)$$

$$\frac{d}{dt}\hat{z} = \lambda_2(-\theta_1 - \hat{y}_2) \qquad (6.27)$$

Clearly, this observer produces an estimate $\hat{y}_3$ of the third-order time derivative of y and also estimates, through the variable z, the disturbances caused by the environment, the unmodeled dynamics, and the nonlinearities neglected by the linearization process. Here, we have opted for a first-order extension in the ESO observer for the estimation of the overall disturbances.

The (full-state) output feedback control is given by

$$u = \frac{\rho(1-\epsilon)}{\epsilon}\left\{\varphi^*(t)^{(4)} - z - \sum_{i=0}^{3} k_i\left(\hat{y}_i - [\varphi^*(t)]^{(i)}\right)\right\} \qquad (6.28)$$

with $\hat{y}_0 = \varphi$ and $\hat{y}_2 = -\theta_1$, whereas $\hat{y}_1$ and $\hat{y}_3$ are estimated using the reduced-order observers arising from the natural cascaded structure and availability of the flat output acceleration of the incremental flat output variable dynamics. The controller gains $k_i, i = 0, 1, \cdots, 4$, are chosen so that the underlying closed-loop characteristic polynomial for the linearized system exhibits stable roots.

6.2.7.1 Simulations

We set the parameter values to be identical to those used in the previous simulation. The initial angular displacement of mass m_1 was taken to be $\theta_1(0) = 1.047$ [rad] with zero angular velocity, whereas $\theta_2(0) = 0$ [rad] and $\dot{\theta}_2 = 0$ [rad/s]. Fig. 6.6 depicts the controlled response of all states of the non-linear system and the generated torque control input.

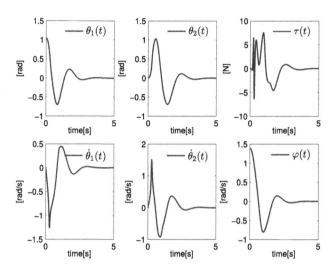

FIGURE 6.6 State responses of ADRC controlled pendubot system.

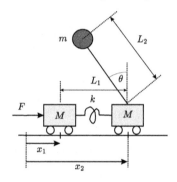

FIGURE 6.7 A two mass spring system with inverted pendulum.

6.3 A TWO-MASS SPRING SYSTEM WITH AN INVERTED PENDULUM

6.3.1 Derivation of the Model

Consider the system constituted by two masses, joined by a spring, of resting length L_1, which can only be displaced horizontally thanks to a force acting on the first mass. The second mass supports a nonactuated inverted pendulum, as shown in Fig. 6.7.

The system Lagrangian is given by

$$\mathcal{L} = \frac{1}{2}M\dot{x}_1^2 + \frac{1}{2}M\dot{x}_2^2 + \frac{1}{2}m\left(\dot{x}_2^2 - 2L_2\dot{x}_2\dot{\theta}\cos\theta + L_2^2\dot{\theta}^2\right)$$
$$- \frac{1}{2}k(x_1 - x_2 + L)^2 - mgL_2(1 + \cos\theta) + Fx_1 \qquad (6.29)$$

Here, we have used the fact that the bob mass in the pendulum is located at

$$x_m = x_2 - L_2 \sin\theta, \quad y_m = -L_2 \cos\theta \qquad (6.30)$$

with the corresponding velocities

$$\dot{x}_m = \dot{x}_2 - L_2\dot{\theta}\cos\theta, \quad \dot{y}_m = L_2\dot{\theta}\sin\theta \qquad (6.31)$$

The potential energy is due to the height of the pendulum mass, the spring, and the forcing input.

The model of the system is just derived from the Euler–Lagrange equations

$$\frac{d}{dt}\frac{\partial L}{\partial \dot{x}_1} - \frac{\partial L}{\partial x_1} = 0, \quad \frac{d}{dt}\frac{\partial L}{\partial \dot{x}_2} - \frac{\partial L}{\partial x_2} = 0, \quad \frac{d}{dt}\frac{\partial L}{\partial \dot{\theta}} - \frac{\partial L}{\partial \theta} = 0 \qquad (6.32)$$

obtaining the following set of nonlinear coupled controlled differential equations:

$$M\ddot{x}_1 + k(x_1 - x_2 + L) = F$$
$$(m + M)\ddot{x}_2 - m_2 L_2 \ddot{\theta}\cos\theta + m L_2\dot{\theta}^2\sin\theta - k(x_1 - x_2 + L) = 0$$
$$m_2 L_2^2\ddot{\theta} - m L_2\ddot{x}_2\cos\theta - mgL_2\sin\theta = 0 \qquad (6.33)$$

6.3.2 Normalization

The following scaling operations and definitions were carried out to normalize the system equations:

$$\chi_1 = \frac{x_1}{L_2}, \quad \chi_2 = \frac{x_2}{L_2}, \quad t' = t\sqrt{\frac{g}{L_2}}, \quad \epsilon = \frac{M}{m}, \quad \beta = \frac{k}{m}\frac{L_2}{g}, \quad u = \frac{F}{mg}, \quad \gamma = \frac{L_1}{L_2} \qquad (6.34)$$

We obtain the following set of normalized equations for the system:

$$\epsilon\ddot{\chi}_1 + \beta(\chi_1 - \chi_2 + \gamma) = u$$
$$(1 + \epsilon)\ddot{\chi}_2 - \ddot{\theta}\cos\theta + \dot{\theta}^2\sin\theta - \beta(\chi_1 - \chi_2 + \gamma) = 0$$
$$\ddot{\theta} - \ddot{\chi}_2\cos\theta - \sin\theta = 0 \qquad (6.35)$$

6.3.3 Tangent Linearization

Assume that the mass spring pendulum system is initially at rest at the equilibrium point

$$\bar{\chi}_1 = 0, \quad \bar{\chi}_2 = \gamma, \quad \dot{\bar{\chi}}_1 = 0, \quad \dot{\bar{\chi}}_2 = 0, \quad \bar{u} = 0, \quad \bar{\theta} = 0, \quad \dot{\bar{\theta}} = 0 \qquad (6.36)$$

and the system is perturbed by a normalized impulsive force f_d applied to the second mass. Device a controller that brings the mass spring pendulum system to its original equilibrium state in a reasonable amount of time.

Define the incremental variables with respect to the initial equilibrium point:

$$\chi_{1\delta} = \chi_1 - 0 = \chi_1, \quad \chi_{2\delta} = \chi_2 - \gamma, \quad \dot{\chi}_{i\delta} = \dot{\chi}_i - 0, \quad i = 1, 2$$
$$\theta_\delta = \theta - 0, \quad \dot{\theta}_\delta = \dot{\theta} - 0, \quad u_\delta = u - 0 = u \tag{6.37}$$

These variables satisfy, up to first order, the following set of linear equations:

$$\epsilon \ddot{\chi}_{1\delta} + \beta(\chi_{1\delta} - \chi_{2\delta}) = u_\delta$$
$$(1 + \epsilon)\ddot{\chi}_{2\delta} - \ddot{\theta}_\delta - \beta(\chi_{1\delta} - \chi_{2\delta}) = 0$$
$$\ddot{\theta}_\delta - \ddot{\chi}_{2\delta} = \theta_\delta \tag{6.38}$$

The horizontal projection χ of the bob position onto the horizontal axis is given by $\chi = x_2 - L_2 \sin\theta$. Its normalized version is just $\chi = \chi_2 - \sin\theta$. The equilibrium value is readily seen to be $\overline{\chi} = \overline{\chi}_2 = \gamma$. The associated normalized velocity is $\dot{\chi} = \dot{\chi}_2 - \dot{\theta}\cos\theta$.

Then consider the dimensionless incremental variable χ_δ and its time derivatives:

$$\chi_\delta = \theta_\delta - \chi_{2\delta}, \quad \dot{\chi}_\delta = \dot{\theta}_\delta - \dot{\chi}_{2\delta}, \quad \ddot{\chi}_\delta = \ddot{\theta}_\delta - \ddot{\chi}_{2\delta} = \theta_\delta, \quad \chi_\delta^{(3)} = \dot{\theta}_\delta$$
$$\chi_\delta^{(4)} = \ddot{\theta}_\delta = \frac{1}{\epsilon}\left[\beta(x_{1\delta} - x_{2\delta}) + (1 + \epsilon)\theta_\delta\right]$$
$$\chi_\delta^{(5)} = \frac{1}{\epsilon}\left[\beta(\dot{\chi}_{1\delta} - \dot{\chi}_{2\delta}) + (1 + \epsilon)\dot{\theta}_\delta\right]$$
$$\chi_\delta^{(6)} = \frac{\beta}{\epsilon^2}u_\delta + \left[\frac{\beta(1 + \epsilon) - 2\beta^2}{\epsilon^2}\right](\chi_{1\delta} - \chi_{2\delta}) + \left[\frac{(1 + \epsilon)^2 - \beta}{\epsilon^2}\right]\theta_\delta \tag{6.39}$$

The inverse relations, i.e., the incremental state variables in terms of the incremental flat output and its time derivatives are easily obtained by inspection using the system of linearized equations:

$$\chi_{1\delta} = -\chi_\delta + \left(1 - \frac{1 + \epsilon}{\beta}\right)\ddot{\chi}_\delta + \frac{\epsilon}{\beta}\chi_\delta^{(4)}$$
$$\dot{\chi}_{1\delta} = -\dot{\chi}_\delta + \left(1 - \frac{1 + \epsilon}{\beta}\right)\chi_\delta^{(3)} + \frac{\epsilon}{\beta}\chi_\delta^{(5)}$$
$$\chi_{2\delta} = \theta_\delta - \chi_\delta = \ddot{\chi}_\delta - \chi_\delta, \quad \dot{\chi}_{2\delta} = \chi_\delta^{(3)} - \dot{\chi}_\delta, \quad \theta_\delta = \ddot{\chi}_\delta, \quad \dot{\theta}_\delta = \chi_\delta^{(3)} \tag{6.40}$$

The incremental control input u_δ is expressed in terms of the flat output χ_δ as

$$u_\delta = \frac{\epsilon^2}{\beta}\chi_\delta^{(6)} + \frac{\epsilon(1 + \beta) + \epsilon^2}{\beta}\chi_\delta^{(4)} - (1 + 2\epsilon)\ddot{\chi}_\delta \tag{6.41}$$

The acceleration and the fourth-order time derivative of the incremental flat output are functions of the incremental position variables alone. The sixth-order time derivative of the incremental flat output is a function of the incremental control input (i.e., the flat output candidate is confirmed to have full relative degree, as expected) and of even-order time derivatives of the flat output. This will allow for the design of three independent observers generating the unmeasured time derivatives of the flat output, necessary for closing the control loop.

6.3.4 A Linear Controller Based on Tangent Linearization

Controlling the value of the incremental projection of the bob χ_δ to zero, together with all its time derivatives, i.e., $(\chi_\delta, \dot{\chi}_\delta, \ddot{\chi}_\delta, \chi_\delta^{(3)}, \chi_\delta^{(4)}, \chi_\delta^{(5)})$, yields the desired equilibrium for all the system state variables.

A linear controller based on the input–output relation is given by

$$u_\delta = \frac{\epsilon^2}{\beta} \left\{ v - \left[\frac{\beta(1+\epsilon) - 2\beta^2}{\epsilon^2} \right] (\chi_{1\delta} - \chi_{2\delta}) - \left[\frac{(1+\epsilon)^2 - \beta}{\epsilon^2} \right] \theta_\delta \right\} \quad (6.42)$$

with

$$v = -\gamma_5 \chi_\delta^{(5)} - \cdots - \gamma_1 \dot{\chi}_\delta - \gamma_0 \chi_\delta \quad (6.43)$$

where, as we have already seen in the previous examples, all the time derivatives of χ_δ are expressible in terms of the system states $\chi_{1\delta}, \dot{\chi}_{1\delta}, \chi_{2\delta}, \dot{\chi}_{2\delta}, \theta_\delta, \dot{\theta}_\delta$. The coefficients $\{\gamma_5, \ldots, \gamma_0\}$ are chosen so that the corresponding characteristic polynomial for the closed-loop linearized system is Hurwitz.

6.3.5 Simulations

An impulsive force is given to the second car, i.e., to the cart carrying the inverted pendulum. The task of the controller is to recuperate the lost equilibrium after the impulsive force has been delivered. A flatness-based controller is shown to efficiently control the perturbed system. However, all six state variables must be accurately measured.

The following set of parameter values were used in the simulations illustrating the performance of a flatness-based controller:

$$M = 0.5 \text{ [kg]}, \quad m = 0.3 \text{ [kg]}, \quad L_1 = 0.1 \text{ [m]}, \quad L_2 = 1 \text{ [m]}$$
$$g = 9.8 \text{ [m/s}^2], \quad k = 30 \text{ [N/m]} \quad (6.44)$$

Fig. 6.8 shows the behavior of the position of, both, the cars and the pendulum as well as the control input in the pendulum stabilization process. Fig. 6.9 shows the robustness of the scheme when an impulsive disturbance force is applied.

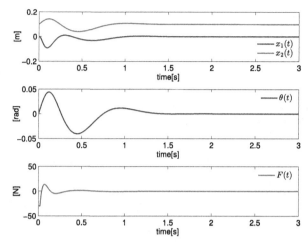

FIGURE 6.8   Closed-loop stabilization responses to linear feedback controller after an impulsive disturbance.[2]

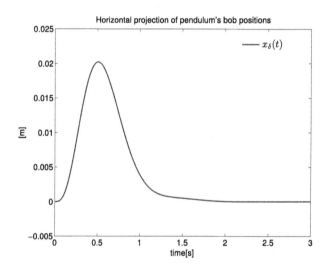

FIGURE 6.9   Stabilization of the flat output after an impulsive disturbance force.

6.3.6   An ADRC Controller Based on Tangent Linearization

The even-order time derivatives of the measurable flat output χ_δ are given by

$$\ddot{\chi}_\delta = \theta_\delta$$

---

2. For interpretation of the references to color in this and the following figures, the reader is referred to the web version of this chapter.

$$\chi_\delta^{(4)} = \frac{1}{\epsilon}\left[\beta(x_{1\delta} - x_{2\delta}) + (1+\epsilon)\theta_\delta\right]$$

$$\chi_\delta^{(6)} = \frac{\beta}{\epsilon^2}u_\delta + \left[\frac{\beta(1+\epsilon) - 2\beta^2}{\epsilon^2}\right](\chi_{1\delta} - \chi_{2\delta}) + \left[\frac{(1+\epsilon)^2 - \beta}{\epsilon^2}\right]\theta_\delta \quad (6.45)$$

Let $\chi_\delta = \theta_\delta - x_{2\delta}$ be denoted by y_0, and its redundant estimated value be $\hat{y}_0$. The first relation prompts the following observer:

$$\dot{\hat{y}}_0 = \hat{y}_1 + \lambda_1(\theta_\delta - x_{2\delta} - \hat{y}_0)$$
$$\dot{\hat{y}}_1 = \theta_\delta + \lambda_0(\theta_\delta - x_{2\delta} - \hat{y}_0) \quad (6.46)$$

This observer produces the estimate $\hat{y}_1$ of $\dot{\chi}_\delta$.

The second relation is translated into the observer

$$\dot{\hat{y}}_2 = \hat{y}_3 + \lambda_3(\theta_\delta - \hat{y}_2)$$
$$\dot{\hat{y}}_3 = \frac{1}{\epsilon}\left[\beta(x_{1\delta} - x_{2\delta}) + (1+\epsilon)\theta_\delta\right] + \lambda_2(\theta_\delta - \hat{y}_2) \quad (6.47)$$

where $\hat{y}_3$ is the estimated value of the third-order time derivative of the incremental flat output χ_δ. Note that the quantity $\ddot{\theta}_\delta = \frac{1}{\epsilon}\left[\beta(x_{1\delta} - x_{2\delta}) + (1+\epsilon)\theta_\delta\right]$ representing $d^2(\theta_\delta)/dt^2$ acts as a "control" input to this observer, only written in terms of the measured positions. This observer produces the third-order time derivative of the incremental flat output $\hat{y}_3$.

Finally, the extended observer

$$\dot{\hat{y}}_4 = \hat{y}_5 + \lambda_6(\frac{1}{\epsilon}\left[\beta(x_{1\delta} - x_{2\delta}) + (1+\epsilon)\theta_\delta\right] - \hat{y}_4)$$
$$\dot{\hat{y}}_5 = \frac{\beta}{\epsilon^2}u_\delta + \left[\frac{\beta(1+\epsilon) - 2\beta^2}{\epsilon^2}\right](\chi_{1\delta} - \chi_{2\delta}) + \left[\frac{(1+\epsilon)^2 - \beta}{\epsilon^2}\right]\theta_\delta$$
$$+ \lambda_5(\frac{1}{\epsilon}\left[\beta(x_{1\delta} - x_{2\delta}) + (1+\epsilon)\theta_\delta\right] - \hat{y}_4) + \hat{z}$$
$$\dot{\hat{z}} = \lambda_4(\frac{1}{\epsilon}\left[\beta(x_{1\delta} - x_{2\delta}) + (1+\epsilon)\theta_\delta\right] - \hat{y}_4) \quad (6.48)$$

produces the fifth-order time derivative of the incremental flat output as the variable $\hat{y}_5$. Also, this observer produces an estimate of the total disturbance by means of the extended variable $\hat{z}$. A block diagram for the mass spring pendulum observers is shown in Fig. 6.10.

Note, however, that the last observer can be considerably simplified by invoking the cancelling feature of ADRC schemes. In fact, all the linear state-dependent additive terms can be considered to be inherited from endogenous disturbance (i.e., state-dependent additive terms acting as disturbances) that affect the simplified linear system $\chi^{(6)} = (\beta/\epsilon^2)u_\delta$. This is precisely one of the

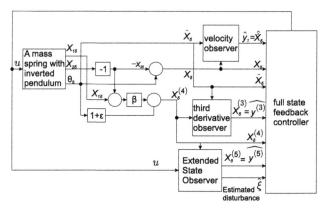

FIGURE 6.10 ADRC scheme for the mass spring system with an inverted pendulum.

remarkable features of ADRC control schemes that make it so similar to Model-Free control strategies. We use the following extended observer for the total disturbance estimation in the simplified linearized system (see Fig. 6.10):

$$\dot{\hat{y}}_4 = \hat{y}_5 + \lambda_6 \left(\frac{1}{\epsilon} \left[\beta(x_{1\delta} - x_{2\delta}) + (1 + \epsilon)\theta_\delta \right] - \hat{y}_4 \right)$$

$$\dot{\hat{y}}_5 = \frac{\beta}{\epsilon^2} u_\delta + \hat{z}$$

$$\dot{\hat{z}} = \lambda_4 \left(\frac{1}{\epsilon} \left[\beta(x_{1\delta} - x_{2\delta}) + (1 + \epsilon)\theta_\delta \right] - \hat{y}_4 \right) \qquad (6.49)$$

6.3.7 Simulations

Assume that the mass spring pendulum system is initially at rest in the following equilibrium point:

$$\bar{\chi}_1 = 0 \quad \bar{\chi}_2 = \gamma \quad \dot{\bar{\chi}}_1 = 0 \quad \dot{\bar{\chi}}_2 = 0 \quad \bar{u} = 0 \quad \bar{\theta} = 0 \quad \dot{\bar{\theta}} = 0 \qquad (6.50)$$

At time $t = 0$, the system is located at the equilibrium point used for the linearization. At this instant, an impulsive force is applied to the second cart (holding the inverted pendulum), and the controller is to restore the initial equilibrium position with pendulum in its vertical unstable position.

The initial conditions for the system were all set to be zero except the initial position for the second cart placed at the position corresponding to the resting spring. Fig. 6.11 shows the ADRC behavior of the position of, both, the cars and the pendulum as well as the control input in the pendulum stabilization process in the presence of an impulsive disturbance. Fig. 6.12 shows the robustness of the scheme when an impulsive disturbance force is applied.

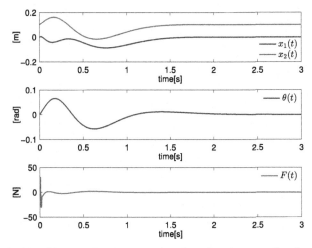

FIGURE 6.11 Closed-loop stabilization responses to linear feedback controller after an impulsive disturbance with ADRC scheme.

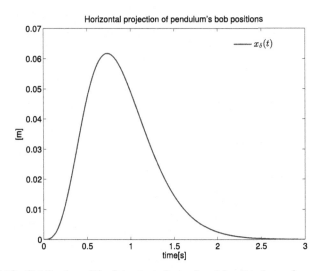

FIGURE 6.12 Stabilization of the flat output after an impulsive disturbance force with ADRC scheme.

6.4 DOUBLE INVERTED PENDULUM ON A CART

Consider a double inverted pendulum placed on a cart shown in Fig. 6.13. Each bob has the same mass m, and the links are assumed to have negligible mass and moment of inertia. The lengths of the links are considered to be different, of values L_1 and L_2. The cart is free to move without friction in a horizontal manner thanks to the action of an external force F. The horizontal position of

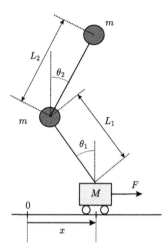

FIGURE 6.13 Double inverted pendulum on a cart system.

the cart x_M is denoted by x, whereas those of the bobs are, respectively, denoted by x_{m1} and x_{m2}. The configuration space is constituted by $S^1 \times S^1 \times \mathbb{R}$ with generalized coordinates given by the position x of the cart and the angular positions θ_1 and θ_2 of the links.

6.4.1 Derivation of the Model via Euler–Lagrange Equations

We have

$$x_M = x$$
$$x_{m1} = x - L_1 \sin(\theta_1) y_{m1} = L_1 \cos(\theta_1)$$
$$x_{m2} = x - L_1 \sin(\theta_1) + L_2 \sin(\theta_2), \ y_{m2} = L_1 \cos(\theta_1) + L_2 \cos(\theta_2) \quad (6.51)$$

The velocities associated with each mass are obtained by straightforward time differentiation of their respective positions:

$$\dot{x}_M = \dot{x}, \quad \dot{x}_{m1} = \dot{x} - L_1 \dot{\theta}_1 \cos(\theta_1), \quad \dot{y}_{m1} = -L_1 \dot{\theta}_1 \sin(\theta_1)$$
$$\dot{x}_{m2} = \dot{x} - L_1 \dot{\theta}_1 \cos(\theta_1) + L_2 \dot{\theta}_2 \cos(\theta_2), \quad \dot{y}_{m2} = -L_1 \dot{\theta}_1 \sin(\theta_1) - L_2 \dot{\theta}_2 \cos(\theta_2)$$
$$(6.52)$$

The kinetic energies are obtained as follows:

$$K_M = \frac{1}{2} M \dot{x}^2, \quad K_{m1} = \frac{1}{2} m (\dot{x}^2 - 2L_1 \dot{x} \dot{\theta}_1 \cos(\theta_1) + L_1^2 \dot{\theta}_1^2)$$
$$K_{m2} = \frac{1}{2} m (\dot{x}^2 + L_1^2 \dot{\theta}_1^2 + L_2^2 \dot{\theta}_2^2 - 2L_1 \dot{x} \dot{\theta}_1 \cos(\theta_1) + 2L_2 \dot{x} \dot{\theta}_2 \cos(\theta_2)$$
$$- 2L_1 L_2 \dot{\theta}_1 \dot{\theta}_2 \cos(\theta_1 + \theta_2) \quad (6.53)$$

The potential energies associated with the bobs masses positions and the work of the external input force are expressed as

$$u = mgL_1 \cos(\theta_1) + mg(L_1 \cos(\theta_1) + L_2 \cos(\theta_2)) - xF \qquad (6.54)$$

The Lagrangian of the system is computed as

$$\mathcal{L} = \frac{1}{2}(M + 2m)\dot{x}^2 - 2mL_1\dot{x}\dot{\theta}_1 \cos(\theta_1) + mL_2\dot{x}\dot{\theta}_2 \cos(\theta_2)$$
$$- mL_1L_2\dot{\theta}_1\dot{\theta}_2 \cos(\theta_1 + \theta_2) + mL_1^2\dot{\theta}_1^2$$
$$+ \frac{1}{2}mL_2^2\dot{\theta}_2^2 - 2mgL_1 \cos(\theta_1) - mgL_2 \cos(\theta_2) + xF \qquad (6.55)$$

6.4.2 Normalization

Define the normalized variables and parameters

$$\epsilon = \frac{M}{m}, \quad \rho = \frac{L_2}{L_1}, \quad \chi = \frac{x}{L_1}, \quad u = \frac{F}{mg}, \quad d\tau = dt\sqrt{\frac{g}{L_1}} \qquad (6.56)$$

We obtain the normalized Lagrangian as

$$\mathcal{L}' = \frac{1}{2}(2 + \epsilon)\dot{\chi}^2 - \dot{\chi}\dot{\theta}_1 \cos(\theta_1) + \rho\dot{\chi}\dot{\theta}_2 - \rho\dot{\theta}_1\dot{\theta}_2 \cos(\theta_1 + \theta_2) + \dot{\theta}_1$$
$$+ \frac{1}{2}\rho^2\dot{\theta}_2^2 - 2\cos(\theta)_1 - \rho\cos(\theta_2) + \chi u \qquad (6.57)$$

Applying the Euler–Lagrange formalism to the three generalized coordinates expressing the degrees of freedom of the system, we obtain the following set of normalized coupled second-order differential equations for the controlled system:

$$(2 + \epsilon)\ddot{\chi} - 2\ddot{\theta}_1 \cos(\theta)_1 + 2\dot{\theta}_1^2 \sin(\theta)_1 + \rho\ddot{\theta}_2 \cos(\theta_2) - \rho\dot{\theta}_2^2 \sin(\theta_2) = u$$
$$-2\ddot{\chi}\cos(\theta)_1 - \rho\ddot{\theta}_2 \cos(\theta_1 + \theta_2) + \rho\dot{\theta}_2^2 \sin(\theta_1 + \theta_2) + 2\ddot{\theta}_1 - 2\sin(\theta_1) = 0$$
$$\rho\ddot{\chi}\cos(\theta_2) - \rho\ddot{\theta}_1 \cos(\theta_1 + \theta_2) + \rho\dot{\theta}_1^2 \sin(\theta_1 + \theta_2) + \rho^2\ddot{\theta}_2 - \rho\sin(\theta_2) = 0$$
$$(6.58)$$

In the customary matrix form

$$M(q)\ddot{q} + C(q, \dot{q})\dot{q} + G(q) = bu \qquad (6.59)$$

with generalized coordinate, $q^T = (\chi, \theta_1, \theta_2)$, we have

$$\begin{bmatrix} 2 + \epsilon & -2\cos(\theta_1) & \rho\cos(\theta_2) \\ -2\cos(\theta_1) & 2 & -\rho\cos(\theta_1 + \theta_2) \\ \rho\cos(\theta_2) & -\rho\cos(\theta_1 + \theta_2) & \rho^2 \end{bmatrix} \begin{bmatrix} \ddot{\chi} \\ \ddot{\theta}_1 \\ \ddot{\theta}_2 \end{bmatrix}$$

$$+ \begin{bmatrix} 0 & 2\dot{\theta}_1 \sin(\theta_1) & -\rho\dot{\theta}_2 \sin(\theta_2) \\ 0 & 0 & \rho\dot{\theta}_2 \sin(\theta_1 + \theta_2) \\ 0 & \rho\dot{\theta}_1 \sin(\theta_1 + \theta_2) & 0 \end{bmatrix} \begin{bmatrix} \dot{\chi} \\ \dot{\theta}_1 \\ \dot{\theta}_2 \end{bmatrix}$$

$$+ \begin{bmatrix} 0 \\ -2\sin(\theta_1) \\ -2\rho s \in (\theta_2) \end{bmatrix} = \begin{bmatrix} 1 \\ 0 \\ 0 \end{bmatrix} u \tag{6.60}$$

The double inverted pendulum on a cart system is not feedback linearizable. Here we propose an ESO observer-based ADRC controller for a trajectory tracking task defined on the basis of the controllable linearization system variables.

6.4.3 Tangent Linearization

Next, we linearize the system around the unstable equilibrium point

$$\bar{\chi} = 0, \quad \dot{\bar{\chi}} = 0, \quad \bar{\theta}_1 = 0, \quad \dot{\bar{\theta}}_1 = 0, \quad \bar{\theta}_2 = 0, \quad \dot{\bar{\theta}}_2 = 0, \quad \bar{u} = 0 \tag{6.61}$$

Then all incremental variables coincide with the actual variables. The linearized system is written as

$$\ddot{\chi} = \frac{2}{\epsilon}\theta_1 + \frac{1}{\epsilon}u$$
$$\ddot{\theta}_1 = \frac{2(1+\epsilon)}{\epsilon}\theta_1 + \theta_2 + \frac{1}{\epsilon}u$$
$$\ddot{\theta}_2 = \frac{2}{\rho}\theta_1 + \frac{2}{\rho}\theta_2 \tag{6.62}$$

The system in the state space form is found to be controllable. The last row of the inverse of the controllability matrix (see [24]), multiplied by the state vector, reveals that the flat output is given by

$$y = x - \theta_1 + \rho\theta_2 \tag{6.63}$$

This output represents the linearized horizontal projection of the position of the second bob mass. In other words, the linearized position of the "shadow" of the second bob on the horizontal line is a system output that, upon approximate linearization, yields a flat output capable of differentially parameterizing all the incremental system variables, including the incremental control input u.

The consecutive time derivatives of the special output y are computed using the observability matrix of the system. This results in

$$y = x - \theta_1 + \rho\theta_2, \quad \dot{y} = \dot{x} - \dot{\theta}_1 + \rho\dot{\theta}_2, \quad \ddot{y} = \theta_2$$
$$y^{(3)} = \dot{\theta}_2, \quad y^{(4)} = \frac{2}{\rho}\theta_1 + \frac{2}{\rho}\theta_2, \quad y^{(5)} = \frac{2}{\rho}\dot{\theta}_1 + \frac{2}{\rho}\dot{\theta}_2 \tag{6.64}$$

On the other hand,

$$y^{(6)} = \frac{2}{\rho}(\ddot{\theta}_1 + \ddot{\theta}_2) = -(\frac{2(2+\epsilon)}{\rho\epsilon})\ddot{y} + (\frac{2(\rho + \rho\epsilon + \epsilon)}{\rho\epsilon})y^{(4)} + (\frac{2}{\rho\epsilon})u \quad (6.65)$$

Note that once again it is verified that the acceleration $\ddot{y}$ of the flat output y and its fourth-order time derivative $y^{(4)}$ are linear combinations of the position variables alone, whereas the first-, third-, and fifth-order time derivatives of the flat output are functions of the generalized velocities alone. Thus, it is not necessary to estimate $\ddot{y}$ and $y^{(4)}$ since they are conformed in terms of the measured positions.

The inverse relation yields the following differential parameterization of all system variables:

$$\theta_1 = \frac{\rho}{2}y^{(4)} - \ddot{y}, \quad \dot{\theta}_1 = \frac{\rho}{2}y^{(5)} - y^{(3)}, \quad \theta_2 = \ddot{y}, \quad \dot{\theta}_2 = y^{(3)}$$

$$x = y - (1+\rho)\ddot{y} + \frac{\rho}{2}y^{(4)}, \quad \dot{x} = \dot{y} - (1+\rho)y^{(3)} + \frac{\rho}{2}y^{(5)} \quad (6.66)$$

Finally, the differential parameterization of the incremental control input u is computed to be

$$u = \frac{\rho\epsilon}{2}y^{(6)} - (\rho + \rho\epsilon + \epsilon)y^{(4)} + (2+\epsilon)\ddot{y} \quad (6.67)$$

The input–flat output dynamics of the linearized system is therefore given by

$$y^{(6)} = \frac{2}{\rho\epsilon}\left[u + (\rho + \rho\epsilon + \epsilon)y^{(4)} - (2+\epsilon)\ddot{y}\right] \quad (6.68)$$

The simplified input-to-flat output dynamics, which is quite useful in simplifying the trajectory tracking ADRC feedback scheme, may be chosen to be

$$y^{(6)} = \frac{2}{\rho\epsilon}u + \xi(t) \quad (6.69)$$

This unique advantage of ADRC feedback schemes makes the approach so reminiscent of Model-Free Control developed by Fliess and his colleagues.

6.4.4 The Method of Cascaded Observers for the ADRC Controller for Trajectory Tracking

In practise, a high-dimensional, extended observer is rather inconvenient due to the noisy estimates of the state variables. We may alternatively exploit the fact that all even-order time derivatives of the measurable incremental flat output are functions of the incremental generalized positions alone, i.e., they can be synthesized as linear combinations of the measured horizontal displacement of the cart and the links angular positions. The observer design and the corresponding controller specification easily extend to the trajectory tracking case thanks to the linearity of the tangent approximation system.

Let $y_\delta^*(t)$ be a given smooth output reference trajectory for the incremental flat output y. Define $e_y = y - y^*(t)$ and $e_u = u - u^*(t)$, where $u^*(t)$ is computed from (6.67); alternatively, we can resort to the nominal version of equation (6.69) if the option of relegating linear additive endogenous terms to the total disturbance term is chosen. In this case, we simply let $u^*(t) = (\rho\epsilon/2)[y^*(t)]^{(6)}$. Similarly, we let $e_{\theta_1} = \theta_1 - \theta_1^*(t)$, $e_{\theta_2} = \theta_2 - \theta_2^*(t)$, and $e_x = x - x^*(t)$. The incremental state variable nominal trajectories $\theta_1^*(t), \theta_2^*(t), x^*(t)$, as we have seen in previous chapters, are easily obtained from the flatness property, which differentially parameterizes all system state variables in terms of the flat output and a finite number of its time derivatives. In our case, the tangent linearized system enjoys such a special property.

Consider the following relations satisfied by the even-order time derivatives of the incremental flat output trajectory tracking errors:

$$\ddot{e}_y = e_{\theta_2}, \quad e_y^{(4)} = \frac{d^2}{dt^2}(\ddot{e}_y) = \ddot{e}_{\theta_2} = \frac{2}{\rho}\left(e_{\theta_1} + e_{\theta_2}\right)$$

$$e_y^{(6)} = \frac{2}{\rho\epsilon}\left[e_u + (\rho + \rho\epsilon + \epsilon)e_y^{(4)} - (2+\epsilon)\ddot{e}_y\right] \tag{6.70}$$

These relations may be viewed as dynamical subsystems with input given by linear combinations of the generalized position tracking error variables. The first two relations are used to build two ESO observers for the unmeasured intermediate odd-degree time derivatives of the incremental flat output tracking error.

Let $\hat{e}_{y_0}$ denote the redundant estimate of the measured flat output tracking error $y - y^*(t)$. The first observer is devised as follows:

$$\frac{d}{dt}\hat{e}_{y_0} = \hat{e}_{y_1} + \lambda_1(e_{y_0} - \hat{e}_{y_0})$$

$$\frac{d}{dt}\hat{e}_{y_1} = e_{\theta_2} + \lambda_0(e_{y_0} - \hat{e}_{y_0}) \tag{6.71}$$

This observer produces an estimate $\hat{e}_{y_1}$ of the first-order time derivative of e_y.

Let the actual value of $\ddot{e}_y$ be expressed as the position error variable $\theta_2 - \theta_2^*(t)$. The redundant estimate of $\ddot{e}_y$ is denoted by $\hat{e}_{y_2}$. The second observer is synthesized as follows:

$$\frac{d}{dt}\hat{e}_{y_2} = \hat{e}_{y_3} + \lambda_3(e_{\theta_2} - \hat{e}_{y_2})$$

$$\frac{d}{dt}\hat{e}_{y_3} = \frac{2}{\rho}\left(e_{\theta_1} + e_{\theta_2}\right) + \lambda_2(e_{\theta_2} - \hat{e}_{y_2}) \tag{6.72}$$

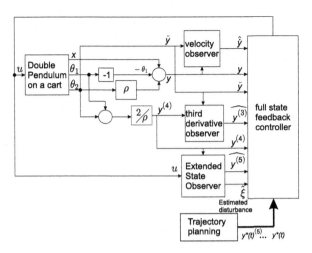

FIGURE 6.14 ADRC scheme for the double inverted pendulum on a cart.

Clearly, this observer produces an estimate $\hat{e}_{y_3}$ of the third derivative of the incremental flat output tracking error $y - y^*(t)$.

The third relation prompts an ESO that simultaneously estimates the fifth-order time derivative of the flat output tracking error, here denoted by $\hat{e}_{y_5}$, and the possible unknown disturbances summarized in the extension variable z due to neglected nonlinearities and exogenous perturbations (unmodeled dynamics, force disturbances, and so on). The actual value of the $e_{y^{(4)}}$ coincides with the linear combination $\frac{2}{\rho}\left(e_{\theta_1} + e_{\theta_2}\right)$, and the redundant estimate of $e_{y^{(4)}}$ is denoted by $\hat{e}_{y_4}$. We have

$$\frac{d}{dt}\hat{e}_{y_4} = \hat{e}_{y_5} + \lambda_6\left[\frac{2}{\rho}\left(e_{\theta_1} + e_{\theta_2}\right) - \hat{e}_{y_4}\right]$$

$$\frac{d}{dt}\hat{e}_{y_5} = \hat{z} + \frac{2}{\rho\epsilon}\left[e_u + \frac{2}{\rho}(\rho + \rho\epsilon + \epsilon)(e_{\theta_1} + e_{\theta_2}) - (2 + \epsilon)e_{\theta_2}\right]$$

$$+ \lambda_5\left[\frac{2}{\rho}\left(e_{\theta_1} + e_{\theta_2}\right) - \hat{e}_{y_4}\right]$$

$$\frac{d}{dt}\hat{z} = \lambda_4\left[\frac{2}{\rho}\left(e_{\theta_1} + e_{\theta_2}\right) - \hat{e}_{y_4}\right] \tag{6.73}$$

This observer produces an estimate $\hat{e}_{y_5}$ of the fifth-order time derivative of the incremental flat output tracking error e_y and of the natural endogenous disturbances caused by the nonlinearities neglected in the linearization of the original system and by some possible additional uncertainties affecting the system.

As in the previous example, the observer may be simplified by neglecting the state terms inherited from the linearized model (see Fig. 6.14). We may safely propose

$$\frac{d}{dt}\hat{e}_{y4} = \hat{e}_{y5} + \lambda_6 \left[\frac{2}{\rho}(\theta_1 + \theta_2) - \hat{y}_4 \right)$$
$$\frac{d}{dt}\hat{y}_5 = \hat{z} + \frac{2}{\rho\epsilon}e_u + \lambda_5 \left[\frac{2}{\rho}(e_{\theta_1} + e_{\theta_2}) - \hat{e}_{y4} \right]$$
$$\frac{d}{dt}\hat{z} = \lambda_4 \left[\frac{2}{\rho}(e_{\theta_1} + e_{\theta_2}) - \hat{e}_{y4} \right] \tag{6.74}$$

The output observation error $e_0 = e_{y0} - \hat{e}_{y0}$ and its time derivatives $e_1 = e_{\theta_2} - \hat{e}_{y0}$ and $e_2 = \frac{2}{\rho}(e_{\theta_1} + e_{\theta_2}) - \hat{e}_{y4}$ generate the following set of reconstruction error dynamics:

$$\ddot{e}_0 + \lambda_1 \dot{e}_0 + \lambda_0 e_0 = 0$$
$$\ddot{e}_1 + \lambda_3 \dot{e}_1 + \lambda_2 e_1 = 0$$
$$e_2^{(3)} + \lambda_6 \ddot{e}_2 + \lambda_5 \dot{e}_e + \lambda_4 e_2 = 0 \tag{6.75}$$

An appropriate choice of the coefficients $(\lambda_6, \ldots, \lambda_0)$ guarantees exponentially decreasing estimation errors e_0, e_1, and e_2. Using, respectively, a second- and third-order Hurwitz polynomials of the forms $(s^2 + 2\varsigma_0\omega_{n0}s + \omega_{n0}^2)$ and $(s^2 + 2\varsigma_1\omega_{n1}s + \omega_{n1}^2)(s + p)$, we obtain

$$\lambda_1 = 2\varsigma_0\omega_{n0}, \quad \lambda_0 = \omega_{n0}^2, \quad \lambda_3 = 2\varsigma_0\omega_{n0}, \quad \lambda_2 = \omega_{n0}^2, \quad \lambda_6 = 2\varsigma_1\omega_{n1} + p$$
$$\lambda_5 = 2p\varsigma_1\omega_{n1} + \omega_{n1}^2, \quad \lambda_4 = p\omega_{n1}^2$$

The tracking controller is synthesized with a canceling strategy in mind based on the estimate of the total disturbance input function $\xi(t)$. The estimate of $\xi(t)$ is denoted by $\hat{z}_1$.

The output feedback control $e_u = u - u^*(t)$ is given by

$$e_u = \frac{\rho\epsilon}{2} \left\{ -\hat{z} - \sum_{i=0}^{5} k_i \left(\hat{e}_{y_i} \right) \right\} \tag{6.76}$$

The linear controller gains $\{k_0, \cdots, k_5\}$ were set by using the coefficients of a desired sixth-order Hurwitz polynomial of the form $(s^2 + 2\zeta\omega_n s + \omega_n^2)^3$. We obtain

$$k_5 = 6\zeta\omega_n, \quad k_4 = (3\omega_n^2 + 12\zeta^2\omega_n^2), \quad k_3 = 8\zeta^3\omega_n^3 + 12\zeta\omega_n^3$$
$$k_2 = 12\zeta^2\omega_n^4 + 3\omega_n^4, \quad k_1 = 6\zeta\omega_n^5, \quad k_0 = \omega_n^6$$

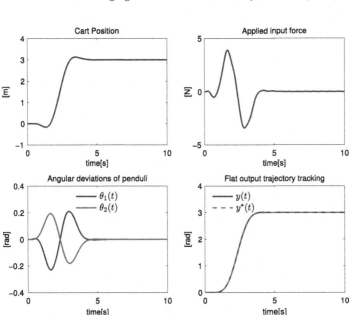

FIGURE 6.15 Performance of double inverted pendulum system to synthesized ADRC control.

6.4.5 Simulations

The following set of parameter values was used in the simulations:

$$M = 0.5 \text{ [kg]}, \quad m_1 = 0.3 \text{ [kg]}, \quad m_2 = 0.3 \text{ [kg]}$$

$$L_1 = 1 \text{ [m]}, \quad L_2 = 0.75 \text{ [m]}, \quad g = 9.8 \text{ [m/s}^2] \qquad (6.77)$$

The incremental flat output trajectory $y_\delta^*(t)$ was specified as a rest-to-rest trajectory to smoothly interpolate between the initial value of zero and a final desired value 3. The initial conditions for the system were all set to be zero. Fig. 6.15 shows the behavior of the double inverted pendulum in a trajectory tracking task on the flat output.

6.5 THE FURUTA PENDULUM

The Furuta pendulum [25], also called the rotational pendulum, is one of the most popular underactuated systems used in academic laboratories around the world. The system is provided with one control input and has two mechanical degrees of freedom. It consists of an actuated arm that rotates in the horizontal plane; the actuated arm is joined to a nonactuated pendulum that rotates loosely in a vertical plane perpendicular, at the tip, to the horizontal rotating arm. The nonlinearities of the system include the gravitational, Coriolis, and centripetal forces and the acceleration couplings [26]. In addition, the system is nonfeedback linearizable [1].

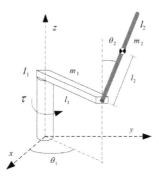

FIGURE 6.16 Schematics of the Furuta Pendulum.

The tangent linearization around the unstable equilibrium point of the Furuta pendulum is flat (i.e., it is controllable) with a physically measurable flat output. Thus, the problem of stabilization and tracking can be tackled from a combined perspective of flatness and ADRC. Here, low-order cascaded observers are used, motivated by the structure revealed by flatness in the context of an ADRC solution to the trajectory tracking problem. In this section, an ADRC scheme is proposed for a trajectory tracking problem associated with the Furuta pendulum (actuated link), whereas the unactuated pendulum is to remain around its unstable vertical position without falling during the entire tracking maneuver.

6.5.1 The Furuta Pendulum Model

The Furuta pendulum is shown in Fig. 6.16. The nonlinear model of the mechanical part of the system can be derived from either Newton equations or from the Euler–Lagrange formalism [27]:

$$\left(I_1 + m_2(l_1^2 + l_2^2 \sin^2(\theta_2))\right)\ddot{\theta}_1 - m_2 l_2 l_1 \cos(\theta_2)\ddot{\theta}_2 + 2m_2 l_2^2 \dot{\theta}_1 \dot{\theta}_2 \sin(\theta_2)\cos(\theta_2)$$
$$+ m_2 l_2 l_1 \dot{\theta}_2^2 \sin(\theta_2) = \tau \tag{6.78}$$

$$\left(I_2 + m_2 l_2^2\right)\ddot{\theta}_2 - m_2 l_2 l_1 \cos(\theta_2)\ddot{\theta}_1 - m_2 l_2^2 \dot{\theta}_2^2 \sin(\theta_2)\cos(\theta_2)$$
$$- m_2 g l_2 \sin(\theta_2) = 0 \tag{6.79}$$

where θ_2 is the angular displacement of the pendulum with respect to the vertical line passing through the pivot of the pendulum, θ_1 is the angle of the horizontal arm measured with respect to an arbitrary but fixed direction in the (x, y) plane, m_2 denotes the mass of the pendulum, I_2 stands for the inertia of the pendulum, I_1 represents the inertia of the horizontal arm, l_1 and l_2 are, respectively, the lengths of the horizontal arm and the distance between the center of mass of the pendulum and the pendulum pivot, and τ is the control input directly applied by the DC motor to the horizontal arm. We specifically assume that only the angular positions θ_1 and θ_2 are measurable. The model (6.78)–(6.79) is not feedback linearizable, i.e., it is nonflat.

Consider the tangent linearization of the system around the following arbitrary unstable equilibrium point:

$$\bar{\theta}_1 = 0,\ \bar{\theta}_2 = 0,\ \bar{\tau} = 0,\ \dot{\bar{\theta}}_1 = 0,\ \dot{\bar{\theta}}_2 = 0 \tag{6.80}$$

We readily obtain

$$\left(I_1 + m_2 l_1^2\right)\ddot{\theta}_{\delta 1} - m_2 l_2 l_1 \ddot{\theta}_{\delta 2} = \tau_\delta \tag{6.81}$$

$$\left(I_2 + m_2 l_2^2\right)\ddot{\theta}_{\delta 2} - m_2 l_2 l_1 \ddot{\theta}_{\delta 1} - m_2 g l_2 \theta_{\delta 2} = 0 \tag{6.82}$$

where $\theta_{\delta 1} = \theta_1 - 0 = \theta_1$, $\theta_{\delta 2} = \theta_2 - 0 = \theta_2$, and $\tau_\delta = \tau - 0 = \tau$ are the incremental states of the linearized system. To simplify the notation, we define

$$\alpha = I_1 + m_2 l_1^2 \quad \epsilon = m_2 l_2 l_1 \quad \rho = I_2 + m_2 l_2^2 \quad \eta = m_2 g l_2$$

We obtain the following implicit incremental description of the system:

$$\alpha \ddot{\theta}_{\delta 1} - \epsilon \ddot{\theta}_{\delta 2} = \tau_\delta \tag{6.83}$$

$$\rho \ddot{\theta}_{\delta 2} - \epsilon \ddot{\theta}_{\delta 1} - \eta \theta_{\delta 2} = 0 \tag{6.84}$$

6.5.2 Flatness of the Linearized Furuta Pendulum Model

The linear model (6.84) is differentially flat with incremental flat output, denoted by F_δ, given in this case by the following expression:

$$F_\delta = \theta_{\delta 1} - \frac{\rho}{\epsilon}\theta_{\delta 2} \tag{6.85}$$

Indeed, all system variables in the linear model, i.e., states and the control input are expressible as differential functions of the incremental flat output. In other words, they are expressible as functions of the flat output F_δ and a finite number of its time derivatives:

$$\theta_{\delta 1} = F_\delta - \frac{\rho}{\eta}\ddot{F}_\delta, \quad \dot{\theta}_{\delta 1} = \dot{F}_\delta - \frac{\rho}{\eta}F_\delta^{(3)}, \quad \theta_{\delta 2} = -\frac{\epsilon}{\eta}\ddot{F}_\delta, \dot{\theta}_{\delta 2} = -\frac{\epsilon}{\eta}F_\delta^{(3)}$$

$$\tau_\delta = \left(\frac{\epsilon^2 - \alpha\rho}{\eta}\right)F_\delta^{(4)} + \alpha\ddot{F}_\delta \tag{6.86}$$

The linearized input to flat output model of the system is given by

$$F_\delta^{(4)} = \left(\frac{\eta}{\epsilon^2 - \alpha\rho}\right)\tau_\delta - \left(\frac{\alpha\eta}{\epsilon^2 - \alpha\rho}\right)\ddot{F}_\delta \tag{6.87}$$

where it is assumed that $\epsilon^2 - \alpha\rho \neq 0$.

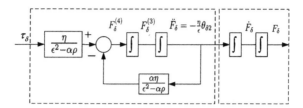

FIGURE 6.17 Cascade structure of the linearized Furuta Pendulum System.

6.5.3 A Set of Low-Order Observers for ADRC

It is immediate from the differential parameterization (6.86) that the tangent system naturally decomposes into a cascade connection of two independent blocks, the first one controlled by the torque input τ_δ with the corresponding output given by the flat output incremental acceleration $\ddot{F}_\delta$. This output coincides, modulo a constant factor $-\frac{\eta}{\epsilon}$, with the vertical arm incremental angular position $\theta_{\delta 2}$, i.e., $\ddot{F}_\delta = -\frac{\eta}{\epsilon}\theta_{\delta 2}$. The signal $-\frac{\eta}{\epsilon}\theta_{\delta 2}$ acts then as an auxiliary known (i.e., measurable) input to the second block, which consists of a chain of two integrators rendering the phase variables $\dot{F}_\delta$ and F_δ. The last variable F_δ is the incremental flat output to be controlled in the overall system (see Fig. 6.17).

This cascading property simplifies and decouples the observer design task in the flatness-based ADRC scheme to be presented next.

6.5.4 Problem Formulation

On the basis of (6.87), we adopt the following simplified perturbed model for the underlying nonlinear Furuta pendulum system (6.87):

$$F_\delta^{(4)} = \left(\frac{\eta}{\epsilon^2 - \alpha\rho}\right)\tau_\delta + \tilde{\xi}(t) \qquad (6.88)$$

where $\tilde{\xi}(t)$ represents state-dependent expressions, all the higher-order terms (h.o.t.) are neglected by linearization, the possibly unmodeled dynamics, and external unknown disturbances affecting the system. We lump all this uncertain terms into a single time-varying function model represented by $\tilde{\xi}(t)$, which in our case is of the form

$$\tilde{\xi}(t) = -\left(\frac{\alpha\eta}{\epsilon^2 - \alpha\rho}\right)\ddot{F}_\delta + \text{h.o.t.} \qquad (6.89)$$

Suppose that we desire to transfer the horizontal arm from the initial position $\theta_{\delta 1}(0) = 0$ toward the final position $\theta_{\delta 1}(t_f) = \Theta$ in a finite prespecified time interval $[0, t_f]$. The maneuver is to be carried out while the vertical arm (pendulum) evolves closely around the unstable position (i.e., $\theta_{\delta 2}(0) = 0$). We desire to accomplish this maneuver without losing the vertical unstable position of the pendulum arm $\theta_{\delta 1}(t_f) = 0$ at the end of, and long thereafter of, the prescribed time interval. Clearly, such a rest-to-rest maneuver is feasible via an adequate planning of the trajectory for the flat output $F_\delta = \theta_{\delta 1} - \frac{\rho}{\epsilon}\theta_{\delta 2}$. The vertical arm is required to start and end at the unstable position, and then $\theta_2(0) = \theta_2(t_f) = 0$. The horizontal arm orientation θ_1 starts at rest at $\theta_1(0) = 0$ and ends the maneuver at a different rest point $\theta_1(t_f) = \Theta_f$ with $\dot{\theta}_1(0) = \dot{\theta}_1(t_f) = 0$. Hence, the initial value of the flat output is $F_\delta(0) = 0$, whereas its final value is $F_\delta(t_f) = \Theta_f$. We could prescribe a smooth rest to rest trajectory for $F_\delta^*(t)$ using, for instance, a Bèzier polynomial with sufficient time derivatives being zero at the initial and final instants.

The flat output trajectory tracking error $e_{F_\delta} = F_\delta - F_\delta^*(t)$ is seen to evolve according to

$$e_{F_\delta}^{(4)} = \left(\frac{\eta}{\epsilon^2 - \alpha\rho}\right)\tau_\delta + \xi(t) \qquad (6.90)$$

where $\xi(t)$ represents the previously defined unknown input term complemented now by the effects arising from the prescribed nominal flat output trajectory and its various time derivatives. Let $e_{F_\delta}^{(i)} = e_i$, $i = 0, 1, 2, 3$, be the flat output tracking error and its first three time derivatives. The flat output trajectory tracking error perturbed state space model is given by

$$\dot{e}_0 = e_1 \qquad (6.91)$$
$$\dot{e}_1 = e_2$$
$$\dot{e}_2 = e_3$$
$$\dot{e}_3 = \left(\frac{\eta}{\epsilon^2 - \alpha\rho}\right)\tau_\delta + \xi(t)$$

At this point, we make use of the cascading property and view the previous system as the connection of two subsystems. Note that

$$e_2 = \ddot{F}_\delta - \ddot{F}_\delta^*(t) \qquad (6.92)$$

is the known input to the second-order pure integration system

$$\dot{e}_0 = e_1 \qquad (6.93)$$
$$\dot{e}_1 = \ddot{F}_\delta - \ddot{F}_\delta^*(t)$$

An observer for this subsystem will give an estimate of the unmeasured first-order time derivative of the flat output tracking error e_1.

The rest of the system is given by

$$\dot{e}_2 = e_3 \qquad\qquad (6.94)$$

$$\dot{e}_3 = \left(\frac{\eta}{\epsilon^2 - \alpha\rho}\right)\tau_\delta + \xi(t)$$

This subsystem prompts an ESO for the simultaneous observation of e_3, the third-order time derivative of the incremental flat output tracking error, and the total disturbance function $\xi(t)$.

6.5.5   A Cascaded GPI Observer-Based ADRC for the Furuta Pendulum

The perturbation term $\xi(t)$ is algebraically observable [28] since it can be expressed algebraically in terms of the input, the output, and their finite time derivatives. Then, we may propose an internal time polynomial model for such an unknown time-varying function $\xi(t)$, here denoted by z_1, and adopt, say, the following sixth-order time polynomial model for $\xi(t)$: $z^{(6)} = 0$. Then, with $z(t) = z_1$, the flat output trajectory tracking error model (6.91) may be rewritten as follows:

$$
\begin{aligned}
\dot{e}_0 &= e_1 \\
\dot{e}_1 &= e_2 \\
\dot{e}_2 &= e_3 \\
\dot{e}_3 &= \left(\frac{\eta}{\epsilon^2 - \alpha\rho}\right)\tau_\delta + z_1 \\
\dot{z}_1 &= z_2 \\
\dot{z}_2 &= z_3 \\
&\;\;\vdots \\
\dot{z}_6 &= 0
\end{aligned}
\qquad\qquad (6.95)
$$

We propose a set of two lower-order linear observers for the simultaneous estimation of the unmeasured phase variables associated with the output tracking error e_{F_δ} and the disturbance signal term $\xi(t)$:

$$
\begin{aligned}
\dot{\hat{e}}_0 &= \hat{e}_1 + \kappa_1(e_0 - \hat{e}_0) \\
\dot{\hat{e}}_1 &= F_\delta - F_\delta^*(t) + \kappa_0(e_0 - \hat{e}_0)
\end{aligned}
\qquad (6.96)
$$

where $F_\delta = \theta_{\delta 1} - \frac{\rho}{\epsilon}\theta_{\delta 2}$. This observer yields $\hat{e}_1$, an estimate of $\dot{e}_{F_\delta}$. The second observer, including a disturbance observer, is given by

$$\dot{\hat{e}}_2 = \hat{e}_3 + \lambda_7(e_2 - \hat{e}_2)$$

$$\dot{\hat{e}}_3 = \left(\frac{\eta}{\epsilon^2 - \alpha\rho}\right)\tau_\delta + \hat{z}_1 + \lambda_6(e_2 - \hat{e}_2)$$
$$\dot{\hat{z}}_1 = \hat{z}_2 + \lambda_5(e_2 - \hat{e}_2)$$
$$\dot{\hat{z}}_2 = \hat{z}_3 + \lambda_4(e_2 - \hat{e}_2)$$
$$\dot{\hat{z}}_3 = \hat{z}_4 + \lambda_3(e_2 - \hat{e}_2)$$
$$\dot{\hat{z}}_4 = \hat{z}_5 + \lambda_2(e_2 - \hat{e}_2)$$
$$\dot{\hat{z}}_5 = \hat{z}_6 + \lambda_1(e_2 - \hat{e}_2)$$
$$\dot{\hat{z}}_6 = \lambda_0(e_2 - \hat{e}_2) \tag{6.97}$$

where $e_2 = \ddot{F}_\delta - \ddot{F}^*(t) = -\frac{\eta}{\epsilon}\theta_{\delta 2} - \ddot{F}^*(t)$. The observation error $\tilde{e}_0 = e_0 - \hat{e}_0$ of the incremental flat output tracking error generates the following linear injected estimation error dynamics:

$$\ddot{\tilde{e}}_0 + \kappa_1\dot{\tilde{e}}_0 + \kappa_0\tilde{e}_0 = 0 \tag{6.98}$$

An appropriate choice of the design coefficients $\{\kappa_1, \kappa_0\}$, placing the roots of the corresponding characteristic polynomial deep into the left half of the complex plane, renders an asymptotically exponentially decreasing estimation error state $\tilde{e}_1$, $\dot{\tilde{e}}_1 = \tilde{e}_2$. The tracking error velocity for the flat output $\dot{e}_{F_\delta}$ is, thus, accurately estimated for feedback purposes.

Similarly, consider the observation error $\tilde{e}_2 = e_2 - \hat{e}_2$ of the flat output acceleration tracking error. It generates the following dominantly linear reconstruction error dynamics:

$$\tilde{e}_3^{(8)} + \lambda_7\tilde{e}_3^{(7)} + \lambda_6\tilde{e}_3^{(6)} + \cdots + \lambda_1\dot{\tilde{e}}_3 + \lambda_0\tilde{e}_3 = \xi^{(6)}(t) \tag{6.99}$$

We can prove that a necessary and sufficient condition for having the incremental flat output acceleration estimation error $\tilde{e}_3$ and its associated phase variables $\dot{\tilde{e}}_3, \ddot{\tilde{e}}_3, \ldots, \tilde{e}_3^{(8)}$ ultimately uniformly converge toward a small as desired neighborhood of the acceleration estimation error phase space is that $\xi^{(6)}(t)$ is uniformly absolutely bounded. An appropriate choice of the design coefficients $\{\lambda_7, \ldots, \lambda_1, \lambda_0\}$, placing the poles of the associated linear homogeneous system sufficiently far into the left half of the complex plane, renders a uniformly asymptotically convergent estimation error $\tilde{e}_3$ toward an arbitrary small vicinity of the origin along with a finite number of its time derivatives.

In the case at hand, the observer gain parameters λ_j for $j = 0, 1, 2, \cdots, 8$ are chosen using the methodology proposed by Kim et al. [29] based on extensive use of the so-called characteristic ratios of the characteristic polynomial. This methodology suitably mitigates the "peaking phenomenon," typical for high gain pole placement injected responses of the observer [30].

Consider a characteristic polynomial $p(s)$ of the form

$$a_n s^n + a_{n-1} s^{n-1} + \cdots + a_2 s^2 + a_1 s + a_0, \qquad a_i > 0 \qquad (6.100)$$

and let α_i be the characteristic ratios of $p(s)$. It has been shown [29] that if the following two conditions are satisfied, then the polynomial (6.100) is Hurwitz:

1) $\alpha_1 > 2$

2) $\alpha_k = \dfrac{\sin\left(\frac{k\pi}{n}\right) + \sin\left(\frac{\pi}{n}\right)}{2\sin\left(\frac{k\pi}{n}\right)} \alpha_1$ \qquad (6.101)

for $k = 2, 3, \ldots, n - 1$. The construction of the all-pole stable characteristic polynomial involves only α_1, which we require to be greater than 2. Thus, this result allows us to characterize the reference all-pole systems by adjusting a single parameter α_1 to achieve the desired damping. Since the generalized time constant can be chosen independently of α_i.

For arbitrary a_0 and $T > 0$,

$$a_1 = T a_0$$

$$a_i = \frac{T^i a_0}{\alpha_{i-1}\alpha_{i-2}^2\alpha_{i-3}^3 \cdots \alpha_1^{i-1}} \quad \text{for } i = 2, 3, \ldots, n$$

$$\lambda_j = \left(\frac{a_j}{a_n}\right), \quad \text{for } j = 0, 1, , 3, \ldots, 8 \qquad (6.102)$$

The smoothness of the error responses and noise rejection properties, bestowed by the scheme, make it a highly recommendable choice for pole placement in practical situations. The reader may find details in the above-cited reference.

6.5.6 The ADRC Controller Design

The control input may then be readily synthesized with an active disturbance canceling strategy for the uncertain input $\xi(t)$ in terms of its estimated value $\hat{z}_1$ and the use, for feedback purposes, of the estimated time derivatives associated with the incremental flat output tracking error e_1 and the measurable incremental flat output acceleration error e_3. We then propose

$$\tau_\delta = -\left(\frac{\epsilon^2 - \alpha\rho}{\eta}\right)\left[\hat{z}_1 + k_3\hat{e}_3 + k_2 e_2 + k_1\hat{e}_1 + k_0 e_0\right] \qquad (6.103)$$

where, naturally, the measured tracking errors e_0 and e_2 are used instead of their redundant estimates. This is due to the fact that these variables are known

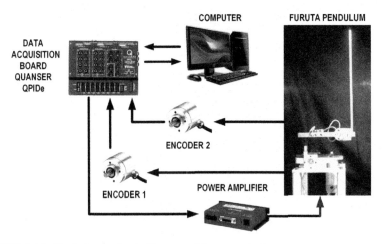

FIGURE 6.18 Block diagram for the Furuta pendulum control scheme.

to be measurable through the original position variables θ and ϕ and their incremental values. Note that the coefficients of the controller should be chosen in accordance with the fact that the incremental flat output tracking error is approximately governed by the differential equation

$$e_0^{(4)} + k_3 e_0^{(3)} + k_2 \ddot{e}_0 + k_1 \dot{e}_0 + k_0 e_0 = \xi(t) - \hat{z}_1 \tag{6.104}$$

and the set of design coefficients $\{k_3, ..., k_1, k_0\}$ should render the underlying characteristic polynomial $p(s) = s^{(4)} + k_3 s^{(3)} + k_2 s^2 + k_1 s + k_0$ to be a Hurwitz polynomial. We propose $k_3 = 4\zeta_c \omega_c$, $k_2 = 2\omega_c^2 + 4\zeta_c^2 \omega_c^2$, $k_1 = 4\zeta_c \omega_c^3$, $k_0 = \omega_c^4$.

6.5.7 Experimental Results

Fig. 6.18 shows a diagram of the experimental platform used for the Furuta pendulum. The experimental device (Fig. 6.19) consists of a Brushed servomotor from Moog, model C34L80W40, which drives the horizontal arm through a synchronous belt with ration 4.5:1. The angles of the pendulum and arm (motor) can be measured with incremental optical encoders of 2500 counts per revolution. A digital amplifier, model Junus 90, working in current mode, is in charge of driving the motor.

The data acquisition is carried out through a data card from Quanser consulting, model QPIDe terminal board. This card reads signals from the optical incremental encoders and supplies control voltages to the power amplifiers. The control strategy was implemented in the Matlab-Simulink platform. Finally, the sampling time was set to be 0.0005 [s]. The Furuta pendulum parame-

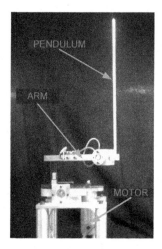

FIGURE 6.19 Furuta pendulum system prototype.

ters were $l_0 = 0.33$ [m], $l_1 = 0.275$ [m], $m_0 = 1.64$ [kg], $m_1 = 0.141$ [kg], $I_0 = 0.0481$ [kgm^2], $I_1 = 0.0036$ [kgm^2], which are used in (6.97) and (6.103) to obtain ρ, η, α, ϵ, and γ respectively.

The initial conditions for the joint variables in the system were [$\theta_1 = 0$, $\theta_2 = 0$]. The observer gain parameters for the observation error $\tilde{e}_1$ were set to be $\zeta_o = 2$, $\omega_o = 15$. The observer gain parameters for the dynamics of the observation error $\tilde{e}_3$ were set to be $n = 8$, $T = 6$, $a_0 = 16$, $\alpha = 4$. The controller design parameters were specified to be $\zeta_c = 1$, $\omega_c = 12$.

Fig. 6.20 shows the performance of the Furuta pendulum experimental platform; while the horizontal arm is rotating, the pendulum lies nearly vertically at the unstable position $\theta_2 = 0$. Fig. 6.21 shows the performance of the horizontal arm from the initial position $\theta_1(2) = 0$ toward the final position $\theta_1(5.4) = 2\pi$ during the time interval $t \in [2, 5.4]$ [s]. The produced control torque is depicted in Fig. 6.22. Note that the lumped disturbance estimation (see Fig. 6.23) determines the waveform of the control input, which tries to cancel out the additive disturbance inputs. Fig. 6.24 shows the behavior of the flat output $F_\delta = \theta_{\delta 1} - \frac{\rho}{\epsilon}\theta_{\delta 2}$ from the initial value $F_\delta(2) = 0$ toward the desired final value $F_\delta(5.4) = 2\pi$, Fig. 6.25 shows the tracking error, which remains restricted to a small bounded zone centered around the origin of the error phase space.

6.6 THE BALL-AND-BEAM SYSTEM

Consider the ball-and-beam system shown in Fig. 6.26. This system consists of a ball placed on a beam, which undergoes an angular displacement around a centered pivot, actuated by a dc motor directly coupled to the beam by means of

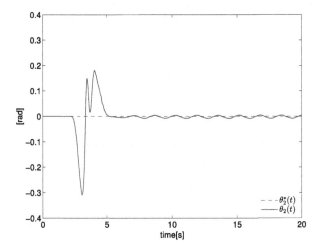

FIGURE 6.20 Vertical angle closed-loop behavior.

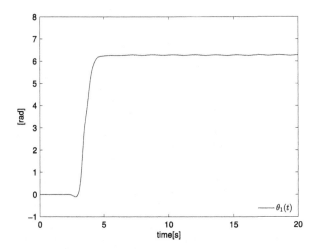

FIGURE 6.21 Horizontal arm angular reference trajectory tracking behavior.

a pulley. The dynamical model is represented as follows [31]:

$$\left(m_2 x^2 + m_1 l_1^2 + I_1 + I_p\right)\ddot{\theta} + 2m_2 x\dot{x}\dot{\theta} + m_2 gx\cos\theta + m_1 gl_1\sin\theta = N\tau$$

(6.105)

$$\left(m_2 + \frac{I_2}{R^2}\right)\ddot{x} - m_2 x\dot{\theta}^2 + m_2 g\sin\theta = 0$$

(6.106)

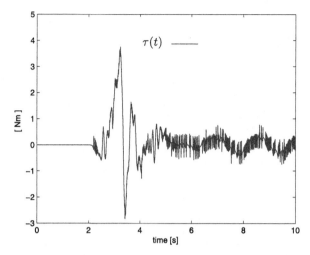

FIGURE 6.22 Torque input trajectory.

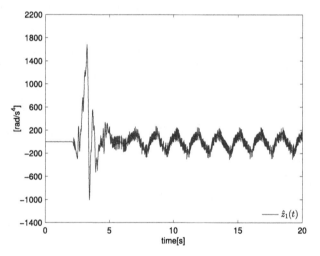

FIGURE 6.23 Lumped online disturbance estimation.

where $x \in \mathbb{R}$ denotes the position of the ball measured from the center of mass of the beam, $\theta \in \mathbb{R}$ is the angular position of the beam, $m_2 \in \mathbb{R}$ and $R \in \mathbb{R}$ denote, respectively, the ball mass and its radius, $I_2 = \frac{2}{5} m_2 R^2 \in \mathbb{R}$ is the ball inertia, $I_1 \in \mathbb{R}$ denotes the beam inertia, $m_1 \in \mathbb{R}$ is the beam mass, $I_p \in \mathbb{R}$ represents the inertia of the pulley system, $N \in \mathbb{R}$ is a ratio of distances concerning the pulley/motor actuator. The control input (motor torque) $\tau \in \mathbb{R}$, can be expressed as a function of the motor voltage through the approximate relation

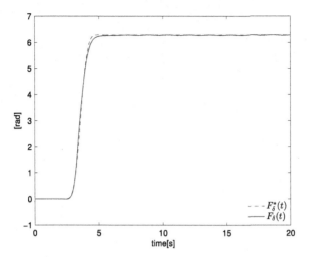

FIGURE 6.24 Flat output reference trajectory tracking.

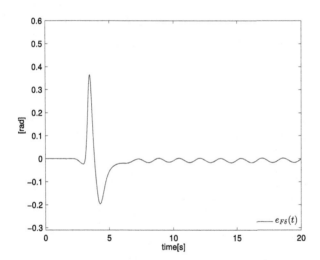

FIGURE 6.25 Tracking error for the flat output.

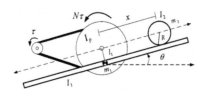

FIGURE 6.26 A schematic of the Ball-and-Beam system.

$\tau(t) = \dfrac{k_\tau}{R_a} V(t)$, where $k_\tau \in \mathbb{R}$ is the motor torque constant, and $R_a \in \mathbb{R}$ is the motor armature electric resistance.

The problem formulation is as follows: *It is desired to transfer the ball position from an initial value $x(0)$ to a given final value $x(t_{final})$ in a prescribed finite interval of time $[0, t_{final}]$ through a smooth rest-to-rest reference trajectory.*

6.6.1 Tangent Linearization

The system desired equilibrium point is described by $\bar{\theta} = 0, \dot{\bar{\theta}} = 0, \bar{x} = 0, \dot{\bar{x}} = 0,$ $\bar{V} = 0$. The approximate linearization of system (6.105)–(6.106) around the equilibrium point is given by

$$\left(m_1 l_1^2 + I_1 + I_p \right) \ddot{\theta}_\delta + m_2 g x_\delta + m_1 g l_1 \theta_\delta = \frac{k_\tau N}{R_a} V_\delta \qquad (6.107)$$

$$\left(m_2 + \frac{I_2}{R^2} \right) \ddot{x}_\delta + m_2 g \theta_\delta = 0 \qquad (6.108)$$

with incremental states $\theta_\delta = \theta - \bar{\theta} = \theta, \dot{\theta}_\delta = \dot{\theta} - \dot{\bar{\theta}} = \dot{\theta}, x_\delta = x - \bar{x} = x, \dot{x}_\delta = \dot{x} - \dot{\bar{x}} = \dot{x}$, and $V_\delta = V - \bar{V} = V$. To simplify the notation, we define:

$$\alpha = m_1 l_1^2 + I_1 + I_p, \quad \beta = m_2 g, \quad \sigma = m_1 g l_1,$$

$$\gamma = m_2 + \frac{I_2}{R^2}, \quad \mu = \frac{k_\tau N}{R_a} \qquad (6.109)$$

The system is reexpressed as

$$\alpha \ddot{\theta}_\delta + \beta x_\delta + \sigma \theta_\delta = \mu V_\delta \qquad (6.110)$$

$$\gamma \ddot{x}_\delta + \beta \theta_\delta = 0 \qquad (6.111)$$

6.6.2 Obtaining a Flat Output

The linearized model (6.110)–(6.111) is controllable and hence flat. The flat output coincides with the incremental position of the ball. We define:

$$x_\delta = x \qquad (6.112)$$

coincident here with the actual ball position x. We obtain the following parameterization of the ball and beam system states in term of flat output and a finite number of its time derivatives:

$$x = x_\delta, \quad \dot{x} = \dot{x}_\delta, \quad \theta_\delta = -\frac{\alpha}{\beta} \ddot{x}_\delta, \quad \dot{\theta}_\delta = -\frac{\alpha}{\beta} x_\delta^{(3)}$$

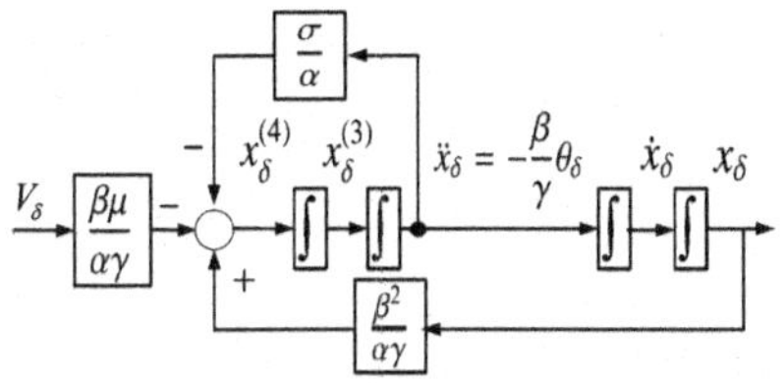

FIGURE 6.27 Cascade structure of ball-and-beam system.

$$V_\delta = -\frac{\alpha\gamma}{\beta\mu}x_\delta^{(4)} - \frac{\sigma\gamma}{\beta\mu}\ddot{x}_\delta + \frac{\beta}{\mu}x_\delta \tag{6.113}$$

6.6.3 On the Cascade Form of the Linearized Ball-and-Beam System

The linearized acceleration of the ball can be expressed in terms of the angular position of the beam through the following relation:

$$\ddot{x}_\delta = -\frac{\beta}{\gamma}\theta_\delta \tag{6.114}$$

The incremental input-to-flat output relation for the ball-and-beam system (6.113) is represented by means of a fourth-order linear time-invariant system

$$x_\delta^{(4)} = -\frac{\beta\mu}{\alpha\gamma}V_\delta - \frac{\sigma}{\alpha}\ddot{x}_\delta + \frac{\beta^2}{\alpha\gamma}x_\delta \tag{6.115}$$

Note that the linearized system naturally decomposes into a cascade connection of two blocks. The first one is controlled by the input voltage $V_\delta(t)$ with the corresponding output given by the flat output incremental acceleration $\ddot{x}_\delta = \ddot{x}$. This output coincides, modulo a constant factor $\frac{\beta}{\gamma}$, with the beam incremental angular position $\theta_\delta = \theta$, i.e., $\ddot{x}_\delta = -\frac{\beta}{\gamma}\theta_\delta$. The signal $-\frac{\beta}{\gamma}\theta_\delta$ then acts as an auxiliary measurable input to the second block, which consists of an elementary chain of two integrators. These facts render, simply enough, the possibility of estimation of the phase variables $\dot{x}_\delta$ and x_δ (Fig. 6.27). This cascade property simplifies and decomposes the Extended State Observer design task for the ADRC scheme into two lower-order observer designs. The net result is that online observer-based phase variable estimations are limited to a simpler estimation of the first- and third-order time derivatives of the flat output. The alternative is to estimate the consecutive time derivatives of the flat output up to third order.

6.6.4 Implementing the Active Disturbance Rejection Controller for the Linearized System

The linearized system can be expressed as follows:

$$x_\delta^{(4)} = -\frac{\beta\mu}{\alpha\gamma}V_\delta - \frac{\sigma}{\alpha}\ddot{x}_\delta + \frac{\beta^2}{\alpha\gamma}x_\delta + \text{h.o.t.} + v(t) \tag{6.116}$$

where h.o.t. stands for high-order terms, and $v(t)$ includes the unmodeled dynamics, external disturbances, exogenous perturbations, etc. The flat output trajectory tracking error is defined as $e_{x_\delta} := x_\delta - x^*(t) = x - \bar{x} - (x^*(t) - \bar{x})$, which coincides with the tracking error $x - x^*(t)$.

Let us treat the tracking error dynamics as the following simplified perturbed model:

$$e_{x_\delta}^{(4)} = -\frac{\beta\mu}{\alpha\gamma}V_\delta + \xi(t) \tag{6.117}$$

where $\xi(t)$ is the total disturbance representing the effects of the neglected incremental state-dependent linearities, un-modeled dynamics, external disturbances, and, especially, the nonlinearities collected in the higher-order terms, eliminated by the linearization, and $V_\delta^*(t) \in \mathbb{R}$ is the feedforward control input. In other words,

$$\xi(t) = \frac{\beta\mu}{\alpha\gamma}V_\delta^* - \frac{\sigma}{\alpha}\ddot{x}_\delta + \frac{\beta^2}{\alpha\gamma}x_\delta + \text{h.o.t.} + v(t) \tag{6.118}$$

6.6.4.1 Extended State Observer Design

Denote, $e_{x_\delta}^{(i)} \in \mathbb{R}$ simply as e_i, $i = 0, 1, \dots, n-1$. The flat output trajectory tracking error perturbed state space model is given by

$$\dot{e}_0 = e_1$$
$$\dot{e}_1 = e_2$$
$$\dot{e}_2 = e_3$$
$$\dot{e}_3 = -\frac{\beta\mu}{\alpha\gamma}V_\delta + \xi(t) \tag{6.119}$$

Note that the variable e_2 may be expressed as

$$e_2 = \ddot{x}_\delta - \ddot{x}_\delta^*(t) = -\frac{\beta}{\gamma}\theta_\delta - \ddot{x}_\delta^*(t) \tag{6.120}$$

which represents a known input to the second-order pure integration system

$$\dot{e}_0 = e_1$$
$$\dot{e}_1 = e_2 = -\frac{\beta}{\gamma}\theta_\delta - \ddot{x}_\delta^*(t) \tag{6.121}$$

whereas the rest of the error system is given by

$$\dot{e}_2 = e_3$$
$$\dot{e}_3 = -\frac{\beta\mu}{\alpha\gamma}V_\delta + \xi(t) \tag{6.122}$$

A set of two second-order linear Extended State Observers can be proposed for the simultaneous estimation of the flat output tracking error phase variables associated with e_{x_δ} and of the perturbation signal $\xi(t) = z_1 \in \mathbb{R}$. Using the cascade property, for the phase variables e_0 and e_1 in Eq. (6.121), we set

$$\dot{\hat{e}}_0 = \hat{e}_1 + \rho_1(e_0 - \hat{e}_0)$$
$$\dot{\hat{e}}_1 = -\frac{\beta}{\gamma}\theta_\delta - \ddot{x}_\delta^*(t) + \rho_0(e_0 - \hat{e}_0) \tag{6.123}$$

whereas for the rest of the dynamics associated with e_2 and e_3, in (6.122), including the disturbance z_1, we synthesize:

$$\dot{\hat{e}}_2 = \hat{e}_3 + \lambda_{\varphi+1}(e_2 - \hat{e}_2)$$
$$\dot{\hat{e}}_3 = -\frac{\beta\mu}{\alpha\gamma}V_\delta + \hat{z}_1 + \lambda_\varphi(e_2 - \hat{e}_2)$$
$$\dot{\hat{z}}_j = \hat{z}_{j+1} + \lambda_{p-j}(e_2 - \hat{e}_2), \quad j = 1, \dots, m$$
$$\dot{z}_\varphi = \lambda_0(e_2 - \hat{e}_2) \tag{6.124}$$

where m is the order of the dynamic state extension for the disturbance observer, which, in this case, it is set to $m = 6$. The observation error $\tilde{e}_0 = e_0 - \hat{e}_0$ of the incremental flat output tracking error satisfies the following perturbed dynamics:

$$\dddot{e}_0 + \gamma_1\ddot{e}_0 + \gamma_0\tilde{e}_0 = \eta(\dot{\tilde{e}}_0, \tilde{e}_0, e_2) \tag{6.125}$$

where $\eta(\dot{\tilde{e}}_0, \tilde{e}_0, e_2)$ is a perturbation term depending on the estimates of the tracking errors and on e_2. An appropriate choice of the design coefficients $\{\gamma_1, \gamma_0\}$, placing the roots of the corresponding characteristic polynomial into the left half of the complex plane, renders an asymptotically exponentially decreasing estimation error phase variables $\tilde{e}_0$, $\dot{\tilde{e}}_0$, and $\ddot{\tilde{e}}_0$ toward previously chosen arbitrarily small neighborhoods of the origin. In this case, the parameters γ_1, γ_0 are selected such that a stable second-order dynamics with characteristic polynomial $s^2 + 2\zeta_o\omega_o s + \omega_o^2$, $\zeta_o, \omega_o \in \mathbb{R}^+$ being matched. The parameters are chosen as follows: $\gamma_1 = 2\zeta_o\omega_o$ and $\gamma_0 = \omega_o^2$, $\omega_0, \zeta_0 \in \mathbb{R}^+$. The tracking error velocity $\dot{e}_{r_\delta} = e_1$ is accurately estimated for feedback purposes by $\hat{e}_1$. The observation error of the flat output acceleration tracking error $\tilde{e}_2 = e_2 - \hat{e}_2$ generates the following linear reconstruction error dynamics:

$$\tilde{e}_2^{(m+2)} + \lambda_{m+1}\tilde{e}_2^{(m+1)} + \ldots + \lambda_1\dot{\tilde{e}}_2 + \lambda_0\tilde{e}_2 = [\xi(t)]^{(m)} \qquad (6.126)$$

A necessary and sufficient condition for having the incremental flat output acceleration estimation error $\tilde{e}_2$ and its associated phase variables $\tilde{e}_2, \dot{\tilde{e}}_2, \ldots, \tilde{e}_2^{(m+1)}$ ultimately uniformly converging toward a sufficiently small neighborhood of the acceleration estimation error phase space consists in choosing the observer gains λ_k $\{k = 0, \ldots, m+1\}$ sufficiently large and such that the linear injected error dynamics becomes Hurwitz. As performed in the last section, we propose the characteristic ratio assignment pole placement method [29].

The controller may be readily synthesized as a disturbance-canceling ADRC, properly removing the total input disturbance, $\xi(t)$ while imposing a desired exponentially dominated tracking error dynamics. This canceling action is achieved by using the total disturbance estimated value $\hat{z}_1$. Simultaneously, the tracking error phase variables estimates $\hat{e}_1$ and $\hat{e}_3$ are obtained from the proposed observers. The controller is thus given by

$$V_\delta = \frac{\alpha\gamma}{\beta\mu}\left[k_0 e_0 + k_1\hat{e}_1 + k_2 e_2 + k_3\hat{e}_3 + \hat{z}_1\right] \qquad (6.127)$$

where, naturally, the tracking errors e_0 and e_2 themselves are used instead of their estimates. The coefficients of the feedback controller are chosen considering the differential equation of the closed-loop flat output tracking error,

$$e_0^{(4)} + k_3 e_0^{(3)} + k_2\ddot{e}_0 + k_1\dot{e}_0 + k_0 e_0 = \xi(t) - \hat{z}_1 \qquad (6.128)$$

The set of coefficients $\{k_3, k_2, k_1, k_0\}$ is chosen so that the associated characteristic polynomial of the dominantly linear closed-loop dynamics $p_c(s) = s^4 + k_3 s^3 + k_2 s^2 + k_1 s + k_0$ becomes a Hurwitz polynomial. In this case, $k_3 = 4\zeta_c\omega_c, k_2 = 2\omega_c^2 + 4\zeta_c^2\omega_c^2, k_1 = 4\zeta_c\omega_c^3, k_0 = \omega_c^4$, with $0 < \zeta_c < 1, \omega_c \in \mathbb{R}^+$. We may force the closed-loop tracking error dynamics to exhibit the desired stable fourth-order dynamics with characteristic polynomial coincident with $(s^2 + 2\zeta_c\omega_c s + \omega_c^2)^2$.

6.6.5 Experimental Results

A diagram of the experimental platform of the ball-and-beam system is shown in Fig. 6.28. The experimental device depicted in Fig. 6.29 consists of a 24 [V] DC motor, which drives an aluminium beam via a synchronous belt and a pulley with ratio $N = 6{:}1$. The angular position of the beam is measured using an incremental optical encoder of 2500 pulses per revolution. A linear sensor, consisting on an etched nickel–chromium wire, measured the position of the ball

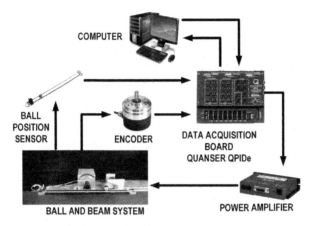

FIGURE 6.28 Block diagram of ball-and-beam implemented control system.

FIGURE 6.29 Ball-and-beam system prototype.

along the beam with resolution of 25 [mm/V]. The data acquisition is carried out through a Quanser data card, model QPIDe. The control strategy described before was implemented in the Matlab-Simulink Quarc platform. The sampling time was 0.001 [s]. The parameters for the beam are $I_1 = 0.0045$ [kgm^2], $m_1 = 0.065$ [kg], $l_1 = 0.015$ [m], $I_p = 0.001$ [kgm^2], $m_2 = 0.065$ [kg], and $I_2 = 0.0045$ [kgm^2]. The radius of the ball is $R = 0.0127$ [m]. The parameters of the motor are $R_a = 2.983$ [Ω] and $k_\tau = 0.0724$ [Nm/A]. The initial conditions for the position variables in the system were $[x = 0, \theta = 0]$. The observer gain parameters for the observation error $\tilde{e}_0$ were $\zeta_o = 1.3$, $\omega_o = 13$. The observer gain parameters for observation error $\tilde{e}_2$ dynamics were set to $T = 6$, $a_0 = 4$, $\alpha_1 = 4.1$. The controller design parameters were $\zeta_c = 1$, $\omega_c = 15$. Fig. 6.30 shows the ball tracking trajectory performance from the initial position $x_\delta(0) = 0$ toward the position $x_\delta(9) = 0.12$, and the ball is moved to the final position $x_\delta(18) = -0.12$. The tracking error is detailed in Fig. 6.31, where it is shown that the position error is restricted to the interval $[-0.004, 0.004]$ [m], Fig. 6.32 shows the ball trajectory tracking performance. The applied control input voltage is depicted in Fig. 6.33. Notice that the lumped disturbance es-

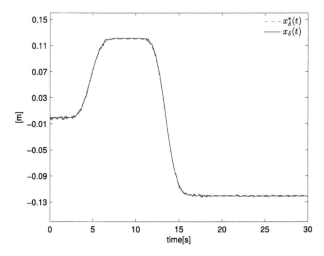

FIGURE 6.30 Trajectory tracking performance of the ADRC controller.

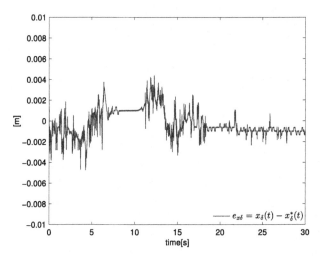

FIGURE 6.31 Output trajectory tracking error.

timation (Fig. 6.34) largely determines the form of the disturbance canceling control input signal.

6.7 REMARKS

This chapter has shown that challenging nonlinear systems, such as the class of underactuated nonlinear systems, which cannot be exactly feedback linearized, constitute excellent examples that may be used to illustrate the effectiveness of

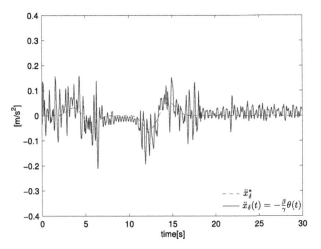

FIGURE 6.32 Tracking of ball acceleration.

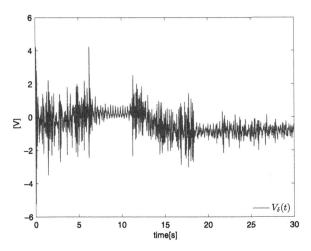

FIGURE 6.33 Voltage control input signal.

ADRC schemes when combined with the flatness property, possibly present in the tangent linearization model of the system. This applies both to extended-observer-based designs or to flat-filtering-based control schemes. The remarkable feature is that if the system exhibits a controllable tangent linearization around a given equilibrium point, then the ADRC control scheme enjoys global features when the design is based on the incremental flat output. Incremental flat output reference trajectories, taking the system operation substantially away from the operating equilibrium, may be imposed and tracked, regardless of the excitation that the faithful controlled tracking may cause to the

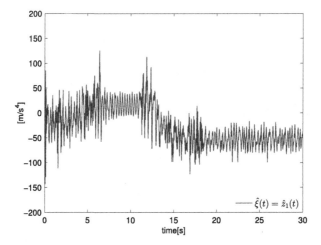

FIGURE 6.34 Total online disturbance estimation.

system nonlinearities neglected by the linearization process. The results show how the combination of incremental flatness and ADRC naturally overcome a traditional obstacle to linearization-based control of nonlinear systems. One of the key issues is to plan the output reference trajectories on the basis of a suitable behavior for the incremental flat output variables. The observer-based ADRC scheme online estimates and cancels the effects of the neglected non-linearities, which may be substantially excited by the velocity and amplitude features of the planned incremental flat output reference trajectory. The ADRC scheme, online, cancels the discrepancies from the linearized dynamics, forcing the controlled plant dynamics to behave as a linear system around the prescribed trajectory. ADRC, in any of its two forms advocated in this book, extended observer-based or flat filter controllers, will efficiently handle the nonlineari-ties neglected by the linearized system for the trajectory tracking task while taking the system operation far from the vicinity of the initial equilibrium point where the linearized model is assumed to be valid. Here, we have also uncovered and exploited a unique property of controllable tangent linearization models of underactuated nonlinear systems. This property is based on the fact that even-order time derivatives of the incremental flat outputs, before hitting the control inputs, are functions of the measurable incremental position variables alone. Similarly, odd-order time derivatives, below the system order, of the flat outputs are functions of the velocity variables alone. This property has been exploited to substantially simplify the observer-based ADRC scheme design, thus obtain-ing a robust controlled performance that only requires position measurements, linear controllers, and rather unrestricted planning of output reference trajecto-ries.

REFERENCES

[1] I. Fantoni, R. Lozano, Non-Linear Control for Underactuated Mechanical Systems, Springer Science & Business Media, 2002.

[2] R. Olfati-Saber, Normal forms for underactuated mechanical systems with symmetry, IEEE Transactions on Automatic Control 47 (2) (2002) 305–308.

[3] M. Reyhanoglu, A. van der Schaft, N.H. McClamroch, I. Kolmanovsky, Dynamics and control of a class of underactuated mechanical systems, IEEE Transactions on Automatic Control 44 (9) (1999) 1663–1671.

[4] M. Fliess, J. Lévine, P. Martin, P. Rouchon, Flatness and defect of non-linear systems: introductory theory and applications, International Journal of Control 61 (1995) 1327–1361.

[5] M. Fliess, J. Lévine, Ph. Martin, P. Rouchon, A Lie–Backlund approach to equivalence and flatness of nonlinear systems, IEEE Transactions on Automatic Control 44 (5) (1999) 922–937.

[6] D. Auckly, L. Kapitanski, W. White, Control of nonlinear underactuated systems, Communications on Pure and Applied Mathematics 53 (3) (2000) 354–369.

[7] D.J.W. Belleter, K.Y. Pettersen, Leader–follower synchronisation for a class of underactuated systems, in: Nonlinear Systems, Springer, 2017, pp. 157–179.

[8] F. Bullo, Stabilization of relative equilibria for underactuated systems on Riemannian manifolds, Automatica 36 (12) (2000) 1819–1834.

[9] S. Finet, L. Praly, Feedback linearization of the transverse dynamics for a class of one degree underactuated systems, in: Proceedings of the 54th IEEE Conference on Decision and Control (CDC), IEEE, 2015, pp. 7802–7807.

[10] K.Y. Pettersen, Underactuated marine control systems, in: Encyclopedia of Systems and Control, 2015, pp. 1499–1503.

[11] M.W. Spong, Underactuated mechanical systems, in: Control Problems in Robotics and Automation, Springer, 1998, pp. 135–150.

[12] R. Xu, U. Özgüner, Sliding mode control of a class of underactuated systems, Automatica 44 (1) (2008) 233–241.

[13] A. Bloch, J. Baillieul, P. Crouch, J. Marsden, D. Zenkov, P. Krishnaprasad, R.M. Murray, Nonholonomic Mechanics and Control, Interdisciplinary Applied Mathematics, vol. 24, Springer, 2003.

[14] R. Ortega, M.W. Spong, F. Gómez-Estern, G. Blankenstein, Stabilization of a class of underactuated mechanical systems via interconnection and damping assignment, IEEE Transactions on Automatic Control 47 (8) (2002) 1218–1233.

[15] A. Chelouah, Extensions of differential flat fields and Liouvillian systems, in: Proceedings of the 36th IEEE Conference on Decision and Control, 1997, pp. 4268–4273.

[16] H. Sira-Ramirez, Soft landing on a planet: a trajectory planning approach for the Liouvillian model, in: American Control Conference, vol. 4, IEEE, 1999, pp. 2936–2940.

[17] H. Sira-Ramirez, R. Castro-Linares, E. Liceaga-Castro, A Liouvillian systems approach for the trajectory planning-based control of helicopter models, Int. J. Robust Nonlinear Control 10 (301) (2000) 320.

[18] X. Lu, S.K. Spurgeon, Asymptotic feedback linearization and control of non-flat systems via sliding modes, in: Proceedings of the 3rd European Control Conference, 1995, pp. 693–698.

[19] V. Hagenmeyer, M. Zeitz, Internal dynamics of flat nonlinear SISO systems with respect to a non-flat output, Systems & Control Letters 52 (3) (2004) 323–327.

[20] M. Ramirez-Neria, H. Sira-Ramírez, R. Garrido-Moctezuma, A. Luviano-Juarez, On the linear active disturbance rejection control of the Furuta pendulum, in: 2014 American Control Conference, IEEE, 2014, pp. 317–322.

[21] M. Cefalo, G. Oriolo, Task-constrained motion planning for underactuated robots, in: 2015 IEEE International Conference on Robotics and Automation (ICRA), IEEE, 2015, pp. 2965–2970.

[22] I. Fantoni, R. Lozano, M.W. Spong, Energy based control of the pendubot, IEEE Transactions on Automatic Control 45 (4) (2000) 725–729.

[23] K. Graichen, M. Treuer, M. Zeitz, Swing-up of the double pendulum on a cart by feedforward and feedback control with experimental validation, Automatica 43 (1) (2007) 63–71.

[24] H. Sira-Ramírez, S.K. Agrawal, Differentially Flat Systems, Marcel Dekker, 2004.

[25] K. Furuta, M. Yamakita, S. Kobayashi, M. Nishimura, A new inverted pendulum apparatus for education, in: IFAC Advances in Control Education Conference, 1991, pp. 191–196.

[26] B.S. Cazzolato, Z. Prime, On the dynamics of the Furuta pendulum, Journal of Control Science and Engineering 2011 (2011), 8 pp.

[27] S. Mori, H. Nishihara, K. Furuta, Control of unstable mechanical system control of pendulum, International Journal of Control 23 (5) (1976) 673–692.

[28] S. Diop, M. Fliess, Nonlinear observability, identifiability, and persistent trajectories, in: Proceedings of the 30th IEEE Conference on Decision and Control, IEEE, 1991, pp. 714–719.

[29] Y.C. Kim, L.H. Keel, S.P. Bhattacharyya, Transient response control via characteristic ratio assignment, IEEE Transactions on Automatic Control 48 (1) (2003) 2238–2244.

[30] H.J. Sussmann, P.V. Kokotovic, The peaking phenomenon and the global stabilization of nonlinear systems, IEEE Transactions on Automatic Control 36 (4) (1991) 424–440.

[31] J. Hauser, S. Sastry, P. Kokotovic, Nonlinear control via approximate input–output linearization: the ball and beam example, IEEE Transactions on Automatic Control 37 (3) (1992) 392–398.

Appendix A

Differential Flatness

A.1 DEFINITION OF FLATNESS

Flatness is a rather ubiquitous property of controlled dynamic systems that offers several unique features that contribute to simplify the controller design task, the trajectory planning problem, and to directly perform off-line analysis related to manifolds of equilibria, effects of inputs and states restrictions, and feasibility of controllers in the context of uncertainties. In the simplest case, that is, linear time-invariant systems (single- or multivariable cases), flatness and controllability are equivalent. In such cases, a flat set of outputs is also observable.

Flatness may be found in linear and nonlinear systems, continuous- and discrete-time systems, single- and multivariable systems, finite-dimensional, and even infinite-dimensional systems (systems with input and state delays, dynamic systems described by fractional-order time derivatives, and systems governed by partial differential equations, usually controlled from their boundaries).

The most general definition we attempt, in the interest of simplicity, is necessarily restricted to the class of finite-dimensional nonlinear multivariable systems. We consider systems of the form

$$\dot{x} = f(x, u), \quad y = h(x, u), \quad x \in \mathbb{R}^n, \quad u \in \mathbb{R}^m, \quad y \in \mathbb{R}^k \qquad (A.1)$$

Let γ, α, β be finite multiindices, i.e., vectors of integer components, respectively, $(\gamma_1, \ldots, \gamma_p)$, $(\alpha_1, \ldots, \alpha_q)$, and $(\beta_1, \ldots, \beta_r)$. System (A.1) is said to be differentially flat if there exists a vector

$$z = \phi(x, \dot{x}, \cdots, x^{(\gamma)}) \in \mathbb{R}^m \qquad (A.2)$$

called the *set of flat outputs*, equal in number to the number of inputs, whose components are differentially independent (i.e., they do not satisfy differential equations by themselves) and such that there are functions ψ and θ for which

$$x = \psi(z, \dot{z}, \cdots, z^{(\alpha)}), \quad u = \theta(z, \dot{z}, \cdots, z^{(\beta)}) \qquad (A.3)$$

Evidently, since the output is a function of the state vector x and, possibly, of the input vector u, the components of the vector y are expressible in terms of the

Active Disturbance Rejection Control of Dynamic Systems. http://dx.doi.org/10.1016/B978-0-12-849868-2.00015-0
Copyright © 2017 Elsevier Inc. All rights reserved.
285

flat outputs z and a finite number of derivatives of such outputs. Flatness implies that all variables in the system (states, inputs, and outputs) can be differentially parameterized in terms of the components of a privileged vector z that is a function of the components of the state and a finite number of their time derivatives. In general, the flat output (or the set of flat outputs) is not unique.

A.2 ILLUSTRATIVE EXAMPLES

We begin by illustrating that a linear single-input–single-output (SISO) time-invariant controllable system is flat in the sense defined before.

A.2.1 Linear Single-Input–Single-Output Time-Invariant Continuous Systems

Consider the linear system

$$\dot{x} = Ax + bu, \quad x \in \mathbb{R}^n, \quad u \in \mathbb{R} \tag{A.4}$$

Let the pair (A, b) be controllable, i.e., the Kalman controllability matrix

$$\mathcal{K} = [b, Ab, \cdots, A^{n-1}b] \tag{A.5}$$

is of full rank n. Suppose that the characteristic polynomial of A in the complex domain is represented by

$$p(s) = s^n + \alpha_{n-1}s^{n-1} + \cdots + \alpha_1 s + \alpha_0 \tag{A.6}$$

Consider the state coordinate transformation $z = \mathcal{K}^{-1}x$. The transformed system is of the form

$$\dot{z} = Fz + gu \tag{A.7}$$

where

$$F = \mathcal{K}^{-1}A\mathcal{K}, \quad g = \mathcal{K}^{-1}b \tag{A.8}$$

Using the definition of $\mathcal{K}$ and the facts that $\mathcal{K}F = A\mathcal{K}$ and $\mathcal{K}b = g$, we obtain in explicit form:

$$\dot{z}_1 = -\alpha_0 z_n + u$$
$$\dot{z}_2 = z_1 - \alpha_1 z_n$$
$$\vdots$$
$$\dot{z}_{n-1} = z_{n-2} - \alpha_{n-2} z_n$$
$$\dot{z}_n = z_{n-1} - \alpha_{n-1} z_n \tag{A.9}$$

It is not difficult to establish that $\zeta = z_n$ is a (scalar) flat output. Indeed, all variables in the transformed system (states and the input) are expressible as a differential function of ζ, i.e., each state variable in the original system is a function of ζ and of a finite number of its time derivatives. Clearly,

$$z_1 = \zeta^{(n-1)} + \alpha_{n-1} z^{(n-2)} + \cdots + \alpha_2 \dot{\zeta} + \alpha_1 \zeta$$

$$\vdots$$

$$z_{n-3} = \zeta^{(3)} + \alpha_{n-1} \ddot{\zeta} + \alpha_{n-2} \dot{\zeta} + \alpha_{n-3} \zeta$$
$$z_{n-2} = \ddot{\zeta} + \alpha_{n-1} \dot{\zeta} + \alpha_{n-2} \zeta$$
$$z_{n-1} = \dot{\zeta} + \alpha_{n-1} \zeta$$
$$z_n = \zeta \qquad\qquad\qquad (A.10)$$

From the first equation in (A.9), the control input u is differentially parameterized as

$$u = \zeta^{(n)} + \alpha_{n-1} z^{(n-1)} + \cdots + \alpha_2 \ddot{\zeta} + \alpha_1 \dot{\zeta} + \alpha_0 \zeta \qquad (A.11)$$

The flat output is $\zeta = z_n = (0, 0, \cdots, 0, 1)z = (0, 0, \cdots, 1)\mathcal{K}^{-1}x$, i.e., the flat output is given by the last row of the inverse of the controllability matrix, multiplied by the state of the system. Any scalar multiple of the flat output is also a flat output.

The transfer function of the flat output ζ is characterized by a constant numerator:

$$\zeta(s) = \left[\frac{1}{s^n + \alpha_{n-1} s^{n-1} + \cdots + \alpha_2 s^2 + \alpha_1 s + \alpha_0} \right] u(s) \qquad (A.12)$$

Eq. (A.10) establishes the following matrix relation:

$$
\begin{bmatrix} z_1 \\ z_2 \\ \vdots \\ z_{n-2} \\ z_{n-1} \\ z_n \end{bmatrix}
=
\begin{bmatrix}
\alpha_1 & \alpha_2 & \alpha_3 & \cdots & \alpha_{n-1} & 1 \\
\alpha_2 & \alpha_3 & \alpha_4 & \cdots & 1 & 0 \\
\vdots & \vdots & \vdots & \vdots & \vdots & \vdots \\
\alpha_{n-2} & \alpha_{n-1} & 1 & \cdots & 0 & 0 \\
\alpha_{n-1} & 1 & 0 & \cdots & 0 & 0 \\
1 & 0 & 0 & \cdots & 0 & 0
\end{bmatrix}
\begin{bmatrix} \zeta \\ \dot{\zeta} \\ \vdots \\ \zeta^{(n-3)} \\ \zeta^{(n-2)} \\ \zeta^{(n-1)} \end{bmatrix}
\qquad (A.13)
$$

Clearly, the matrix relating the transformed state z to the phase variables associated with the flat output ζ is invertible. This means that the flat output ζ is an observable output and, hence, ζ cannot satisfy a differential equation by itself without relation to the system variables. The above relation also establishes that state and input variable equilibria are directly parameterizable in terms of the flat output equilibrium value. This is particularly useful in trajectory planning tasks involving desirable rest-to-rest maneuvers.

A flat output usually enjoys an important physical meaning. If a requirement on the system operation is established in terms of the original state and input, or output, variables, then this can be immediately translated into an operation requirement for the flat output. A desired trajectory for the flat output, consistent with the requirement, allows us, thanks to the differential parameterization characterizing flatness, to off-line compute all the system state variable trajectories and the nominal input trajectory. This is a unique advantage of flatness, which provides the designer with a unique insight into trajectory planning tasks, addressing the ways to avoid control input saturations, how to stay away from state restrictions, and how to impose some quality features to the nominal responses via smart planning.

A.2.1.1 Example

Consider the following system, representing a set of three carts, with identical masses joined by identical springs. The set of carts is acted upon by a single force moving the train along a horizontal line from the first car. Suppose that there is no friction affecting the motions of the carts. The normalized model of the system is given by

$$\ddot{z}_1 = u + (z_2 - z_1 - 1)$$
$$\ddot{z}_2 = z_1 - 2z_2 + z_3$$
$$\ddot{z}_3 = z_2 - z_3 + 1 \tag{A.14}$$

By inspection a flat output is given by the third cart (normalized) position $\zeta = z_3$. All variables can be expressed as differential functions of ζ. Indeed,

$$z_3 = \zeta$$
$$z_2 = \ddot{\zeta} + \zeta - 1$$
$$z_1 = \zeta^{(4)} + 3\ddot{\zeta} + \zeta - 2$$
$$u = \zeta^{(6)} + 4\zeta^{(4)} + 3\ddot{\zeta} \tag{A.15}$$

A.2.2 Linear Single-Input–Single-Output Time-Invariant Discrete-Time Systems

This case is closely related to the previous one, and details will not be provided. We summarize the case as follows.

Given a controllable linear discrete-time SISO system of the form

$$x_{k+1} = Ax_k + bu_k, \quad x_k \in \mathbb{R}^n, \quad u_k \in \mathbb{R}, \quad \forall k \qquad (A.16)$$

with

$$\mathcal{K} = \left[b, Ab, \cdots, A^{n-1}B \right], \quad \text{rank}\mathcal{K} = n \qquad (A.17)$$

The flat output $\zeta_k \in \mathbb{R}$ is given by the variable obtained from the multiplication of the last row of the inverse of the controllability matrix by the state vector:

$$\zeta_k = [\, 0 \quad 0 \quad \cdots \quad 1 \,]\mathcal{K}^{-1}x_k \qquad (A.18)$$

In the discrete-time case, the flat output parameterizes all system variables in terms of the flat output and a finite number of advances (i.e., future values) of such an output. This, seemingly, introduces a lack of causality in the definition of the flat output and all the consequences that might be derived from it. From the definition itself, however, we can always proceed with a flatness-based controller design, or an ADRC design, in which causality is fully respected (see Appendix B).

A.2.2.1 Example

Consider the following example, representing the Newton heating model of a slab of three compartments. The temperatures of the slabs at time k are denoted by a_k, b_k, and c_k. The control input is the temperature u_k applied to the first slab whose temperature at time k is identified by a_k:

$$c_{k+1} = (1 - 2p)c_k + pb_k$$
$$b_{k+1} = pa_k + (1 - 2p)b_k + pc_k$$
$$a_{k+1} = (1 - 2p)a_k + pb_k + pu_k \qquad (A.19)$$

In this case,

$$A = \begin{bmatrix} 1-2p & p & 0 \\ p & 1-2p & p \\ 0 & p & 1-2p \end{bmatrix}, \quad b = \begin{bmatrix} 0 \\ 0 \\ p \end{bmatrix}$$

$$\mathcal{K} = \begin{bmatrix} 0 & 0 & p^3 \\ 0 & p^2 & 2(1-2p)p^2 \\ p & (1-2p)p & (1-4p+5p^2)p \end{bmatrix} \qquad (A.20)$$

The last row of the inverse of the controllability matrix multiplied by the state vector $(a_k, b_k, c_k)^T$ results in the flat output

$$\zeta_k = \frac{c_k}{p^3} \tag{A.21}$$

The flat output may be taken to be the state c_k itself, or any scalar multiple of c_k, as before. In fact, a flat output $\zeta_k = c_k$ is easily found by inspection in this case.

Note that all variables in the system are parameterizable in terms of $\zeta_k = c_k$ and a finite number of advances of ζ_k. Clearly,

$$
\begin{aligned}
c_k &= \zeta_k \\
b_k &= \frac{1}{p}(\zeta_{k+1} - (1 - 2p)\zeta_k) \\
a_k &= \frac{1}{p^2}\left[\zeta_{k+2} - 2(1 - 2p)\zeta_{k+1} + ((1 - 2p)^2 - p^2)\zeta_k\right] \\
u_k &= \frac{1}{p^3}\zeta_{k+3} + \cdots
\end{aligned}
\tag{A.22}
$$

A.2.3 Linear Multiple-Input–Multiple-Output Time-Invariant Continuous Systems

We follow a similar route to establish the characterization of flat systems in the linear multivariable case (MIMO). Since there exist two alternative procedures for the construction of the Kalman controllability matrix in the MIMO case, we will use one of the methods, known as "filling the crate by columns."

Consider the linear time-invariant system

$$\dot{x} = Ax + Bu, \quad x \in \mathbb{R}^n, \quad u \in \mathbb{R}^m \tag{A.23}$$

with $B = [b_1, b_2, \cdots, b_m] \in \mathbb{R}^{n \times m}$ being a full-rank m matrix.

The Kalman controllability matrix is given by the nonsquare $n \times nm$ matrix

$$\mathcal{K} = \left[B; AB; A^2B; \cdots; A^{n-1}B\right] \tag{A.24}$$

Since controllability demands that the rectangular matrix, $\mathcal{K}$ be of full rank n, there are $n(m - 1)$ linearly dependent column vectors in $\mathcal{K}$ that must be excluded. We assume that the n-column full-rank controllability matrix $\mathcal{K}$ is of the form

$$
\begin{aligned}
\mathcal{K}_c = \Big[&b_1; Ab_1; \cdots; A^{\gamma_1-1}b_1; b_2; Ab_2; \cdots; A^{\gamma_2-1}b_2; \cdots; b_m; \\
&Ab_m; \cdots; A^{\gamma_m-1}b_m\Big]
\end{aligned}
\tag{A.25}
$$

where $\{\gamma_1, \cdots, \gamma_m\}$ is the set of Kronecker controllability indices satisfying $\sum_i \gamma_i = n$.

The state coordinate transformation $z = \mathcal{K}_c^{-1} x$ leads to a system of the form

$$\dot{z} = Fz + Gu \qquad (A.26)$$

with $F = \mathcal{K}_c^{-1} A \mathcal{K}_c$ and $G = \mathcal{K}_c^{-1} B$. In the transformed state variables

$$z = [z_{11}, z_{12} \cdots z_{1\gamma_1}, z_{21}, z_{22}, \cdots, z_{2\gamma_2}, \cdots z_{m1}, \cdots, z_{m\gamma_m}]^T \qquad (A.27)$$

the privileged set of state components, capable of parameterizing the remaining state variables and m control inputs, is constituted by the set of m states

$$\{z_{\gamma_1}, z_{2\gamma_2}, \cdots, z_{m\gamma_m}\} \qquad (A.28)$$

These states are the last components of the m blocks in which the state vector z can be partitioned according to the values of the Kronecker controllability indices $\{\gamma_1, \cdots \gamma_m\}$. We call these blocks the Kronecker blocks. These are then easy to establish, in terms of the original x variables, with the help of a matrix of zeroes and ones whose rows depict the position, in the transformed z vector, of the last state variables $z_{j\gamma_j}$ in each Kronecker block. The flat outputs are then given by

$$Z = [z_{\gamma_1}, z_{2\gamma_2}, \cdots, z_{m\gamma_m}]^T = \begin{bmatrix} \phi_1 \\ \phi_2 \\ \vdots \\ \phi_m \end{bmatrix} z = \begin{bmatrix} \phi_1 \\ \phi_2 \\ \vdots \\ \phi_m \end{bmatrix} \mathcal{K}_c^{-1} x \qquad (A.29)$$

where ϕ_j, $j = 1, 2, \ldots, m$, is of the form

$$\phi_j = [\underbrace{0, \cdots, 0}_{\gamma_1}, \cdots, \underbrace{0, \cdots, 0, 1}_{\gamma_j}, 0, \cdots, 0] \qquad (A.30)$$

with the 1 occupying the position $\sum_{i=1}^{j} \gamma_i$ in the n-dimensional row vector.

A.2.4   Linear Multiple-Input–Multiple-Output Time-Invariant Discrete-Time Systems

The MIMO linear time-invariant case follows similarly to the previous case. We summarize the flat output expression in terms of the original state vector of the system

$$x_{k+1} = Ax_k + Bu_k, \quad x_k \in \mathbb{R}^n, \quad u \in \mathbb{R}^m, \quad B = [b_1, b_2, \cdots b_m], \quad \mathrm{rank}\, B = m \qquad (A.31)$$

with the Kalman controllability matrix given by

$$\mathcal{K}_c = \left[b_1, Ab_1, \cdots, A^{\gamma_1-1}b_1, b_2, Ab_2, \cdots, A^{\gamma_2-1}b_2, \cdots, b_m, \right.$$
$$\left. Ab_m \cdots, A^{\gamma_m}b_m \right] \qquad (A.32)$$

The vector of flat outputs ζ is given by

$$\zeta = (\zeta_1, \cdots, \zeta_m)^T = \Phi\mathcal{K}_c^{-1}x, \quad \Phi = \begin{bmatrix} \phi_1 \\ \phi_2 \\ \vdots \\ \phi_m \end{bmatrix}$$

$$\phi_j = [\underbrace{0, \cdots, 0}_{\gamma_1}, \cdots, \underbrace{0, \cdots, 0}_{\gamma_j}, 1, 0, \cdots, 0] \qquad (A.33)$$

A.2.5   Nonlinear Single-Input–Single-Output Time-Invariant Continuous Systems

The nonlinear system characterized by n-dimensional smooth vector fields f and g,

$$\dot{x} = f(x) + g(x)u, \quad x \in \mathbb{R}^n, \quad u \in \mathbb{R} \qquad (A.34)$$

is said to be feedback linearizable (i.e., flat) at the point x if the following two conditions are satisfied (see [1]):

- The distribution $\left[g, \mathrm{ad}_f g, \cdots \mathrm{ad}_f^{n-2} g \right]$ is involutive in an open neighborhood containing x;
- The set of vectors $\{ g, \mathrm{ad}_f g, \cdots \mathrm{ad}_f^{n-2} g, \mathrm{ad}_f^{n-1} g \}$ is linearly independent at the point x,

where $\mathrm{ad}_f^j g$ are recursively defined via the following Lie bracket operation:

$$\mathrm{ad}_f^j g = [f, \mathrm{ad}_f^{j-1} g], \quad [f, g] = \mathrm{ad}_f g = \left[\left(\frac{\partial g(x)}{\partial x^T} \right) f(x) - \left(\frac{\partial f(x)}{\partial x^T} g(x) \right) \right] \qquad (A.35)$$

Define the invertible matrix $\mathcal{K}(x)$ as follows:

$$\mathcal{K}(x) = \left[g(x), \mathrm{ad}_f g(x), \cdots, \mathrm{ad}_f^{n-2} g(x), \mathrm{ad}_f^{n-1} g(x) \right] \qquad (A.36)$$

The flat output ζ is determined, up to a nonzero smooth factor function $\alpha(x)$, via integration of its gradient function. Such a gradient is given by

$$\frac{\partial \zeta(x)}{\partial x^T} = \alpha(x)[0, 0, \cdots, 0, 1]\mathcal{K}^{-1}(x) \tag{A.37}$$

A.2.5.1 A Normalized Boost Converter Example

Let Q be a constant scalar. Consider the nonlinear smooth system,

$$\dot{x}_1 = -ux_2 + 1, \quad \dot{x}_2 = ux_1 - \frac{x_2}{Q}, \quad u \in \mathbb{R} \tag{A.38}$$

Here $f(x) = (1, -\frac{x_2}{Q})^T$ and $g(x) = (-x_2, x_1)^T$. The vector field $g(x)$ is in itself a one-dimensional distribution, which is trivially involutive. The set of vectors

$$\{g(x), \text{ad}_f g(x)\} = \left\{ \begin{bmatrix} -x_2 \\ x_1 \end{bmatrix}, \begin{bmatrix} \frac{x_2}{Q} \\ \frac{1}{Q} \end{bmatrix} \right\} \tag{A.39}$$

is an everywhere linearly independent set, except on the manifolds $x = -1$, $x_2 = 0$. The gradient of the flat output ζ, modulo a smooth nonzero factor function, is the row vector (x_1, x_2). A flat output is given by the function $\zeta(x) = \frac{1}{2}(x_1^2 + x_2^2)$.

A.2.5.2 Soft Landing Model for a Spacecraft

Consider the following system:

$$\dot{x}_1 = x_2, \quad \dot{x}_2 = g - \frac{\sigma\alpha}{x_3}u, \quad \dot{x}_3 = -\sigma u \tag{A.40}$$

where x_1 is the height of the aircraft, measured negatively upwards, x_2 is the corresponding velocity, and x_3 represents the mass of the vehicle. The parameters σ and α are, respectively, mass rate and velocity of combustion gases. The control u may be assumed to smoothly take values in the open interval $(0, 1)$.

This system is nonflat since the second condition for feedback linearization fails.

The system, however, enjoys a closely related property to flatness establishing that there exists a largest linearizable subsystem by means of a state coordinate transformation and partial feedback. Such a subsystem is of dimension 2 and represented by the variables (x_1, x_2). The defect of the system, given in this case by the variable x_3, is integrable via quadratures. Indeed, the partial feedback, with auxiliary input v,

$$u = \frac{x_3}{\sigma\alpha}(g - v), \quad x_1 = \zeta, \quad x_2 = \dot{\zeta}, \quad v = \ddot{\zeta} \tag{A.41}$$

renders the system partially linear, and the mass differential equation becomes a linear time-varying equation dependent upon the output $\zeta = z_1$, which represents the variable with the largest relative degree in the system. We have

$$\dot{x}_3 = -\frac{1}{\alpha}\left(g - \ddot{\zeta}\right)x_3 \tag{A.42}$$

Such systems with integrable defect are addressed as Liouvillian (see [2]).

A.3 ABOUT THE ADVANTAGES AND DISADVANTAGES OF FLATNESS

Several disadvantages are usually claimed regarding the use of flatness in feedback controller design. A prevailing one is that there exists no general method for finding the flat output. As we have seen, for the most commonly used classes of systems, there exist specific formulae that give the flat output or the gradient of the flat output function. Whereas it is rather elaborate and complicated to give a formula for the set of flat outputs in the multivariable nonlinear system case requiring dynamic feedback for their feedback linearization (see [3]), it is also a fact that flatness may be assessed by inspection in the vast majority of engineering examples.

Whereas the problem of characterizing flatness in the most general case still challenges the mathematically inclined part of the control systems community, it is also true that significant advances have been provided in the literature in that respect [4]; see also the book by Lévine [5].

In some cases, the flat outputs cannot be measured. This is certainly a fact for a quite particular class of nonlinear systems. The induction motor example is perhaps the most significant example, given that one of the flat outputs is related to the argument of the unmeasured complex flux variables. In the majority of interesting engineering cases and applications, the flat outputs enjoy a distinctive physical nature.

In this book, we emphasize the fundamental importance of the flatness property in the ADRC control of dynamical systems of various kinds (single-variable, multivariable, linear or nonlinear, continuous or discrete). It is our conviction, through many successful experimental results and applications, that flatness is a property seemingly taylored, in a special manner, around ADRC schemes. The combination of flatness and ADRC provides the designer with a systematic procedure to obtain an efficient robust controller with minimum of complexity. This vision is even applicable to nondifferentially flat systems, provided that they exhibit a controllable linearization.

Another frequent complaint about flatness resides in the singularities and lack of regularity of some mathematical models. Clearly, flatness in this respect cannot prevent singularities from occurring since they are properties of the system rather than properties induced by the method used for analysis and controller design. Wherever such singularities do exist, at least the flatness property and its state and inputs parameterizations explicitly depict them in a fashion that is usually related to the physics of the system.

A definite advantage of flatness resides in the fact that it is relatively easy to assess, even if the system is nonlinear in the control inputs, as suggested by Eq. (A.1).

A.4 DIFFERENTIAL FLATNESS AND UNCERTAINTY

Disturbance, plant and measurement noises, unmodeled dynamics, unknown external influences, time-varying unknown parameters constitute an inescapable reality in control system engineering. Satisfactory performance of a designed control system, for a given plant operating under such uncertainty conditions, is difficult to guarantee from the outset. Controller design methods are usually confronted with this reality and the need for certain choices regarding how to proceed with a design that performs reasonably well.

No controller design method is so general as to be able to accept as hypothesis the active presence of all possible sources of uncertainty and imperfections, affecting a given plant and then proceed to synthesize a remarkable controller that successfully withstands stringent measures of quality performance. At the outset, one must compromise, to the best of knowledge, regarding what will be relevant and what will not. Surprising results can only be casted either as new discoveries or as a gross underestimation of some of the prevailing conditions.

In the design of feedback controllers, flatness is of great help, but it is not the miracle option. It goes a long way in producing quite smart controllers with perhaps more advantages than disadvantages. Naturally, it does have limitations of its own, and one should be aware of them. The issue of uncertain signals plaguing the behavior of the plant system and of the measurements of the available signals constitutes a delicate feature that should be analyzed with caution and care. To begin with, flatness is a system structural property, i.e., one that should not be part of the description of the system surrounding environment or of our interaction with it. As a result, the system is to be cleaned up of all uncertainties before even attempting to analyze it for flat outputs candidates. However, it is highly instructive and recommendable to include and trace the effects of such initially neglected uncertainties once one has clearly identified the best candidates for flat outputs (usually related to measurability of such variables).

A.4.1 A DC Motor Example

Consider, just for the sake of argument in our preceding discussion, the equations of a linear DC motor:

$$L\frac{di}{dt} = -Ri - k\omega + V$$

$$J\frac{d\omega}{dt} = -B\omega + ki - \tau$$

$$\frac{d\theta}{dt} = \omega \qquad\qquad\qquad (A.43)$$

where i is the armature current, θ is the motor shaft angular position, and ω is the angular velocity of the shaft. The control input to the system is the armature voltage V, whereas τ represents the load torque.

Here, every parameter in the system may be a source of uncertainty. The load torque τ is, generally speaking, unknown, and we are certain that the mathematical model constitutes a quite nice approximation of reality, but, nevertheless, it is relatively far from the actual phenomenon. Suppose that J and τ are unknown; moreover, we may have reasons to believe that k is time-varying.

The flat output is the angular position of the motor shaft, since it clearly allows for a differential parameterization of ω, i, and V in terms of θ including the effects of the disturbance load torque input τ, and of its first order time derivative as well as the presence of the uncertain parameter k and of its time derivative, $\dot{k}$. Indeed, we have

$$\omega = \dot{\theta}$$

$$i = \frac{1}{k}\left[J\ddot{\theta} + B\dot{\theta} + \tau\right]$$

$$V = \frac{JL}{k}\theta^{(3)} + \left(\frac{BL}{k} + \frac{RJ}{k} - \frac{LJ}{k^2}\dot{k}\right)\ddot{\theta} + \left(\frac{RB+k^2}{k} - \frac{BL}{k^2}\dot{k}\right)\dot{\theta}$$

$$+ \left(\frac{R}{k}\tau + \frac{L}{k}\dot{\tau}\right) \qquad\qquad (A.44)$$

The last equation is plagued by uncertain terms: $k, \dot{k}, J, \tau$ and $\dot{\tau}$. This results in an uncertain input gain $(k/(JL))$ which is a fundamental quantity for the controller design. The load torque may give rise to impulses in this input–output model if such a load is modeled as a step function starting somewhere along the time frame of the system operation.

In the literature, there are many options to deal with the controller design issue on an input–output system like that. It is the distinctive advantage of ADRC that one may get away with an effective controller computed on the basis of a simplified model of the form

$$\theta^{(3)} = \frac{k_0}{J_0 L} V + \xi(t) \tag{A.45}$$

where k_0 and J_0 are nominal constant values of $k(t)$ and J chosen on the basis of "engineering judgement." The signal $\xi(t)$ may be estimated with the help of a linear extended state disturbance observer. A more reasonable option, but surely a bit more complex to implement, may be based on obtaining an online estimate of the time-varying uncertain gain $k(t)/(JL)$, using algebraic identification methods (see [6]), and proceed to use the still simplified pure integration dynamics in an extended observer-based scheme using the online identified gain.

Uncertainty does affect the final working description of the plant. Flatness at least gives you a clear indication related to where exactly does the obstacle to be surmounted resides. The rest is pretty much your choice.

REFERENCES

[1] A. Isidori, Nonlinear Control Systems, Springer Science & Business Media, 1995.
[2] A. Chelouah, Extensions of differential flat fields and Liouvillian systems, in: Proceedings of the 36th IEEE Conference on Decision and Control, 1997, pp. 4268–4273.
[3] H. Sira-Ramirez, On the sliding mode control of multivariable nonlinear systems, International Journal of Control 64 (4) (1996) 745–765.
[4] Ph. Martin, R.M. Murray, P. Rouchon, Flat systems, in: Proceedings of the 4th European Control Conference, 1997, pp. 211–264.
[5] J. Levine, Analysis and Control of Nonlinear Systems: A Flatness-Based Approach, Springer Science & Business Media, 2009.
[6] H. Sira-Ramírez, C. García-Rodríguez, J. Cortés-Romero, A. Luviano-Juárez, Algebraic Identification and Estimation Methods in Feedback Control Systems, John Wiley & Sons, 2014.

Appendix B

Generalized Proportional Integral Control

B.1 DEFINITIONS AND GENERALITIES

The use of efficient controllers, with the least possible amount of information about the complex system to be controlled, represents a challenging problem in controller design for dynamical systems. Robust strategies that involve the use of observers for unknown dynamics and external disturbance have been the basis of the so-called Active Disturbance Rejection Control (ADRC) scheme [1]. The scheme relies on the use of Extended State Observers, of linear or nonlinear nature, which use the input–output information to reconstruct an online cancelation of the effects of the lumped (i.e., total) disturbance inputs. The effectiveness of this scheme has been shown through many academic and industrial applications (see [2], [3], [4], and references therein). Another interesting scheme is the Model-Free Control (MFC) proposed by Fliess and Join [5]. This technique takes advantage of algebraic estimation techniques in the context of ultramodels, which are used to "replace" the original system by another one having the same response with the help of the control input (see [6] for a detailed analysis). The approach does not require a detailed mathematical description of the system thanks to the online estimation and rejection of the unmodeled elements of the dynamics [7]. In ADRC, as pioneered by Prof. Han, one of the original ideas was to establish a direct connection between the classical PID control and other modern control theory techniques. In this sense, an alternative approach and some equivalences are found in the domain of Classical Compensation Networks by means of the Generalized Proportional Integral Control (GPI).

Classical Compensation Networks controlling linear systems constitute dynamical control systems that are naturally differentially flat [8], [9], [10] (see Appendix A) although flatness in this new context is not to be taken in a strict control oriented sense, but in a new filtering sense. (The input of the compensation network is in fact constituted by the plant output signal, and the compensation network output is in fact the system input.) All variables in the compensator or filter are indeed expressible in terms of the filtered output signal and a finite number of its time derivatives, a general defining property of linear flat (input–output) systems.

Active Disturbance Rejection Control of Dynamic Systems. http://dx.doi.org/10.1016/B978-0-12-849868-2.00016-2
Copyright © 2017 Elsevier Inc. All rights reserved.
299

In this Appendix, a reinterpretation of GPIC is proposed in terms of CCN. This allows, in turn, the conception of a "flat filtering" device, which constitutes a tool for output feedback control design in linear controllable systems. A controllable linear system exhibits a natural flat output (Brunovsky's output) from which the system is also trivially observable. Flat filtering is based on the fact that a GPI controller is viewed as a dynamical linear system that exhibits as a natural flat output a filtered version of the plant output signal. This property is particularly helpful in the design of efficient output feedback stabilization schemes and in solving output reference trajectory tracking tasks. The approach is naturally extended, using Active Disturbance Rejection Control (ADRC) ideas [1,11], to the control of significantly perturbed SISO nonlinear systems affected by unknown endogenous nonlinearities in the presence of exogenous disturbances and unmodeled dynamics.

This Appendix is organized as follows. Section B.2 introduces, in a tutorial fashion, the parallel between GPI control, based on integral reconstructors, and CCN for linear systems. In this section, flat filtering is exhibited as a natural reinterpretation of GPI control. Section B.2 also contains the robust version of flat-filtering-based control schemes, which allows one to treat endogenous and exogenous disturbances as joint unstructured total disturbance signals. These joint disturbances will be locally representable as time polynomial models (also called ultramodels) efficiently rejected by the designed robust flat filtering compensator via attenuation. The scheme is similar (in fact, dual) to extended observer-based Active Disturbance Rejection Control (ADRC). In Section B.3, an application example, including simulations, is provided. Section B.4 presents the counterpart of GPI control to the case of discrete-time linear systems. Section B.5 contains some generalizations of discrete GPI control to the case of state space representations. A heat equation example is presented along with computer simulations.

B.2 GENERALIZED PROPORTIONAL INTEGRAL CONTROL AND CLASSICAL COMPENSATION NETWORKS

Flat filtering constitutes a reinterpretation in the form of Classical Compensation Networks (CCN) of GPI controllers introduced by Fliess and his colleagues [12]). Roughly speaking, any classically perturbed[1] linear controllable system whose output of the unperturbed system model is the Brunovsky output can be regulated with the help of a well-tuned proper linear filter and a suitable linear combination of the available internal states of such a filter. Here we show that such a classical tool is also capable of efficiently handling control

---

1. Additively perturbed by classical disturbances: constant, linear, quadratic, etc. functions of time, i.e., polynomial functions of time.

tasks on perturbed linearizable nonlinear systems (i.e., flat systems), such as the boost converter, including unknown, or neglected, nonlinearities, exogenous disturbances, and unmodeled dynamics; in a fashion similar as these uncertainties are handled in ADRC schemes (see Han [1], Zheng et al. [13,14], and Guo et al. [15–17], and a recent survey by Madoński and Herman [18]).

B.2.1 A Second-Order Pure Integration Example

Here we review, in a tutorial fashion, the intimate links existing between GPIC and CCN. For simplicity, we address pure integration linear SISO systems and state that the results are easily extended to even nonlinear uncertain SISO or MIMO systems under reasonable modifications.

Consider the second-order unperturbed system

$$\ddot{y} = u \tag{B.1}$$

Suppose that we want to asymptotically track a given smooth output reference signal $y^*(t)$. The nominal control input $u^*(t)$ is just given by $u^*(t) = \ddot{y}^*(t)$. The output tracking error dynamics is given by

$$\ddot{e}_y = e_u, \quad e_y = y - y^*(t), \quad e_u = u - u^*(t) \tag{B.2}$$

GPI control evades the need for asymptotic state observers using integral reconstructions of the state variables in the form of uncompensated integrals of inputs and outputs. In this instance, an integral reconstructor of $\dot{e}_y$, which is off by an unknown (constant) initial condition, is simply obtained as

$$\widehat{\dot{e}}_y = \int_0^t e_u(\sigma)d\sigma, \quad \dot{e}_y = \widehat{\dot{e}}_y + \dot{e}_y(0) \tag{B.3}$$

The use of such an integral reconstructor of $\dot{e}_y$ in a linear feedback control scheme entitles the use of an output error integral compensation term to asymptotically overcome the effects of the constant error committed in the tracking error derivative estimation. We propose the following linear feedback controller:

$$e_u = -k_2\widehat{\dot{e}}_y - k_1 e_y - k_0 \int_0^t e_y(\sigma)d\sigma \tag{B.4}$$

where the set of constant coefficients $\{k_2, k_1, k_0\}$ is to be chosen to guarantee the asymptotic stability of the tracking error e_y.

The closed-loop system tracking error dynamics is given by

$$\ddot{e}_y = -k_2\dot{e}_y - k_1 e_y - k_0 \int_0^t e_y(\sigma)d\sigma + k_2\dot{e}_y(0) \tag{B.5}$$

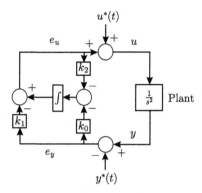

FIGURE B.1   GPI control of a second order system via integral reconstructor.

Note that (B.5) may be rewritten, after defining $z = \int_0^t e_y(\sigma)d\sigma - (k_2/k_0)\dot{e}_y(0)$, as

$$\ddot{e}_y = -k_2\dot{e}_y - k_1 e_y - k_0 z$$
$$\dot{z} = e_y, \quad z(0) = -(k_2/k_0)\dot{e}_y(0) \tag{B.6}$$

The characteristic polynomial of system (B.6) is readily found to be

$$p(s) = s^3 + k_2 s^2 + k_1 s + k_0 \tag{B.7}$$

whose roots are guaranteed to lie in the left half of the complex plane as long as $k_0 > 0$, $k_2 > 0$, $k_2 > k_0/k_1$. We also obtain, using (B.3), the following implicit expression for the linear feedback controller (B.4):

$$
e_u = -k_2 \int_0^t e_u(\sigma)d\sigma - k_1 e_y - k_0 \int_0^t e_y(\sigma)d\sigma
$$
$$
= -k_1 e_y - \int_0^t [k_0 e_y(\sigma) + k_2 e_u(\sigma)]d\sigma \tag{B.8}
$$

Fig. B.1 is a block diagram representation of the second-order plant-GPI controller system.

Taking the Laplace transform of (B.8), after some rearrangement and simplifications, yields the following mixed time-domain–frequency-domain expression for the linear feedback output tracking controller (see Fig. B.2):

$$
u(t) = u^*(t) - \left[\frac{k_1 s + k_0}{s + k_2}\right](y - y^*(t)) \tag{B.9}
$$

The stability requirement on the design coefficients $\{k_2, k_1, k_0\}$ indicates that the controller (B.9) is a lead compensator network since the stable pole $s = -k_2$ is

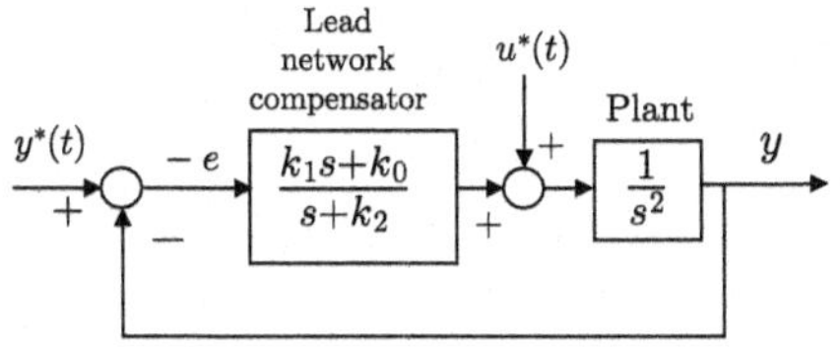

FIGURE B.2 Lead network realization of a GPI controller for second-order plant.

located to the left of the stable zero $s = -k_0/k_1$ on the real line of the complex plane. The classical compensation network (B.9) admits a low-pass filtering interpretation. Define the filtered tracking error e_{y_f} as follows:

$$e_{y_f}(s) = \left[\frac{1}{s + k_2}\right] e_y(s), \quad \text{i.e.,} \quad \dot{e}_{y_f} = -k_2 e_{y_f} + e_y \qquad (B.10)$$

Letting $\zeta := e_{y_f}$, the feedback controller (B.9) is synthesized in the time domain as

$$u = u^*(t) - (k_1 k_2 - k_0)\zeta + k_2(y - y^*(t))$$
$$\dot{\zeta} = -k_2\zeta + (y - y^*(t)) \qquad (B.11)$$

We address the representation of this controller as the flat-filtered-based controller. The reasons for such a name areas follows: 1) Regarding the controller as a dynamical system with "input" represented by the output error of the system e_y, and with "output" represented by the system control input error e_u, the filtered output e_{y_f} is a flat output for such a linear dynamical system; indeed, all variables in the compensator can be expressed in terms of the flat output e_{y_f} and a finite number (just one in this case) of its time derivatives. 2) A crucial property of flat linear time-invariant systems is that their transfer functions exhibit no zero dynamics. Such is the case of the transfer function of the compensator network in (B.10).

B.2.2 Dirty Derivatives in PD Schemes and Lead Networks

A common practise in classical control systems design, as applied particularly to the field of robotics, is using PD controllers for stabilization and tracking. To avoid measurements of the time derivative term for the controller, an alternative approximate synthesis approach is taken, known as *the dirty derivative* (see Kelly, Santibañez, and Loria [19]).

Consider a second-order system stated in terms of an underlying output reference trajectory tracking task:

$$\ddot{e}_y = e_u, \quad e_y = y - y^*(t), \quad e_u = u - u^*(t) = u - \ddot{y}^*(t) \qquad (B.12)$$

We propose a controller mixing time signals with frequency domain signals of the form:

$$e_u = -\left[k_0 + \frac{k_1 s}{s + k_2}\right] e_y = -G(s) e_y \tag{B.13}$$

based on a low-pass filtering approximation of the derivative term $s e_y$. Clearly, $G(s)$ is a stable lead network, as the following manipulations demonstrate. We can rewrite $G(s)$ as

$$G(s) = \frac{(k_0 + k_1)s + k_0 k_2}{s + k_2} \tag{B.14}$$

The closed-loop system reads

$$\left[s^3 + k_2 s^2 + (k_0 + k_1) + k_2 k_0\right] e_y = 0 \tag{B.15}$$

The Routh–Hurwitz stability test reveals the following restrictions for the set of design parameters $\{k_0, k_1, k_2\}$:

$$k_2 > 0, \quad k_1 > 0, \quad k_0 > 0$$

The pole of the transfer function $G(s)$ is located on the real line at $s = -k_2$, whereas the zero of $G(s)$ is located at $s = -k_0 k_2/(k_0 + k_1) = -k_2/(1 + (k_1/k_0))$. The zero of $G(s)$ is located to the right of its pole. Hence, $G(s)$ is a lead compensator.

B.2.3 A Third-Order Pure Integration Example

Consider the linear time-invariant system

$$y^{(3)} = u \tag{B.16}$$

Suppose that we desire to stabilize the output variable and its time derivatives to the origin of the phase space.

Integrating the system expression once, we obtain an estimate of the output acceleration $\ddot{y}$ modulo a constant initial condition term $\ddot{y}(0)$:

$$\ddot{y} = \ddot{y}(0) + \int_0^t u(\sigma_0) d\sigma_0 \tag{B.17}$$

Integrating (B.16) twice, we obtain an expression for the unknown $\dot{y}$ as

$$\dot{y} = [\ddot{y}(0)]t + \dot{y}(0) + \int_0^t \int_0^{\sigma_0} u(\sigma_1) d\sigma_1 d\sigma_0 \tag{B.18}$$

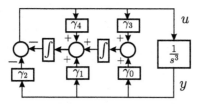

FIGURE B.3   GPI stabilizing control scheme for a third-order system, using integral reconstructors.

We address these faulty estimates as the structural estimates or the integral re-constructed estimates and denote them, respectively, by $\widehat{\dot{y}}$ and $\widehat{\ddot{y}}$ ([12]).

$$\widehat{\dot{y}} = \int_0^t u(\sigma_1)d\sigma_1, \quad \widehat{\ddot{y}} = \int_0^t \int_0^{\sigma_1} u(\sigma_2)d\sigma_2 d\sigma_1 \tag{B.19}$$

The integral reconstructors of the phase variables $\dot{y}$, $\ddot{y}$ are, respectively, off by a first-degree time polynomial and by a constant term whose coefficients depend on the unknown initial conditions.[2]

The use of these faulty estimates in any stabilizing linear feedback scheme demands additive integral output compensation, including, at least, up to a double (iterated) integral of the available output signal y.

Then we propose the stabilizing controller

$$u = \left[-\gamma_4 \widehat{\ddot{y}} - \gamma_3 \widehat{\dot{y}} - \gamma_2 y - \gamma_1 \left(\int y \right) - \gamma_0 \left(\int^{(2)} y \right) \right] \tag{B.20}$$

After substitution of the expressions for the integral reconstructors for $\dot{y}$ and $\ddot{y}$, we obtain the following implicit expression for the controller:

$$u = \left[-\gamma_4 \left(\int u \right) - \gamma_3 \left(\int^{(2)} u \right) - \gamma_2 y - \gamma_1 \left(\int y \right) - \gamma_0 \left(\int^{(2)} y \right) \right] \tag{B.21}$$

Simple association, or nesting, of the iterated integrals leads to the following equivalent expressions, which fully explain the controller representation depicted in Fig. B.3:

$$u = -\gamma_2 y - \left(\int^{(2)} \gamma_0 y + \gamma_3 u \right) + \left(\int^{(1)} \gamma_1 y + \gamma_4 u \right)$$

---

2. We adopt, henceforth, the following notation for multiple integrations on a given time function $\phi(t)$:

$$\int_0^t \int_0^{\sigma_1} \cdots \int_0^{\sigma_{i-1}} \phi(\sigma_i)d\sigma_i \cdots d\sigma_1 = \left(\int^{(i)} \phi(t) \right), \quad \int_0^t \phi(\sigma_1)d\sigma_1 = \left(\int \phi \right)$$

$$= -\gamma_2 y - \left[\int^{(1)} \left(\gamma_1 y + \gamma_4 u + \left(\int^{(1)} \gamma_0 y + \gamma_3 u \right) \right) \right] \qquad (B.22)$$

Taking the Laplace transforms in (B.22), after rearrangement, yields:

$$u(s) = - \left[\frac{\gamma_2 s^2 + \gamma_1 s + \gamma_0}{s^2 + \gamma_4 s + \gamma_3} \right] y(s) \qquad (B.23)$$

The characteristic polynomial of the closed-loop system is clearly given by

$$p(s) = s^5 + \gamma_4 s^4 + \gamma_3 s^3 + \gamma_2 s^2 + \gamma_1 s + \gamma_0 \qquad (B.24)$$

with freely assignable parameters for stability achievement. It is tedious but not difficult to show that the classical requirements for stable roots of $p(s)$ in Eq. (B.24) imply stable locations of both poles and zeroes of the filter in (B.23).

In frequency domain terms, the integral reconstructor-based controller yields a dynamical classical compensation network of the form (B.23). This subsystem can be regarded as a combination of a stable filter, smoothing the available output of the system, and a feed-forward term synthesizing the required plant input in terms of a linear combination of the internal states of such a filter. We address the above classical controller as the flat-filter-based controller. The reasons for this terminology are as follows: 1) Regarding the controller as a dynamical system with "input" represented by the output of the system y and with "output" represented by the system control input, u, the filtered output y_f qualifies as a flat output for such a linear dynamical system; indeed, all variables in the dynamic compensator can be expressed in terms of the flat output y_f and a finite number of its time derivatives. 2) A crucial property of flat linear time-invariant systems is that their transfer functions exhibit no zero dynamics. Such is the case of the transfer function of the filtered output y_f considering y as an input. The flat filtered output is defined as

$$y_f(s) = \left[\frac{1}{s^2 + \gamma_4 s + \gamma_3} \right] y(s), \quad \text{or} \quad \ddot{y}_f = -\gamma_4 \dot{y}_f - \gamma_3 y_f + y \qquad (B.25)$$

The state space representation of the flat filtering controller follows immediately from the controller expression written in compensation network form and expressed back in the time domain. Defining $y_f = \zeta_1$ and $\dot{y}_f = \zeta_2$, we have:

$$\begin{aligned} \dot{\zeta}_1 &= \zeta_2 \\ \dot{\zeta}_2 &= -\gamma_4 \zeta_2 - \gamma_3 \zeta_1 + y \\ u &= -(\gamma_0 - \gamma_2 \gamma_3)\zeta_1 - (\gamma_1 - \gamma_2 \gamma_4)\zeta_2 - \gamma_2 y \end{aligned} \qquad (B.26)$$

B.2.4 The Robustness Issue

Suppose that we desire to control the same system as (B.16) in a perturbed version of the form

$$y^{(3)} = u + \xi(t) \tag{B.27}$$

where $\xi(t)$ is only known to be an absolutely bounded signal. A compensator would try to overcome the unknown disturbance with as many integrations in the compensator as reasonably possible in the hope of facing a classical disturbance of polynomial type or to handle a time-polynomial ultralocal model of such a perturbation with enough time differentiations. Notice, however, that any smooth bounded time-varying perturbation is ultralocally efficiently approximated by a time polynomial of arbitrary degree (piecewise constant, piecewise linear, parabolic, etc.). Any finite-degree time polynomial approximation, taken as internal model of the unstructured additive perturbation, may then be also locally approximately canceled by a sufficient number of differentiations. These differentiations, acting as perturbation annihilators, are easily realized as powers of the complex variable s appearing as factors in the denominator of the compensation network.

Recall that, respectively, in extended state observers and in GPI observers, a single extra integrator or a finite number of integrations extending the observer's state space suffice to have an arbitrarily close estimation to the actual disturbance. Similarly and dually, in GPI-based Flat Filters, the suitable addition of one or a finite number of integrators in the compensation network denominator will result, under closed-loop conditions, in at least the same number of time differentiations of the additive disturbance. This simple duality is at the heart of regarding nonlinear state-dependent, and even input dependent, disturbances as unstructured time polynomial models [1] whose effects can be online identified [5], estimated [6], or cancelled in an approximate manner. That this philosophy works even for nonlinear state- and input-dependent additive nonlinearities rests on the fact that, ultimately, while the system is operating on line, such disturbances are, indeed, time-varying signals. Efforts to generally assess closed-loop stability of the existing control schemes adopting this modeling philosophy for disturbances may be found in [16], [14], [20], [5], and [21]. See also the excellent survey by Madoński and Herman [18].

The preceding paragraph justifies the use of a robust flat-filter-based compensator for the perturbed third-order pure integration system. Suppose that a piecewise constant internal model is adopted for the additive perturbation (hence, $m = 1$ is needed). A trajectory tracking task toward a given smooth output reference trajectory $y^*(t)$ is imposed on the system. We then propose

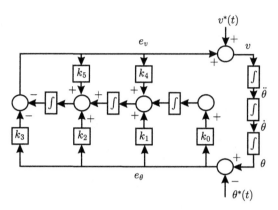

FIGURE B.4 Block diagram of closed-loop system controlled by a GPI compensation network with an extra integrator.

(see Fig. B.4)

$$u = u^*(t) - \left[\frac{k_3 s^3 + k_2 s^2 + k_1 s + k_0}{s(s^2 + k_5 s + k_4)} \right] (y - y^*(t)) \qquad \text{(B.28)}$$

Letting $e_y = (y - y^*(t))$, the closed-loop system evolves in accordance with

$$(s^6 + k_5 s^5 + \cdots + k_1 s + k_0)e_y = (s^3 + k_5 s^2 + k_4 s)\xi(s) \qquad \text{(B.29)}$$

The disturbance signal ξ, ultralocally modeled by a piecewise constant function, is differentiated at least once in the closed-loop system (see Fliess and Join [5] and also [6]).

The state representation of the linear dynamical controller is achieved by defining

$$x = \frac{e_y}{s^3 + k_5 s^2 + k_4 s} \qquad \text{(B.30)}$$

The control signal u, written in terms of x and its time derivatives, in fact constitutes the controller system output

$$u = u^*(t) - k_3 x^{(3)} - k_2 \ddot{x} - k_1 \dot{x} - k_0 x \qquad \text{(B.31)}$$

Defining $x_1 = x$, $x_2 = \dot{x}$, etc., by the definition of x we have

$$\dot{x}_1 = x_2$$
$$\dot{x}_2 = x_3$$
$$\dot{x}_3 = -k_5 x_3 - k_4 x_2 + (y - y^*(t)) \qquad \text{(B.32)}$$

$$u = u^*(t) - (k_2 - k_3 k_5)x_3 - (k_1 - k_3 k_4)x_2 - k_0 x_1 - k_3(y - y^*(t)) \qquad \text{(B.33)}$$

B.2.5 The Dirty Derivative Case

Regarding PID controllers using the dirty derivative expression, we obtain a similar effect of having a pole at the origin of the compensator transfer function. Indeed, consider, in the context of Subsection B.2.2, the following PID controller for a second-order pure integration system:

$$s^2 e_y = e_u = -\left[k_2 + \frac{k_1 s}{s + k_3} + \frac{k_0}{s}\right] e_y = -G(s) e_y \tag{B.34}$$

where $G(s)$ may be rewritten as

$$G(s) = \frac{(k_1 + k_2)s^2 + (k_0 + k_2 k_3)s + k_0 k_3}{s(s + k_3)} \tag{B.35}$$

which has the same form as the corresponding GPI, or flat filtering, controller for a second-order system. The closed-loop system is just

$$\left[s^4 + k_3 s^3 + (k_2 + k_1)s^2 + (k_0 + k_2 k_3)s + k_0 k_3\right] e_y = 0 \tag{B.36}$$

whose characteristic polynomial in the complex variable s can be made into a Hurwitz polynomial without any obstacles.

B.2.6 A Generalization

The previous development can be reproduced for a smooth output reference trajectory tracking problem, defined on an nth-order pure integration system of the form

$$y^{(n)} = u \tag{B.37}$$

The tracking error dynamics corresponding to a smooth output reference trajectory $y^*(t)$ is simply given by

$$e_y^{(n)} = e_u, \quad e_y = y - y^*, \quad e_u = u - u^*, \quad u^* = [y^*]^{(n)} \tag{B.38}$$

The integral reconstructors of the variables $\dot{e}_y, \ldots, e_y^{(n-1)}$, along with their (unstable) reconstruction errors, are proposed to be

$$\widehat{e_y^{(n-i)}} = \int_0^t \int_0^{\sigma_1} \cdots \int_0^{\sigma_{i-1}} e_u(\sigma_i) d\sigma_i \cdots d\sigma_1$$

$$e_y^{(n-i)} = \widehat{e_y^{(n-i)}} + \sum_{j=1}^i e_y^{(n-i+j-1)}(0) \frac{t^{j-1}}{(j-1)!}$$

$$i = 1, \ldots, n-1 \tag{B.39}$$

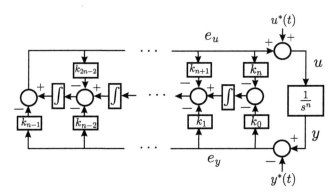

FIGURE B.5 GPI control of an nth-order system via integral reconstructors.

An implicit linear feedback controller, based on the integral reconstructors (B.39), requires additive compensation via multiple, or iterated, integrations performed on the output tracking error signal $e_y(t)$:

$$
\begin{aligned}
e_u &= -\sum_{i=1}^{n-1} k_{2n-i-1} \widehat{e_y^{(n-i)}} - \sum_{i=1}^{n} k_{n-i} \left(\int^{(i-1)} e_y(t) \right) \\
&= - \left[\sum_{i=1}^{n-1} k_{2n-i-1} \left(\int^{(i)} e_u(t) \right) \right] - \sum_{i=1}^{n} k_{n-i} \left(\int^{(i-1)} e_y(t) \right) \\
&= -\sum_{i=1}^{n-1} \left(\int^{(i)} k_{2n-i-1} e_u(t) + k_{n-i-1} e_y(t) \right) - k_{n-1} e_y \qquad \text{(B.40)}
\end{aligned}
$$

A block diagram in Fig. B.5 depicts the GPI trajectory tracking controller along with the pure integration plant. No output signal derivatives are gathered directly from the plant for closing the loop.

From the above expressions, the closed-loop system is also found to be

$$
\begin{aligned}
e_y^{(n)} &= -k_{2n-2} \widehat{e_y^{(n-1)}} - k_{2n-3} \widehat{e_y^{(n-2)}} - \cdots - k_n \widehat{\dot{e}_y} \\
&\quad - k_{n-1} e_y - k_{n-2} \left(\int^{(1)} e_y \right) - \cdots - k_1 \left(\int^{(n-2)} e_y \right) - k_0 \left(\int^{(n-1)} e_y \right)
\end{aligned}
$$
$$\text{(B.41)}$$

Substituting the expressions of the estimated phase variables found in (B.39) into Eq. (B.41), we obtain the following perturbed dynamics including the effects of the neglected initial conditions:

$$
e_y^{(n)} = -k_{2n-2} e_y^{(n-1)} - k_{2n-3} e_y^{(n-2)} - \cdots - k_n \dot{e}_y - k_{n-1} e_y
$$

$$- k_{n-2} \left(\int^{(1)} e_y \right) - \cdots - k_0 \left(\int^{(n-1)} e_y \right)$$

$$+ \left(\sum_{i=1}^{(n-1)} k_{2n-i-1} \sum_{j=1}^{i} e_y^{(n-i+j-1)}(0) \frac{t^{j-1}}{(j-1)!} \right) \tag{B.42}$$

Define

$$z_1 = \left(\int^{(1)} e_y \right) + \frac{k_{n-3}}{k_{n-2}} \left(\int^{(2)} e_y \right) + \cdots$$

$$+ \frac{k_1}{k_{n-2}} \left(\int^{(n-1)} e_y \right) + \frac{k_0}{k_{n-1}} \left(\int^{(n-2)} e_y \right)$$

$$- \left[\left(\sum_{i=1}^{(n-1)} \frac{k_{2n-i-1}}{k_{n-2}} \sum_{j=1}^{i} e_y^{(n-i+j-1)}(0) \frac{t^{j-1}}{(j-1)!} \right) \right]$$

$$z_1(0) = \frac{k_{2n-2}}{k_{n-2}} e_y^{(n-1)}(0)$$

$$z_2 = \left(\int^{(1)} e_y \right) + \frac{k_{n-4}}{k_{n-3}} \left(\int^{(2)} e_y \right) + \cdots$$

$$+ \frac{k_1}{k_{n-2}} \left(\int^{(n-3)} e_y \right) + \frac{k_0}{k_{n-3}} \left(\int^{(n-2)} e_y \right)$$

$$- \left[\left(\sum_{i=1}^{(n-1)} \frac{k_{2n-i-1}}{k_{n-3}} \sum_{j=2}^{i} e_y^{(n-i+j-1)}(0) \frac{t^{j-2}}{(j-2)!} \right) \right]$$

$$z_2(0) = \frac{k_{2n-2}}{k_{n-3}} e_y^{(n-2)}(0) + \frac{k_{2n-3}}{k_{n-3}} e_y^{(n-1)}(0)$$

$$\vdots$$

$$z_{n-2} = \left(\int^{(1)} e_y \right) + \frac{k_0}{k_1} \left(\int^{(2)} e_y \right)$$

$$- \left[\left(\sum_{i=1}^{(n-1)} \frac{k_{2n-i-1}}{k_1} \sum_{j=n-2}^{i} e_y^{(n-i+j-1)}(0) \frac{t^{j-n+2}}{(j-n+2)!} \right) \right]$$

$$z_{n-2}(0) = \frac{k_{2n-2}}{k_1} e_y^{(2)}(0) + \frac{k_{2n-3}}{k_1} e_y^{(3)}(0) + \cdots + \frac{k_{n+1}}{k_1} e_y^{(n-1)}(0)$$

$$z_{n-1} = \left(\int^{(1)} e_y \right)$$

$$z_{n-1}(0) = \frac{k_{2n-2}}{k_0}e_y^{(1)}(0) + \frac{k_{2n-3}}{k_0}e_y^{(2)}(0) + \cdots + \frac{k_n}{k_1}e_y^{(n-1)}(0) \qquad \text{(B.43)}$$

The closed-loop system is then obtained as

$$e_y^{(n)} = -k_{2n-2}e_y^{(n-1)} - \cdots - k_n\dot{e}_y - k_{n-1}e_y - k_{n-2}z_1$$

$$\dot{z}_i = e_y + \frac{k_{n-i-2}}{k_{n-i-1}}z_{i+1}, \quad i = 1, \cdots, n-2$$

$$\dot{z}_{n-1} = e_y \qquad \text{(B.44)}$$

Eliminating the z variables via direct differentiation of the first equation $n-1$ times, we obtain the closed-loop differential equation for the tracking error signal e_y as

$$e_y^{(2n-1)} + k_{2n-2}e_y^{(2n-2)} + \cdots + k_1\dot{e}_y + k_0e_y = 0 \qquad \text{(B.45)}$$

whose characteristic polynomial is simply

$$p(s) = s^{2n-1} + k_{2n-2}s^{2n-2} + \cdots + k_1s + k_0 \qquad \text{(B.46)}$$

Thus, a suitable set of design coefficients $\{k_{2n-2}, k_{2n-1}, \ldots, k_1, k_0\}$ may be prescribed to guarantee that the location of all the roots of the polynomial in Eq. (B.46) are strictly located in the left half portion of the complex plane, guaranteeing the asymptotic stability of the closed-loop output tracking error trajectory.

Note that the Laplace transform of the iterated integral $\left(\int^{(j)} \phi(t)\right)$ is given by $\phi(s)/s^j$. From the expression

$$e_y^{(n)} = e_u = -\left[\sum_{i=1}^{n-1}k_{2n-i-1}\left(\int^{(i)}e_u(t)\right)\right] - \sum_{i=1}^{n}k_{n-i}\left(\int^{(i-1)}e_y(t)\right) \qquad \text{(B.47)}$$

we obtain, after resorting to the frequency domain and rearranging, that the controller may be described by the following classical compensation network (see Fig. B.6):

$$e_u(s) = -\left[\frac{k_{n-1}s^{n-1} + k_{n-2}s^{n-2} + \cdots + k_0}{s^{n-1} + k_{2n-2}s^{2n-2} + \cdots + k_n}\right]e_y(s) \qquad \text{(B.48)}$$

A reinterpretation of the GPI controller leads to a flat-filtering-based control scheme. Indeed, define, in the frequency domain, the filtered output tracking error

$$e_{y_f}(s) = \left[\frac{1}{s^{n-1} + k_{2n-2}s^{2n-2} + \cdots + k_n}\right]e_y(s) \qquad \text{(B.49)}$$

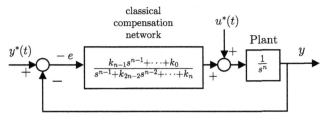

FIGURE B.6 GPI controller for an nth-order pure integration plant as a classical compensation network.

Clearly, a time domain realization of the filtered output tracking error and the controller is just

$$e_{y_f}^{(n-1)} = -k_{2n-2}e_{y_f}^{(n-2)} - \cdots - k_n e_{y_f} + (y - y^*)$$

$$e_u = -k_{n-1}e_{y_f}^{(n-1)} - k_{n-2}e_{y_f}^{(n-2)} - \cdots - k_0 e_{y_f} \qquad (B.50)$$

In other words, the output tracking error $e_y = y - y^*(t)$ and the input error $e_u = u - u^*(t)$ are expressible in terms of the filtered output e_{y_f} and a finite number of its time derivatives (i.e., the basic property of a flat output is valid for e_{y_f}); in state space form,

$$\dot{\zeta}_i = \zeta_{i+1}, \quad i = 1, \ldots n - 2$$

$$\dot{\zeta}_{n-1} = -k_{2n-2}\zeta_{n-1} - \cdots - k_{n+1}\zeta_2 - k_n\zeta_1 + (y - y^*(t))$$

$$e_{y_f} = \zeta_1$$

$$e_u = (k_{n-1}k_{2n-2} - k_{n-2})\zeta_{n-1} + (k_{n-1}k_{2n-3} - k_{n-3})\zeta_{n-2} + \cdots$$

$$+ (k_{n-1}k_{n+1} - k_1)\zeta_2 + (k_{n-1}k_n - k_0)\zeta_1 - k_{n-1}(y - y^*(t)) \qquad (B.51)$$

B.2.7 A Further Generalization

A further generalization, pertaining the output reference trajectory tracking control task without using observers, for a chain of n integrators of the form

$$y^{(n)} = \alpha u + \xi(t), \quad \alpha \in \mathbb{R} \qquad (B.52)$$

and undergoing the effects of a uniformly absolutely bounded additive perturbation input $\xi(t)$ is as follows. The disturbance $\xi(t)$ is assumed to be ultralocally modeled as an $(m - 1)$th-degree time polynomial. A flat filtering controller, or classical compensation network controller, is given by

$$u = u^*(t) - \frac{1}{\alpha} \left[\frac{k_{n+m-1}s^{n+m-1} + k_{n+m-2}s^{n+m-2} + \cdots + k_0}{s^m \left(s^{n-1} + k_{2n+m-2}s^{n-2} + \cdots + k_{n+m} \right)} \right] (y - y^*(t))$$

$$(B.53)$$

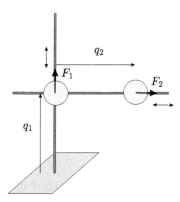

FIGURE B.7 Two-degree-of-freedom Cartesian robot.

with $u^*(t) = (1/\alpha)[y^*(t)]^{(n)}$. The characteristic polynomial of the closed-loop system is given by

$$p(s) = s^{2n+m-1} + k_{2n+m-2}s^{2n+m-2} + k_{2n+m-3}s^{2n+m-3} + \cdots$$
$$+ k_2 s^2 + k_1 s + k_0 \tag{B.54}$$

Since the least order of differentiation, induced by the proposed controller on the additive disturbance term $\xi(t)$, is m, the additive perturbation model is ultralocally annihilated.

B.3 AN ILLUSTRATIVE APPLICATION EXAMPLE

In this section, we present an illustrative application example dealing with the trajectory tracking of a two-degree-of-freedom Cartesian robot. The system, besides being linear, can be controlled using flat filtering without measuring the link velocities.

Consider the two-mass system representing orthogonal links in a Cartesian configuration with each link provided by an independent actuating force, as shown in Fig. B.7.

The mathematical model of the system is given by

$$\begin{cases} \ddot{q}_1 = -g + \dfrac{F_1}{m_1+m_2} \\ \ddot{q}_2 = \dfrac{F_2}{m_2} \end{cases} \tag{B.55}$$

We assume that $|F_1(t)| \le U_1$ and $|F_2(t)| \le U_2$.

A simple change of coordinates in the control input space (F_1, F_2),

$$F_1 = (m_1 + m_2)(g + v_1), \quad F_2 = m_2 v_2 \tag{B.56}$$

reduces the system to a pair of pure second-order integration systems with the auxiliary control inputs v_1 and v_2, i.e.,

$$\ddot{q}_1 = v_1 \quad \ddot{q}_2 = v_2 \tag{B.57}$$

The auxiliary inputs are also constrained by the amplitude limitation on the original forces. In fact,

$$|v_1(t)| \le \left(\frac{U_1}{m_1 + m_2} \right) - g, \quad |v_2(t)| \le \frac{U_2}{m_2} \tag{B.58}$$

Given a smooth trajectory on the plane $(x, y) = (q_2, q_1)$, specified in para-metrics

$$x^* = q_2^*(t), \quad y^* = q_1^*(t) \tag{B.59}$$

we desire that the position coordinates of the end effector (q_2, q_1) track a given smooth trajectory in the plane and, after a reasonable transient regime, avoid saturation of the control input forces F_1 and F_2.

We can control this system by means of flat filtering without need for velocity measurements $\dot{q}_1$ and $\dot{q}_2$:

$$
\begin{aligned}
F_1 &= (m_1 + m_2)\left\{ g + \ddot{q}_1^*(t) - \left[\frac{b_1 s + c_1}{s + a_1} \right] (q_1 - q_1^*(t)) \right\} \\
F_2 &= m_2 \left\{ \ddot{q}_2^*(t) - \left[\frac{b_2 s + c_2}{s + a_2} \right] (q_2 - q_2^*(t)) \right\}
\end{aligned} \tag{B.60}
$$

The nominal values of F_1 and F_2 in permanent regime satisfy

$$
\begin{aligned}
F_1^*(t) &= (m_1 + m_2)(g + \ddot{q}_1^*(t)) \\
F_2^*(t) &= m_2 \ddot{q}_2^*(t)
\end{aligned} \tag{B.61}
$$

The end effector trajectory components must be such that

$$|\ddot{q}_1^*(t)| \le \frac{U_1}{m_1 + m_2} - g, \quad |\ddot{q}_2^*(t)| \le \frac{U_2}{m_2}, \ \forall \, t \tag{B.62}$$

As an example, we take the tracking the following position reference trajectories, given in parametrics

$$
\begin{aligned}
x^* &= (a + R\cos(n\omega t))\cos(\omega t) \\
y^* &= (a + R\cos(n\omega t))\sin(\omega t)
\end{aligned} \tag{B.63}
$$

with the following assigned parameters:

$$m_1 = m_2 = 1 \ [\text{Kg}], \quad g = 9.8 \ [\text{m/s}^2] \tag{B.64}$$

Let

$$\rho(t) = a + R\cos(n\omega t),$$
$$x^*(t) = \rho(t)\cos(\omega t), \quad y^*(t) = \rho(t)\sin(\omega t) \tag{B.65}$$

The second-order time derivatives of the desired trajectories are given by

$$\ddot{x}^* = (\ddot{\rho} - \omega^2\rho)\cos(\omega t) - 2\dot{\rho}\omega\sin(\omega t)$$
$$\ddot{y}^* = (\ddot{\rho} - \omega^2\rho)\sin(\omega t) + 2\dot{\rho}\omega\cos(\omega t) \tag{B.66}$$

with,

$$\dot{\rho} = -(nR)\omega\sin(\omega t), \quad \ddot{\rho} = -R(n\omega)^2\cos(\omega t) \tag{B.67}$$

The estimated maximum value of the control input forces $\ddot{q}_1^*(t) = \ddot{y}^*(t)$ and $\ddot{q}_2^*(t) = \ddot{x}^*(t)$ under nominal tracking conditions are just computed as

$$\sup_t |F_1^*(t)| \leq (m_1 + m_2)(g + (R(n+1)^2 + a)\omega^2)$$

$$= \sup_t |F_2^*(t)| \leq m_2(R(n+1)^2 + a)\omega^2 \tag{B.68}$$

These relations are useful for determining reasonable values of U_1 and U_2 for which, at least in principle, there are no permanent saturations.

We have, for the given robot parameters, that

$$\sup_t |F_1^*(t)| \leq 2(9.8 + (R(n+1)^2 + a)\omega^2) \leq U_1,$$

$$\sup_t |F_2^*(t)| \leq (R(n+1)^2 + a)\omega^2 \leq U_2$$

$$F_{1\,max} = 31.28\ [\text{N}], \quad F_{2\,max} = 5.84$$

B.3.1 Simulations

Simulations were performed for assessing the behavior of the closed loop system under the proposed controller. The following parameters were used:

$$a = 0.5\ [\text{m}], \quad R = 1\ [\text{m}], \quad n = 5, \quad w = 0.4\ [\text{rad/s}]$$

Fig. B.8 depicts the closed loop performance of the flat filtering controller.

B.4 GPI CONTROL FOR DISCRETE-TIME SYSTEMS

B.4.1 The Counterpart of Integral Compensation

Consider the first-order discrete-time system

$$y_{k+1} = u_k + \xi \tag{B.69}$$

where ξ is a constant perturbation. Suppose that we desire to track a given reference output signal $y_k^*(t)$. A proportional plus "adder" controller, counter-

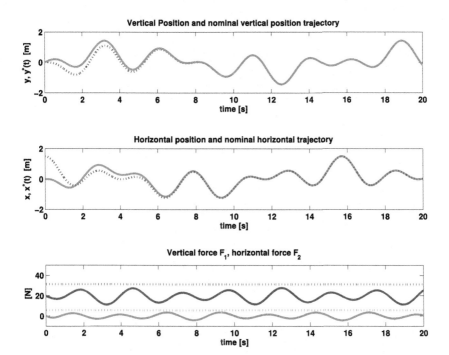

FIGURE B.8 Trajectory tracking performance of flat filtering controller.

part of the "proportional plus integral" controller in continuous time, is given by

$$u_k = y_{k+1}^* + k_1(y_k - y_k^*) + k_0\rho_k$$
$$\rho_{k+1} = \rho_k + (y_k - y_k^*) \tag{B.70}$$

The closed-loop tracking error $e_k = y_k - y_k^*$ evolves according to

$$e_{k+1} = k_1 e_k + k_0\rho_k + \xi$$
$$\rho_{k+1} = \rho_k + e_k \tag{B.71}$$

Thus, taking an advance (time shift) in the error equation (B.71), we have:

$$e_{k+2} = k_1 e_{k+1} + k_0(\rho_k + e_k) + \xi$$
$$= k_1 e_{k+1} + k_0\rho_k + k_0 e_k + \xi \tag{B.72}$$

but from the original error dynamics

$$k_0\rho_k = e_{k+1} - k_1 e_k - \xi \tag{B.73}$$

using Eq. (B.73) in (B.72), we find that the second-order error dynamics is given by the expression

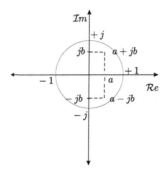

FIGURE B.9   Pole location for GPI controlled discrete-time system.

$$e_{k+2} - (1+k_1)e_{k+1} - (k_0 - k_1)e_k = 0 \qquad (\text{B.74})$$

whose characteristic polynomial in the $\mathcal{Z}$ transform domain is just

$$p(z) = z^2 - (1+k_1)z - (k_0 - k_1) \qquad (\text{B.75})$$

To obtain an asymptotically stable feedback controller design, we chose the coefficients of the characteristic polynomial to equal those of a polynomial $p_d(z)$ with roots known to be strictly inside the unit circle of the complex plane $(a^2 + b^2 < 1)$, as depicted by Fig. B.9

$$p_d(z) = (z - a - jb)(z - a + jb) = z^2 - 2az + a^2 + b^2 \qquad (\text{B.76})$$

i.e.,

$$k_1 = (2a - 1), \quad k_0 = -a^2 - b^2 + (2a - 1) \qquad (\text{B.77})$$

The feedback controller is obtained in terms of the control error $e_{u_k} = u_k - u_k^*$, where u_k^* is found from the unperturbed system dynamics as $u_k^* = y_{k+1}^*$. Since

$$e_{u_k} = k_1 e_k + k_0 \rho_k, \quad \rho_{k+1} = \rho_k + e_k \qquad (\text{B.78})$$

we have, as shown in Fig. B.10, that

$$u_k = u_k^* + \left[k_1 + \frac{k_0}{z-1} \right] (y_k - y_k^*) \qquad (\text{B.79})$$

Rewriting the controller in the form

$$e_u(z) = \left[\frac{k_1 z + (k_0 - k_1)}{z - 1} \right] e_y(z) = \frac{z}{z-1} \left(k_1 + (k_0 - k_1)z^{-1} \right) e_y(z)$$

$$= \left(\frac{1}{1 - z^{-1}} \right) \left(k_1 + (k_0 - k_1)z^{-1} \right) e_y(z) \qquad (\text{B.80})$$

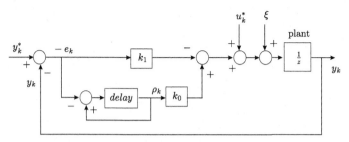

FIGURE B.10 Closed loop system tracking performance of a ramp perturbed first order discrete time system.

it is easy to see that the factor $(1 + z^{-1})^{-1}$ plays a crucial role in the robustness property of the proportional plus adder controller with respect to unknown constant disturbances. Indeed, the factor $(1 - z^{-1})$ is an annihilator of constant additive terms in the system dynamics.

The controller produces the following closed-loop system in the time domain:

$$e_{k+1} - e_k = k_1 e_k + (k_0 - k_1)e_{k-1} + (1 - z^{-1})\xi \qquad (B.81)$$

or

$$e_{k+2} = (1 + k_1)e_{k+1} + (k_0 - k_1)e_k \qquad (B.82)$$

which is the originally found closed-loop dynamics.

B.4.2 Simulations

We set the following output reference trajectory:

$$y_k^* = \cos(\omega k), \quad \omega = 5 \qquad (B.83)$$

The closed-loop characteristic polynomial parameters and the corresponding controller gains were given by

$$a = -0.5, \quad b = 0.5, \quad k_0 = -2.5, \quad k_1 = -2 \qquad (B.84)$$

The constant perturbation ξ was set to

$$\xi = 0.5$$

The sampling time T was taken to be 0.2, i.e., $kT = 0.2k$ with $k = 0, 1, 2, \ldots$. Recall that $kT + T$ is represented as $k + 1$.

Fig. B.11 shows the output reference trajectory tracking features of the proposed controller as well as the nominal and actually applied control input signal.

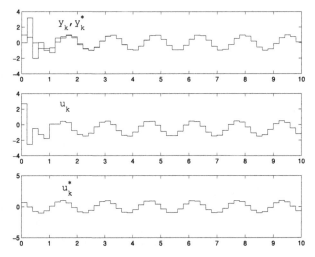

FIGURE B.11   Closed loop response of second order system with "adder" action.

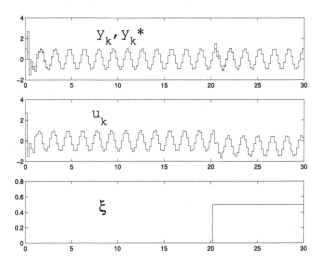

FIGURE B.12   Effect of a step perturbation input ξ on the system tracking performance.

Fig. B.12 depicts the quick recovery of the tracking task in the controlled system, when an unforseen step input perturbation $\xi(t)$ is applied to the system right after $k = 20$.

B.4.3   A Ramp-Perturbed System

Consider now the perturbed first-order system

$$y_{k+1} = u_k + k\xi \tag{B.85}$$

with a perturbation that linearly grows with time, i.e., with a "ramp" perturbation. We first show that a proportional plus adder controller does not exhibit robustness with respect to this type of perturbation. Indeed, suppose that we used the controller

$$u_k = y_{k+1}^* + k_1(y_k - y_k^*) + k_0\rho_k$$
$$\rho_{k+1} = \rho_k + (y_k - y_k^*) \tag{B.86}$$

The closed-loop tracking error dynamics is found to satisfy

$$e_{k+2} - (1 + k_1)e_{k+1} - (k_0 - k_1)e_k = \xi \tag{B.87}$$

whose equilibrium, given by $\bar{e} = -\xi/k_0$, clearly is nonzero and depends upon the unknown value of ξ. We state that the proposed control strategy with a single adder compensation term does not stabilize the tracking error to zero.

We then propose a proportional adder plus "double adder" compensation as follows:

$$u_k = y_{k+1}^* + k_2(y_k - y_k^*) + k_1\rho_k + k_0\zeta_k$$
$$\rho_{k+1} = \rho_k + (y_k - y_k^*)$$
$$\zeta_{k+1} = \zeta_k + v_k$$
$$v_{k+1} = v_k + (y_k - y_k^*) \tag{B.88}$$

Letting $u_k^* = y_{k+1}^*$, we substitute into the controller equation (B.88) the following expressions for ρ_k and ζ_k, mixing time domain signals with transformed signals:

$$e_{u_k} = \left[k_2 + k_1\frac{1}{z-1} + k_0\frac{1}{(z-1)^2} \right] e_k \tag{B.89}$$

Rearranging, we obtain

$$e_{u_k} = \left[\frac{z}{z-1} \right]^2 \left(k_2 + (k_1 - 2k_2)z^{-1} + (k_2 + k_0 - k_1)z^{-2} \right) e_k$$
$$= \left[\frac{1}{(1 - z^{-1})^2} \right] \left(k_2 + (k_1 - 2k_2)z^{-1} + (k_2 + k_0 - k_1)z^{-2} \right) e_k \tag{B.90}$$

The closed-loop system obeys

$$(1 - z^{-1})^2 e_k = \left[k_2 + (k_1 - 2k_2)z^{-1} + (k_2 + k_0 - k_1)z^{-2} \right] e_k + \xi[(1 - z^{-1})^2 k]$$
$$\tag{B.91}$$

Note that $(1 - z^{-1})^2 k = (1 - 2z^{-1} + z^{-2})k = k - 2(k - 1) + (k - 2) = 0$. Using these facts, we find the closed-loop system expression

$$e_{k+3} - (2 + k_2)e_{k+2} + (2k_2 - k_1 + 1)e_{k+1} - (k_0 - k_1 + k_2)e_k = 0 \quad \text{(B.92)}$$

which is totally independent of ξ. The characteristic polynomial $p(z)$ of the closed-loop system (B.92) is, thus, given by

$$z^3 - (2 + k_2)z^2 + (2k_2 - k_1 + 1)z - (k_0 - k_1 + k_2) = 0 \quad \text{(B.93)}$$

Once again, a suitable factor, which in this case is a second-order power of the binomial $1 - z^{-1}$, annihilates the perturbation input and makes the system prone to asymptotically stable behavior by suitable choice of the controller design coefficients.

The controller design is completed by choosing the set of gains $\{k_2, k_1, k_0\}$ so that the characteristic polynomial $p(z)$ coincides with a desired one, denoted by $p_d(z)$, exhibiting all its roots inside the unit circle in the complex plane.

By setting $(a^2 + b^2 < 1, |p| < 1)$ we consider

$$\begin{aligned} p_d(z) &= (z^2 - 2az + a^2 + b^2)(z - p) \\ &= z^3 - (p + 2a)z^2 + (a^2 + b^2 + 2ap)z - p(a^2 + b^2) \end{aligned} \quad \text{(B.94)}$$

Hence, a suitable choice of the gains is represented by

$$\begin{aligned} k_2 &= p + 2a - 2 \\ k_1 &= -\left(a^2 + b^2 + 2ap - 2(p + 2a) + 3\right) \\ k_0 &= -\left(a^2 + b^2 + 2ap - 2(p + 2a) + 3\right) - (p + 2a - 2) + p(a^2 + b^2) \end{aligned}$$
$$\text{(B.95)}$$

B.4.4 GPI Control of a Pure Chain of Delays

Consider the following second-order discrete-time system:

$$y_{k+2} = u_k \quad \text{(B.96)}$$

A feedback controller may be proposed in the form

$$u_k = y_{k+2}^* - k_2(y_{k+1} - y_{k+1}^*) - k_1(y_k - y_k^*) \quad \text{(B.97)}$$

whose corresponding closed-loop dynamics for the tracking error $e_k = y_{k+1} - y_k$ is given by

$$e_{k+2} + k_2 e_{k+1} + k_1 e_k = 0 \quad \text{(B.98)}$$

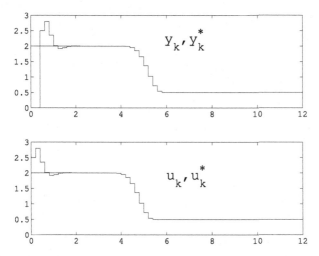

FIGURE B.13 Closed loop response of GPI controlled second order discrete time system.

The appropriate choice of the parameters k_2 and k_1 yields a closed-loop system with zero as an asymptotically stable equilibrium point in the tracking error space.

However, since y_{k+1} is not available at time k, we may exactly estimate it as

$$y_{k+1} = u_{k-1} \tag{B.99}$$

We have, defining $e_{u_k} = u_k - u_k^*$ with $u_k^* = y_{k+2}^*$,

$$e_{u_k} = -k_2 e_{u_{k-1}} - k_1 e_k \tag{B.100}$$

The feedback controller is given, in an abusive combination of notation mixing the time domain and the $\mathcal{Z}$-transform domain, by

$$u_k = u_k^* - \left[\frac{k_1}{1 + k_2 z^{-1}} \right] (y_k - y_k^*), \tag{B.101}$$

This controller, based on an exact prediction estimate of the output y_{k+1} in terms of the control (B.99), is the counterpart of the GPI control in continuous time for a second-order plant. Recall that such a control is synthesized using an integral reconstructor of the first-order time derivative of the output. In that case, a constant error, due to the unknown initial condition of the output derivative, had to be compensated via additional integral control. In the unperturbed discrete-time case, the corresponding prediction estimate is devoid of errors, and therefore no compensation is necessary for that reason.

Simulations of the closed-loop trajectory tracking error performance are shown in Fig. B.13.

B.4.5 A Perturbed Second-Order System

Consider the following simple perturbed linear time-invariant discrete-time system:

$$x_{1,k+1} = x_{2,k}$$
$$x_{2,k+1} = u_k + \xi_k$$
$$y_k = x_{1,k} \tag{B.102}$$

where the external perturbation input is constant, i.e., $\xi_k = \xi$ for all k. We assume that only $y_k = x_{1,k}$ and u_k are available for measurement whereas $x_{2,k} = y_{k+1}$ is not.

A controller with "adder" action (the counterpart of "integral action") for the tracking of a given trajectory y_k^* is given by

$$u_k = y_{k+2}^* - k_2(x_{2,k} - y_{k+1}^*) - k_1(y_k - y_k^*) - k_0\zeta_k$$
$$\zeta_{k+1} = \zeta_k + (y_k - y_k^*) \tag{B.103}$$

The closed-loop system tracking error $e_{1,k} = y_k - y_k^*$ is seen to satisfy the following state equation:

$$\begin{bmatrix} e_{1,k+1} \\ e_{2,k+1} \\ \zeta_{k+1} \end{bmatrix} = \begin{bmatrix} 0 & 1 & 0 \\ -k_1 & -k_2 & -k_0 \\ 1 & 0 & 1 \end{bmatrix} \begin{bmatrix} e_{1,k} \\ e_{2,k} \\ \zeta_k \end{bmatrix} + \begin{bmatrix} 0 \\ \xi \\ 0 \end{bmatrix} \tag{B.104}$$

The closed-loop system has the equilibrium point at

$$e_{1,\infty} = e_{2,\infty} = 0, \quad \zeta_\infty = \xi/k_0 \tag{B.105}$$

and the characteristic polynomial, associated with the closed-loop system, is given in $\mathcal{Z}$-transform terms by

$$z^3 + (k_2 - 1)z^2 + (k_1 - k_2)z + k_0 - k_1 \tag{B.106}$$

whose roots are completely assignable and may be chosen to be stable with the help of the Jury stability test.

Since $x_{2,k} = y_{k+1}$ is not readily available for measurement, we can use the following "delayed reconstructor" for $x_{2,k}$, as obtained directly from the unperturbed system model:

$$\widehat{x_{2,k}} = u_{k-1} \tag{B.107}$$

The proposed delayed reconstructor has the following relation with the actual value of $x_{2,k}$:

$$x_{2,k} = \widehat{x_{2,k}} + \xi \tag{B.108}$$

i.e., the estimate $\widehat{x_{2,k}}$ exhibits a constant unknown reconstruction error with respect to the actual value of the state $x_{2,k}$. Hence, a feasible controller, compensating this error in the feedback loop, is reformulated as

$$u_k = y_{k+2}^* - k_2(\widehat{x_{2,k}} - y_{k+1}^*) - k_1(y_k - y_k^*) - k_0\zeta_k$$
$$\zeta_{k+1} = \zeta_k + (y_k - y_k^*), \quad \widehat{x_{2,k+1}} = u_k \tag{B.109}$$

This feedback produces the following tracking error dynamics of $e_k = y_k - y_k^*$:

$$e_{k+2} + k_2 e_{k+1} + k_1 e_k + k_0\zeta_k = (1 + k_2)\xi$$
$$\zeta_{k+1} = \zeta_k + e_k \tag{B.110}$$

Note that we can express, from the first equation in (B.110), the quantity $k_0\zeta_k$ as

$$k_0\zeta_k = (1 + k_2)\xi - e_{k+2} - k_2 e_{k+1} - k_1 e_k \tag{B.111}$$

Since the tracking error e_k also satisfies

$$e_{k+3} + k_2 e_{k+2} + k_1 e_{k+1} + k_0\zeta_{k+1} = (1 + k_2)\xi$$
$$\zeta_{k+1} = \zeta_k + e_k \tag{B.112}$$

we have:

$$e_{k+3} + k_2 e_{k+2} + k_1 e_{k+1} + k_0[\zeta_k + e_k] = (1 + k_2)\xi \tag{B.113}$$

i.e.,

$$e_{k+3} + k_2 e_{k+2} + k_1 e_{k+1} + k_0\zeta_k + k_0 e_k = (1 + k_2)\xi$$
$$k_0\zeta_k = (1 + k_2)\xi - e_{k+2} - k_2 e_{k+1} - k_1 e_k \tag{B.114}$$

The closed-loop tracking error dynamics then satisfies the following equivalent dynamics:

$$e_{k+3} + (k_2 - 1)e_{k+2} + (k_1 - k_2)e_{k+1} + (k_0 - k_1)e_k = 0 \tag{B.115}$$

whose eigenvalues are completely assignable by appropriate choice of the design parameters $\{k_0, k_1, k_2\}$.

Fig. B.14 depicts the closed loop response of the system under a rest-to-rest trajectory tracking task. Fig. B.15 shows the effect of a constant perturbation $\xi(t)$ on the reconstructed state x_2 and the control input u_k.

The corresponding characteristic equation in $\mathcal{Z}$-transform terms is simply given by

$$z^3 + (k_2 - 1)z^2 + (k_1 - k_2)z + (k_0 - k_1) = 0 \tag{B.116}$$

Recall that the Möbius transformation

$$w = \frac{z+1}{z-1}, \quad z = \frac{w+1}{w-1}, \tag{B.117}$$

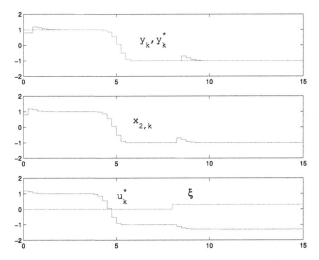

FIGURE B.14   Closed-loop response of second-order system with "adder" action.

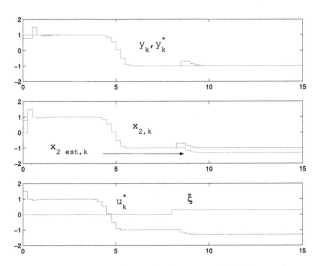

FIGURE B.15   Closed-loop response of second-order system with GPI controller.

allows us to use the traditional Routh–Hurwitz test for assessing the location of the roots of the $\mathcal{Z}$-transform characteristic equation relative to the unit circle in the complex plane.

The proposed feedback controller was given by

$$u_k = y^*_{k+2} - k_2(u_{k-1} - y^*_{k+1}) - k_1(y_k - y^*_k) - k_0\zeta_k$$
$$\zeta_{k+1} = \zeta_k + (y_k - y^*_k) \tag{B.118}$$

The nominal input signal u_k^* is obtained from the unperturbed model by straightforward inversion as $u_k^* = y_{k+2}^*$, and, therefore, it follows that, $u_{k-1}^* = y_{k+1}^*$. By defining $e_{u,k} = u_k - u_k^*$ we have:

$$e_{u,k} + k_2 e_{u,k-1} = -k_1 e_k - k_0 \zeta_k, \quad \zeta_{k+1} = \zeta_k + e_k \tag{B.119}$$

From (B.119) we obtain

$$-k_0 \zeta_k = e_{u,k} + k_2 e_{u,k-1} + k_1 e_k \tag{B.120}$$

Hence,

$$e_{u,k+1} + k_2 e_{u,k} = -k_1 e_{k+1} + (e_{u,k} + k_2 e_{u,k-1} + k_1 e_k) - k_0 e_k$$
$$e_{u,k+2} + (k_2 - 1)e_{u,k+1} - k_2 e_{u,k} = -\left[k_1 e_{k+2} + (k_0 - k_1)e_{k+1} \right] \tag{B.121}$$

In $\mathcal{Z}$-transform representation, the dynamical output feedback controller is expressed as

$$e_u(z) = -\left[\frac{k_1 z^2 + (k_0 - k_1)z}{z^2 + (k_2 - 1)z - k_2} \right] e(z) \tag{B.122}$$

In other words, combining time domain and transform domain expressions, we get:

$$u_k = u_k^* - \left[\frac{k_1 z^2 + (k_0 - k_1)z}{z^2 + (k_2 - 1)z - k_2} \right] (y_k - y_k^*) \tag{B.123}$$

We rewrite the dynamic feedback controller (B.123) as follows:

$$u_k = u_k^* - \frac{z}{z - 1} \left[\frac{k_1 z + (k_0 - k_1)}{z + k_2} \right] e_k$$
$$= u_k^* - \frac{1}{1 - z^{-1}} \left[\frac{k_1 + (k_0 - k_1)z^{-1}}{1 + k_2 z^{-1}} \right] e_k \tag{B.124}$$

Thanks to the denominator of the term $1/(1 - z^{-1})$, the constant additive perturbation ξ is annihilated in the closed-loop dynamics. We obtain the following expression:

$$e_{k+3} + (k_2 - 1)e_{k+2} + (k_1 - k_2)e_{k+1} + (k_0 - k_1)e_k = 0 \tag{B.125}$$

which coincides with Eq. (B.115).

B.4.6 A Third-Order System

Consider now the unperturbed system

$$y_{k+3} = u_k \tag{B.126}$$

with the feedback controller

$$u_k = y^*_{k+3} - k_2(y_{k+2} - y^*_{k+2}) - k_1(y_{k+1} - y^*_{k+1}) - k_0(y_k - y^*_k) \quad \text{(B.127)}$$

resulting in the following closed-loop system for the tracking error $e_k = y_k - y^*_k$:

$$e_{k+3} + k_2 e_{k+2} + k_1 e_{k+1} + k_0 e_k = 0 \quad \text{(B.128)}$$

The characteristic polynomial $p(z)$ of the closed-loop system, given by

$$p(z) = z^3 + k_2 z^2 + k_1 z + k_0 \quad \text{(B.129)}$$

can be made to exhibit all its roots strictly inside the unit circle in the complex plane by appropriate choice of $\{k_2, k_1, k_0\}$, in compliance with Jury's test.

Define $e_{u_k} = u_k - u^*_k$. Replacing in the proposed controller the unavailable signals y_{k+2} and y_{k+1}, respectively, by u_{k-1} and u_{k-2}, along with the required nominal values, we obtain

$$e_{u_k} = -k_2 e_{u_{k-1}} - k_1 e_{u_{k-2}} - k_0 e_k \quad \text{(B.130)}$$

In other words,

$$e_u(z) = -\left[\frac{k_0 z^2}{z^2 + k_2 z + k_1} \right] e(z) \quad \text{(B.131)}$$

The feedback controller, in combination of time and frequency domain notation, is thus given by

$$u_k = u^*_k - \left[\frac{k_0 z^2}{z^2 + k_2 z + k_1} \right] (y_k - y^*_k) \quad \text{(B.132)}$$

or

$$u_k = u^*_k - \left[\frac{k_0}{1 + k_2 z^{-1} + k_1 z^{-2}} \right] (y_k - y^*_k) \quad \text{(B.133)}$$

Consider now a perturbed third-order system with constant additive perturbation input ξ,

$$y_{k+3} = u_k + \xi \quad \text{(B.134)}$$

Let us solve the feedback stabilization problem.

The values of the advances of y_k are simply

$$y_{k+2} = u_{k-1} + \xi, \quad y_{k+1} = u_{k-2} + \xi, \quad y_k \text{ (is measurable)} \quad \text{(B.135)}$$

We take as "retarded reconstructors" of y_{k+2} and y_{k+1} the following:

$$\widehat{y}_{k+2} = u_{k-1}, \quad \widehat{y}_{k+1} = u_{k-2} \qquad \text{(B.136)}$$

Then consider the controller with adder compensation

$$u_k = -k_3\widehat{y}_{k+2} - k_2\widehat{y}_{k+1} - k_1 y_k - k_0\rho_k, \quad \rho_{k+1} = \rho_k + y_k \qquad \text{(B.137)}$$

i.e.,

$$u_k = -k_3 u_{k-1} - k_2 u_{k-2} - k_1 y_k - k_0\rho_k, \quad \rho_{k+1} = \rho_k + y_k \qquad \text{(B.138)}$$

From this we have

$$u_{k+1} = -k_3 u_k - k_2 u_{k-1} - k_1 y_{k+1} - k_0\rho_k - k_0 y_k \qquad \text{(B.139)}$$

Subtracting the last two expressions yields, after rearranging and taking a one step delay,

$$\left[1 + (k_3 - 1)z^{-1} + (k_2 - k_3)z^{-2} - k_2 z^{-3}\right]u_k = -\left[k_1 + (k_0 - k_1)z^{-1}\right]y_k \qquad \text{(B.140)}$$

or, in causal transfer function form $G(z^{-1})$,

$$u_k = -\left[\frac{k_1 + (k_0 - k_1)z^{-1}}{1 + (k_3 - 1)z^{-1} + (k_2 - k_3)z^{-2} - k_2 z^{-3}}\right]y_k \qquad \text{(B.141)}$$

This is immediately written in a form that explicitly exhibits the annihilator factor $(1 - z^{-1})$ in the denominator of the controller transfer function:

$$u_k = -\left[\frac{1}{1 - z^{-1}}\right]\left[\frac{k_1 + (k_0 - k_1)z^{-1}}{1 + k_3 z^{-1} + k_2 z^{-2}}\right]y_k \qquad \text{(B.142)}$$

B.4.7 nth-Order Pure Delay Systems

It is not difficult to see that the unperturbed nth-order delay system

$$y_{k+n} = u_k \qquad \text{(B.143)}$$

admits a compensator of the form

$$u_k = u_k^* - \left[\frac{k_0 z^{n-1}}{z^{n-1} + k_{n-1}z^{n-2} + \cdots + k_1}\right](y_k - y_k^*) \qquad \text{(B.144)}$$

In other words, the stabilizing controller, written in a causal manner, is given by the following expression, where time domain signals are combined with frequency domain signals:

$$u_k = u_k^* - \left[\frac{k_0}{1 + k_{n-1}z^{-1} + \cdots + k_1 z^{-(n-1)}} \right] (y_k - y_k^*) \qquad (B.145)$$

B.4.8   An nth-Order Perturbed System

Consider now the output stabilization to zero of the perturbed system

$$y_{k+n} = u_k + \xi \qquad (B.146)$$

with an unknown constant ξ.

The string of relations between the advanced outputs and the retarded inputs for the perturbed system is just

$$y_{k+n-1} = u_{k-1} + \xi, \quad y_{k+n-2} = u_{k-2} + \xi, \quad \cdots$$
$$y_{k+1} = u_{k-n+1} + \xi, \quad y_k \text{ (measured)} \qquad (B.147)$$

The estimates of the retarded output, off by a constant disturbance term, are

$$\hat{y}_{k+n-1} = u_{k-1}, \quad \hat{y}_{k+n-2} = u_{k-2}, \quad \cdots, \hat{y}_{k+1} = u_{k-n+1}, \quad \hat{y}_k = y_k \qquad (B.148)$$

A feedback controller, based on the retarded output estimates, with an adder compensator is obtained as

$$u_k = -\alpha_{k+n-1}\hat{y}_{k+n-1} - \alpha_{k+n-2}\hat{y}_{k+n-2} - \cdots - \alpha_{k+1}\hat{y}_{k+1} - \alpha_k y_k - \frac{\alpha_{k-1}y_k}{z-1}$$
$$= -\alpha_{k+n-1}u_{k-1} - \alpha_{k+n-2}u_{k-2} - \cdots - \alpha_{k+1}u_{k-(n-1)} - \alpha_k y_k$$
$$- \left[\frac{\alpha_{k-1}z^{-1}}{1 - z^{-1}} \right] y_k \qquad (B.149)$$

This controller, in transfer function form, is just

$$u_k = - \left[\frac{1}{1 - z^{-1}} \right] \left[\frac{\alpha_k + (\alpha_{k-1} - \alpha_k)z^{-1}}{1 + \alpha_{k+n-1}z^{-1} + \alpha_{k+n-2}z^{-2} + \cdots + \alpha_{k+1}z^{-n-1)}} \right] y_k \qquad (B.150)$$

Note that the factor $(1 - z^{-1})^m$ is an annihilator to the transform of any time polynomial in the z domain of degree $m - 1$. Let y_k^* be a given output reference trajectory. Define the tracking error $e_k = y_k - y_k^*$. A trajectory tracking controller, for a perturbed system of the form

$$y_{k+n} = \beta u_k + \xi(k) \qquad (B.151)$$

with $e_{u_k} = u_k - u_k^* = u_k - \frac{1}{\beta} y_{k+n}^*$ is given by

$$
u_k = u_k^* - \left[\frac{1}{1 - z^{-1}} \right]^m
$$
$$
\times \left[\frac{\sum_{j=k-m}^{j=k} \alpha_j (1 - z^{-1})^{k-m-j} z^{j-k}}{1 + \alpha_{k+n-1} z^{-1} + \alpha_{k+n-2} z^{-2} + \cdots + \alpha_{k+1} z^{-(n-1)}} \right] (y_k - y_k^*)
$$

$$(B.152)$$

The success in using a robust flat filtering controller for a discrete-time trajectory tracking task on a perturbed system lies in the proximity of the additive closed-loop term $(1 - z^{-1})^m \xi(k)$ to zero in the time domain.

B.5   A GENERAL RESULT FOR MULTIVARIABLE LINEAR SYSTEMS

In this section, we show that, for linear time-invariant discrete-time unperturbed systems, the state can be exactly obtained by means of a backward shifts parameterization, provided that the system is observable. The result is also valid in the case of perturbed systems modulo classical disturbances represented by time polynomial signals. The actual value of the states is not particularly required as an objective. Nevertheless, the use of such observer-free reconstructors, in a flatness-based linear state feedback scheme, can be adequately incorporated by adding a suitable linear combination of iterated adders of the flat output tracking error.

Since the result equally applies to single- or multivariable systems, we make our considerations in the context of linear multivariable systems. Also, we further assume that the output y_k is not the flat output. Recall that the flat output is observable. Hence, the result particularly applies to systems with measurable flat outputs.

Consider the linear time-invariant unperturbed system

$$
x_{k+1} = A x_k + B u_k, \quad x_k \in R^n, \quad u_k \in R^m
$$
$$
y_k = C x_k, \quad y_k \in R^p
$$

$$(B.153)$$

The system may be equivalently written as

$$
x_k = A x_{k-1} + B u_{k-1}
$$

$$(B.154)$$

Iterating once on the states of the system, we get

$$
x_k = A[A x_{k-2} + B u_{k-2}] + B u_{k-1} = A^2 x_{k-2} + A B u_{k-2} + B u_{k-1} \quad (B.155)
$$

Then $n - 2$ iterations yield

$$x_k = A^{n-1} x_{k-(n-1)} + \sum_{j=0}^{n-2} A^j B u_{k-(j+1)} \tag{B.156}$$

The forward shifts of the output equation result in

$$\begin{bmatrix} y_k \\ y_{k+1} \\ \vdots \\ y_{k+(n-1)} \end{bmatrix} = \begin{bmatrix} C \\ CA \\ CA^2 \\ \vdots \\ CA^{n-1} \end{bmatrix} x_k$$

$$+ \begin{bmatrix} 0 & 0 & \cdots & 0 \\ CB & 0 & \cdots & 0 \\ CAB & CB & \cdots & 0 \\ \vdots & \vdots & \cdots & \vdots \\ CA^{n-2}B & CA^{n-3}B & \cdots & CB \end{bmatrix} \begin{bmatrix} u_k \\ u_{k+1} \\ \vdots \\ u_{k+(n-2)} \end{bmatrix} \tag{B.157}$$

Taking $n - 1$ backward shifts on every equation, we obtain:

$$\begin{bmatrix} y_{k-(n-1)} \\ y_{k-(n-2)} \\ \vdots \\ y_k \end{bmatrix} = \begin{bmatrix} C \\ CA \\ CA^2 \\ \vdots \\ CA^{n-1} \end{bmatrix} x_{k-(n-1)}$$

$$+ \begin{bmatrix} 0 & 0 & \cdots & 0 \\ CB & 0 & \cdots & 0 \\ CAB & CB & \cdots & 0 \\ \vdots & \vdots & \cdots & \vdots \\ CA^{n-2}B & CA^{n-3}B & \cdots & CB \end{bmatrix} \begin{bmatrix} u_{k-(n-1)} \\ u_{k-(n-2)} \\ \vdots \\ u_{k-1} \end{bmatrix} \tag{B.158}$$

which we rewrite as

$$\begin{bmatrix} y_{k-(n-1)} \\ y_{k-(n-2)} \\ \vdots \\ y_k \end{bmatrix} = \mathcal{O} x_{k-(n-1)} + \mathcal{M} \begin{bmatrix} u_{k-(n-1)} \\ u_{k-(n-2)} \\ \vdots \\ u_{k-1} \end{bmatrix} \tag{B.159}$$

FIGURE B.16 Newton's model of the heating of a slab of material.

From the two relations we have:

$$x_k = A^{n-1} x_{k-(n-1)} + \sum_{j=0}^{n-2} A^j B u_{k-(j+1)} \tag{B.160}$$

$$\begin{bmatrix} y_{k-(n-1)} \\ y_{k-(n-2)} \\ \vdots \\ y_k \end{bmatrix} = \mathcal{O} x_{k-(n-1)} + \mathcal{M} \begin{bmatrix} u_{k-(n-1)} \\ u_{k-(n-2)} \\ \vdots \\ u_{k-1} \end{bmatrix} \tag{B.161}$$

We obtain

$$x_k = A^{n-1} (\mathcal{O}^T \mathcal{O})^{-1} \mathcal{O}^T \left\{ \begin{bmatrix} y_{k-(n-1)} \\ y_{k-(n-2)} \\ \vdots \\ y_k \end{bmatrix} - \mathcal{M} \begin{bmatrix} u_{k-(n-1)} \\ u_{k-(n-2)} \\ \vdots \\ u_{k-1} \end{bmatrix} \right\}$$

$$+ \sum_{j=0}^{n-2} A^j B u_{k-(j+1)} \tag{B.162}$$

The state vector can then be exactly computed from inputs, outputs, and a finite number of backward shifts of these vector signals. As a consequence, any linear state feedback can be equivalently synthesized in terms of linear combinations of inputs, outputs, and a finite number of their backward shifts.

In the case of classically perturbed systems, a linear combination of finitely iterated adders of the output components tracking errors completes the feedback scheme, known as GPI control. This is the discrete-time counterpart of Generalized Proportional Integral control of linear time-invariant continuous systems.

B.5.1 The Heat Equation

Here we consider a rather simple space-discretized model of the heating of a slab of material from one of its boundaries (see Fig. B.16). The model is known as the Newton law of heating, and we assume that all its defining parameters are known;

$$c_{k+1} = (1 - 2p)c_k + p b_k$$
$$b_{k+1} = p a_k + (1 - 2p)b_k + p c_k$$

$$a_{k+1} = (1 - 2p)a_k + pb_k + pu_k \tag{B.163}$$
$$y_k = c_k$$

where a_k is the temperature of point a at time k, b_k is the temperature of point b at time k, c_k stands for the temperature of point c at time k, and u_k is the external input temperature.

Here, we have assumed that the temperature to the right of the point c is zero.

From the system equations we obtain

$$c_k = (1 - 2p)y_{k-1} + pb_{k-1}$$
$$b_k = pa_{k-1} + (1 - 2p)b_{k-1} + py_{k-1}$$
$$a_k = (1 - 2p)a_{k-1} + pb_{k-1} + pu_{k-1} \tag{B.164}$$

Iterating once on the unmeasured states, we obtain

$$b_k = 2p(1 - 2p)a_{k-2} + [p^2 + (1 - 2p)^2]b_{k-2} + p^2 u_{k-2}$$
$$+ (1 - 2p)py_{k-2} + py_{k-1}$$
$$a_k = [p^2 + (1 - 2p)^2]a_{k-2} + 2p(1 - 2p)b_{k-2} + p(1 - p)u_{k-2}$$
$$+ p^2 y_{k-2} + pu_{k-1} \tag{B.165}$$

which are still functions of the second backward shifts of the unmeasured states a_{k-2} and b_{k-2}.

The system is flat, with flat output given by c_k, which also happens to be the system output

$$c_k = y_k$$
$$b_k = \frac{1}{p}\left[y_{k+1} - (1 - 2p)y_k\right]$$
$$a_k = \frac{1}{p^2}y_{k+2} - \frac{2}{p^2}(1 - 2p)y_{k+1} + \frac{1 - 4p + 3p^2}{p^2}y_k$$
$$u_k = \frac{1}{p^3}y_{k+3} - \frac{3(1 - 2p)}{p^3}y_{k+2} + \left[\frac{3 - 12p + 10p^2}{p^2}\right]y_{k+1}$$
$$- \left[\frac{(1 - 4p + 2p^2)(1 - 2p)}{p^3}\right]y_k \tag{B.166}$$

From the difference parameterization provided by flatness we obtain

$$b_{k-1} = \frac{1}{p}\left[y_k - (1 - 2p)y_{k-1}\right]$$
$$a_{k-1} = \frac{1}{p^2}y_{k+1} - \frac{2}{p^2}(1 - 2p)y_k + \frac{1 - 4p + 3p^2}{p^2}y_{k-1} \tag{B.167}$$

and, hence,

$$b_{k-2} = \frac{1}{p}\left[y_{k-1} - (1-2p)y_{k-2}\right]$$

$$a_{k-2} = \frac{1}{p^2}y_k - \frac{2}{p^2}(1-2p)y_{k-1} + \frac{1-4p+3p^2}{p^2}y_{k-2} \qquad \text{(B.168)}$$

Substituting the expressions for b_{k-2} and a_{k-2} into the expressions for b_k and a_k obtained by iteration of the state equations, we have the following delayed reconstructor for the unmeasured states:

$$b_k = 2p(1-2p)\left[\frac{1}{p^2}y_k - \frac{2}{p^2}(1-2p)y_{k-1} + \frac{1-4p+3p^2}{p^2}y_{k-2}\right]$$
$$+ \frac{1}{p}[p^2 + (1-2p)^2]\left[y_{k-1} - (1-2p)y_{k-2}\right] + p^2 u_{k-2}$$
$$+ (1-2p)py_{k-2} + py_{k-1}$$

$$a_k = [p^2 + (1-2p)^2]\left[\frac{1}{p^2}y_k - \frac{2}{p^2}(1-2p)y_{k-1}\right.$$
$$\left. + \frac{1-4p+3p^2}{p^2}y_{k-2}\right] + 2(1-2p)\left[y_{k-1} - (1-2p)y_{k-2}\right]$$
$$+ p(1-p)u_{k-2} + p^2 y_{k-2} + pu_{k-1} \qquad \text{(B.169)}$$

which constitute the exact delayed input–output parameterization of the unmeasured state variables.

The consecutive advances of the flat output $y_k = c_k$ may be placed in terms of the unmeasured states as

$$y_k = c_k$$
$$y_{k+1} = (1-2p)y_k + pb_k$$
$$y_{k+2} = [(1-2p)^2 + p^2]y_k + 2p(1-2p)b_k + p^2 a_k \qquad \text{(B.170)}$$

The reconstructors of the advances of the flat output are then obtained as follows:

$$\widehat{y_k} = y_k$$
$$\widehat{y_{k+1}} = (1-2p)y_k + p\widehat{b_k}$$
$$\widehat{y_{k+2}} = [(1-2p)^2 + p^2]y_k + 2p(1-2p)\widehat{b_k} + p^2 \widehat{a_k} \qquad \text{(B.171)}$$

with $\widehat{b_k}$, $\widehat{a_k}$ obtained as in the previous equation.

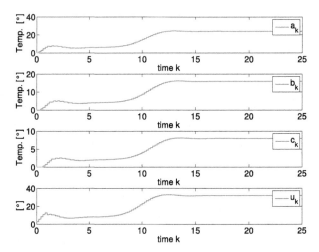

FIGURE B.17 Performance of a GPI controller for a trajectory tracking task on a heat equation system.

Using the difference parameterization of the control input, we have

$$u_k = \frac{1}{p^3} y_{k+3} - \frac{3(1-2p)}{p^3} y_{k+2} + \left[\frac{3-12p+10p^2}{p^2} \right] y_{k+1}$$
$$- \left[\frac{(1-4p+2p^2)(1-2p)}{p^3} \right] y_k \qquad (B.172)$$

We propose the following controller with "adder" corrective action:

$$u_k = \frac{1}{p^3} v_k - \frac{3(1-2p)}{p^3} \widehat{y_{k+2}} + \left[\frac{3-12p+10p^2}{p^2} \right] \widehat{y_{k+1}}$$
$$- \left[\frac{(1-4p+2p^2)(1-2p)}{p^3} \right] y_k$$
$$v_k = p^3 \left[y^*_{k+3} - \sum_{i=0}^{2} k_{i+1}(\widehat{y_{k+i}} - y^*_{k+i}) - k_0 \zeta_k \right]$$
$$\zeta_{k+1} = \zeta_k + y_k - y^*_k \qquad (B.173)$$

Fig. B.17 depicts the performance of the proposed GPI control in an output reference trajectory tracking task implying a rest-to-rest trajectory for the flat output.

REFERENCES

[1] J. Han, From PID to active disturbance rejection control, IEEE Transactions on Industrial Electronics 56 (3) (2009) 900–906.

[2] B.Z. Guo, F.F. Jin, The active disturbance rejection and sliding mode control approach to the stabilization of the Euler–Bernoulli beam equation with boundary input disturbance, Automatica 49 (9) (2013) 2911–2918.

[3] S. Zhao, Z. Gao, An active disturbance rejection based approach to vibration suppression in two-inertia systems, Asian Journal of Control 15 (2) (2013) 350–362.

[4] L. Dong, Y. Zhang, Z. Gao, A robust decentralized load frequency controller for interconnected power systems, ISA Transactions 51 (3) (2012) 410–419.

[5] M. Fliess, C. Join, Model-free control, International Journal of Control 86 (12) (2013) 2228–2252.

[6] H. Sira-Ramírez, J. Linares-Flores, Alberto Luviano-Juarez, J. Cortés-Romero, Ultramodelos Globales y el Control por Rechazo Activo de Perturbaciones en Sistemas No lineales Diferencialmente Planos, Revista Iberoamericana de Automática e Informática Industrial RIAI 12 (2) (2015) 133–144.

[7] R. Madoński, P. Herman, Model-free control or active disturbance rejection control? On different approaches for attenuating the perturbation, in: Control & Automation (MED), 2012 20th Mediterranean Conference on, IEEE, 2012, pp. 60–65.

[8] M. Fliess, J. Lévine, P. Martin, P. Rouchon, Flatness and defect of non-linear systems: introductory theory and applications, International Journal of Control 61 (1995) 1327–1361.

[9] M. Fliess, R. Marquez, E. Delaleau, H. Sira-Ramírez, Correcteurs proportionnels-intégraux généralisés, ESAIM: Control, Optimisation and Calculus of Variations 7 (2002) 23–41.

[10] Hebertt Sira-Ramírez, Sunil K. Agrawal, Differentially Flat Systems, Marcel Dekker, 2004.

[11] Z. Gao, Active disturbance rejection control: a paradigm shift in feedback control system design, in: 2006 American Control Conference, IEEE, 2006, pp. 2399–2405.

[12] M. Fliess, R. Marquez, Continuous-time linear predictive control and flatness: a module-theoretic setting with examples, International Journal of Control 73 (2000) 606–623.

[13] Q. Zheng, Z. Chen, Z. Gao, A practical approach to disturbance decoupling control, Control Engineering Practice 17 (9) (2009) 1016–1025.

[14] Q. Zheng, L.Q. Gao, Z. Gao, On validation of extended state observer through analysis and experimentation, Journal of Dynamic Systems, Measurement, and Control 134 (2) (2012) 024505.

[15] Bao-Zhu Guo, Zhi-liang Zhao, On the convergence of an extended state observer for nonlinear systems with uncertainty, Systems & Control Letters 60 (6) (2011) 420–430.

[16] Bao-Zhu Guo, Zhi-Liang Zhao, On convergence of the nonlinear active disturbance rejection control for MIMO systems, SIAM Journal on Control and Optimization 51 (2) (2013) 1727–1757.

[17] B.Z. Guo, J.J. Liu, Sliding mode control and active disturbance rejection control to the stabilization of one-dimensional Schrödinger equation subject to boundary control matched disturbance, International Journal of Robust and Nonlinear Control 24 (16) (2014) 2194–2212.

[18] R. Madoński, P. Herman, Survey on methods of increasing the efficiency of extended state disturbance observers, ISA Transactions 56 (2015) 18–27.

[19] R. Kelly, V. Santibáñez, A. Loria, Control of Robot Manipulators in Joint Space, Springer Science & Business Media, 2006.

[20] Michel Fliess, Cédric Join, Intelligent PID controllers, in: 16th Mediterranean Conference on Control and Automation, 2008.

[21] H. Sira-Ramirez, J. Linares-Flores, C. Garcia-Rodriguez, M.A. Contreras-Ordaz, On the control of the permanent magnet synchronous motor: an active disturbance rejection control approach, IEEE Transactions on Control Systems Technology (ISSN 1063-6536) 22 (5) (2014) 2056–2063, http://dx.doi.org/10.1109/TCST.2014.2298238.

Index

Printed in the United States
By Bookmasters